Tourism and Sustainable Community Development

As the tourist industry becomes increasingly important to communities around the world, the need to develop tourism sustainably has also become a primary concern. This collection of international case studies addresses this crucial issue by asking what local communities can contribute to sustainable tourism, and what sustainability can offer local communities.

The role of the community in environmental, cultural and economic sustainability is highlighted in an extraordinary variety of contexts, ranging from inner city Edinburgh to rural northern Portugal and the beaches of Indonesia. Individually, through the breadth of their coverage, these investigations present a wealth of original research and source material. Collectively, these studies illuminate the term 'community', the meaning of which, it is argued, is vital to understanding how sustainable tourism development can be implemented in practice.

Derek Hall is head of the Leisure and Tourism Department at the Scottish Agricultural College, and has a personal chair in regional development. He has particular interests in welfare issues in tourism and in the role of tourism in restructuring processes in central and eastern Europe.

Greg Richards is a lecturer in Tourism Management at Tilburg University and is co-ordinator of the European Association for Tourism and Leisure Education (ATLAS). He has directed a number of ATLAS projects for the European Commission on topics including sustainable tourism.

Routledge Advances in Tourism
Series editors: Brian Goodall and Gregory Ashworth

1 **The Sociology of Tourism**
Theoretical and empirical investigations
Edited by Yiorgos Apostolopoulos, Stella Leivadi and Andrew Yiannakis

2 **Creating Island Resorts**
Brian King

3 **Destinations**
Cultural landscapes of tourism
Greg Ringer

4 **Mediterranean Tourism**
Facets of socioeconomic development and cultural change
Edited by Yiorgos Apostolopoulos, Lila Leontidou and Philippos Loukissas

5 **Outdoor Recreation Management**
John Pigram and John Jenkins

6 **Tourism Development**
Contemporary issues
Edited by Douglas G. Pearce and Richard W. Butler

7 **Tourism and Sustainable Community Development**
Edited by Greg Richards and Derek Hall

Tourism and Sustainable Community Development

Edited by Greg Richards and Derek Hall

London and New York

First published 2000
by Routledge
11 New Fetter Lane, London EC4P 4EE

Simultaneously published in the USA and Canada
by Routledge
29 West 35th Street, New York, NY 10001

Routledge is an imprint of the Taylor & Francis Group

Typeset in Garamond by
MHL Typesetting Ltd, Coventry
Printed and bound in Great Britain by
T J International Ltd, Padstow, Cornwall

British Library Cataloguing in Publication Data
A catalogue record for this book is available
from the British Library

Library of Congress Cataloging in Publication Data
Richards, Greg
Tourism and sustainable community development/Greg Richards and Derek Hall.
p. cm.
Includes bibliographical references and index.
1. Tourism. 2. Sustainable development. I. Hall, Derek R. II. Title.

G155.A1 R52 2000
338.4'791–dc21 99-042599

ISBN 0-415-22462-4

Contents

List of figures viii
List of tables ix
List of contributors xi
Foreword xvii

1 The community: a sustainable concept in tourism development? 1
GREG RICHARDS AND DEREK HALL

PART 1
Community participation and identity 15

2 Approaches to sustainable tourism planning and community participation: the case of the Hope Valley 17
BILL BRAMWELL AND ANGELA SHARMAN

3 Residents' perceptions of the socio-cultural impacts of tourism at Lake Balaton, Hungary 36
TAMARA RÁTZ

4 Identity, community and sustainability: prospects for rural tourism in Albania 48
DEREK HALL

PART 2
Sustainable tourism and the community 61

5 Environmental standards and performance measurement in tourism destination development 63
BRIAN GOODALL AND MIKE STABLER

6 Developing sustainable tourism in the Trossachs, Scotland 83
ALISON CAFFYN

7 Establishing the common ground: tourism, ordinary places, grey-areas and environmental quality in Edinburgh, Scotland 101
FRANK HOWIE

8 Local Agenda 21: reclaiming community ownership in tourism or stalled process? 119
GUY JACKSON AND NIGEL MORPETH

PART 3
Developing community enterprise 135

9 Fair trade in tourism – community development or marketing tool? 137
GRAEME EVANS AND ROBERT CLEVERDON

10 Tourism, small enterprises and community development 154
HEIDI DAHLES

11 Gili Trawangan – from desert island to 'marginal' paradise: local participation, small-scale entrepreneurs and outside investors in an Indonesian tourist destination 170
THEO KAMSMA AND KARIN BRAS

12 Tourism in Friesland: a network approach 185
JANINE CAALDERS

13 Understanding community tourism entrepreneurism: some evidence from Texas 205
KHOON Y. KOH

PART 4
Rural communities and tourism development 219

14 Can sustainable tourism positively influence rural regions? 221
JAN VAN DER STRAATEN

15 Cultural tourism and the community in rural Ireland 233
JAYNE STOCKS

16 Agritourism – a path to community development?: the case of Bangunkerto, Indonesia 242
DAVID J. TELFER

17 Community and rural development in Northern Portugal 258
JOACHIM KAPPERT

18 The market for rural tourism in North and Central Portugal: a benefit-segmentation approach 268
ELISABETH KASTENHOLZ

19 Tourism brand attributes of the Alto Minho, Portugal 285
JONATHAN EDWARDS, CARLOS FERNANDES, JULIAN FOX AND ROGER VAUGHAN

20 Conclusions 297
GREG RICHARDS AND DEREK HALL

Index 307

Figures

1.1	Principles behind the approach to sustainable tourism management.	6
2.1	The Hope Valley parishes of Castleton, Edale and Hope.	19
3.1	Tourism's socio-cultural impacts within the framework of wider social change.	38
3.2	The study area.	39
5.1	Tracing the environmental impact of tourism in destinations.	66
5.2	Environmental performance and quality – alternative links between a tourist destination and its tourism businesses.	68
6.1	Location of the Trossachs area.	86
6.2	The Trossachs Trail.	87
9.1	Models and approaches to a fair trade relationship.	144
9.2	Open trading: options for effective monitoring of corporate codes of conduct.	148
12.1	Location of the Frisian WCL area.	195
13.1	Conceptual model of community tourism entrepreneurism.	208
13.2	Quadrant positions of Athens and Mount Pleasant.	212
15.1	The Gaeltacht regions in Ireland.	234
15.2	The Donegal Gaeltacht region.	239
16.1	Salak tourism plantation.	247
19.1	Cognitive model: component parts.	287
19.2	Affective model: overall evaluation.	287
19.3	Applied cognitive model.	292
19.4	Applied affective model.	293

Tables

2.1 Four possible approaches to planning for tourism as a contribution to sustainable development 22
2.2 General aims and specific objectives of the Hope Valley visitor-management plan 25
2.3 General aims and related practical measures in the Hope Valley visitor-management plan 25
2.4 Some issues affecting approaches to community participation in tourism planning 28
3.1 Feelings about the presence of tourists in the Keszthely–Hévíz region 40
3.2 The impact of tourism on the image of the Keszthely–Hévíz region 41
3.3 Residents' opinions on the number of tourists coming to their settlement 41
3.4 Differences perceived by residents between tourists and themselves 41
3.5 Types of difference perceived by residents 42
3.6 Residents' mean response to tourism's positive effects on the region 43
3.7 Residents' mean response to tourism's negative effects on the region 44
3.8 Host's attitudinal/behavioural responses to tourist activities in the region 45
5.1 What environmental standards can be set for tourism? 72
5.2 Quantitative and qualitative measures for evaluating an hotel's environmental performance 75
9.1 Developing countries' services by sector (percentage of countries offering 'free trade') 139
10.1 Number of foreign visitors and revenues 157
11.1 Number of visitors and hotel rooms in Nusa Tenggara Barat 1988–95 177
13.1 Athens' P-factor 211
13.2 Athens' Q-factor 211

13.3 Mount Pleasant's P-factor 211
13.4 Mount Pleasant's Q-factor 212
13.5 Tourism enterprise births in Athens 213
13.6 Tourism enterprise births in Mount Pleasant 213
13.7 Tourism enterprise size and ownership at time of birth in Athens 214
13.8 Tourism enterprise size and ownership at time of birth in Mount Pleasant 214
16.1 Occupation/has agritourism helped salak marketing? 252
16.2 Do you work on the agritourism site/has salak production increased due to tourism? 252
16.3 Have you been involved in planning the agritourism site/do you want a full-time job in tourism? 252
18.1 Ranking of holiday factors 272
18.2 Summary of PCA-solution 275
18.3 Benefits and activities sought 277
18.4 Profile of segments 278
19.1 Variables influencing the attractiveness of a region 289
19.2 Cognitive perceptions of the Alto Minho 291

communities' in LDC destinations. Robert is adviser to the World Tourism Organisation and undertakes the WTO's Global Tourism Forecasts ('2020 Vision') as well as master planning and training for developing countries and regions.

Heidi Dahles received her PhD in cultural anthropology from the University of Nijmegen, the Netherlands. She worked in the field of tourism and leisure studies at Tilburg University (1990–98) and was senior fellow at the International Institute for Asian Studies in Leiden in 1998. She currently teaches in the Department of Business Anthropology, Vrije Universiteit Amsterdam, and holds a postdoctoral fellowship at the Centre for Asian Studies Amsterdam (CASA). Her research interests include entrepreneurial culture, brokerage and patronage in Southeast Asian countries, in particular in the tourism industry.

Jonathan Edwards is a Reader, and Head of the Centre for Land Based Enterprises, in the School of Service Industries, Bournemouth University. His main current research interests focus on the development of rural areas and, in particular, development through tourism.

Graeme Evans is Director of the Centre for Leisure and Tourism Studies at the University of North London. Together with Robert Cleverdon he has been working with Tourism Concern and Voluntary Services Overseas (VSO) on a joint Fair Trade and Tourism action research project with 'southern communities' in LDC destinations. Graeme has worked with small crafts and cultural producers in central Mexico and elsewhere, developing crafts and cultural tourism promotion, and is leading a comparative study of world heritage sites and local community and economic development. He currently advises the World Bank on their 'Culture and Sustainable Development' and 'Culture and Cities' Programmes.

Carlos Fernandes is a Lecturer in the tourism studies course at the Polytechnic Institute of Viana do Castelo. His main research interests are rural development and cultural tourism.

Julian Fox is a Senior Lecturer at Bournemouth University, School of Conservation Science. His research interests are change in cultural landscapes, industrial archaeology, environmental conservation, rural and coastal tourism with particular reference to the Alto Minho, Northern Portugal.

Brian Goodall is Professor of Geography and Dean of the Faculty of Urban and Regional Studies at the University of Reading. His research interests focus on environmental performance in tourism, especially the integration of environmental management into tourism business practices.

Contributors

Bill Bramwell is Reader in Tourism Management in Sheffield Hallam University's Centre for Tourism. He has a PhD in Geography and has led research for many government organisations. He is co-editor of the *Journal of Sustainable Tourism* and has written numerous articles in tourism journals.

Karin Bras is a cultural anthropologist who graduated from the University of Amsterdam, the Netherlands in 1991. Her MA thesis deals with rural integrated tourism in the Basse-Casamance, Senegal. At the moment she is finishing her PhD thesis at the Department of Leisure Studies, Tilburg University and teaching at the Department of Business Anthropology, Vrije Universiteit Amsterdam. Her thesis analyses the role of local tourist guides in the social construction of tourist attractions on the island of Lombok, Indonesia.

Janine Caalders is carrying out PhD research on rural tourism development at Wageningen Agricultural University. In her thesis she approaches this subject from a theoretical and methodological point of view, integrating theories on the nature of (rural) space, innovation and (organisational) networks. Case studies have been carried out in the Netherlands and in France. In addition, she is a partner in a consultancy which specialises in tourism and leisure-related research and advice.

Alison Caffyn is Lecturer in tourism and leisure at the Centre for Urban and Regional Studies, University of Birmingham. She previously worked in development and research for regional tourist boards in Cumbria and Scotland. Her research interests include tourism policy, rural tourism development, partnership approaches, community participation and outdoor recreation management.

Robert Cleverdon is Senior Lecturer in Tourism Development Studies at the University of North London. Together with Graeme Evans he has been working with Tourism Concern and Voluntary Services Overseas (VSO) on a joint Fair Trade and Tourism action research project with 'southern

Khoon Y. Koh's teaching and research interests are community tourism development and tourism education. Presently, he is an Assistant Professor in the Department of Recreation, Tourism and Hospitality Management, Mesa State College, Colorado, USA.

Nigel Morpeth is a Lecturer in tourism within the Department of Tourism at the University of Lincolnshire and Humberside. Additionally he is completing his PhD research into the role that cycle tourism initiatives can play within the implementation of sustainable development policies for tourism and transport.

Tamara Rátz is a Lecturer and PhD candidate at the Tourism Research Centre, Budapest University of Economic Sciences, Hungary, and guest lecturer at Hame Polytechnic, Finland. Her current research interests include the socio-cultural impacts of tourism, sustainability in rural tourism development and heritage tourism.

Greg Richards is a Lecturer in Tourism Management at Tilburg University in the Netherlands. He is currently Co-ordinator of the European Association for Tourism and Leisure Education (ATLAS) and has directed a number of ATLAS projects for the European Commission on topics including cultural tourism, sustainable tourism and tourism education.

Angela Sharman is Teaching and Research Associate in the Centre for Tourism at Sheffield Hallam University. She has an MSc in Tourism Management and has conducted research for various projects on environmental management and sustainable tourism.

Mike Stabler is honorary senior research fellow in the department of economics at The University of Reading and is joint director of the Tourism Research and Policy Unit. His research interests are the economics of conservation, the environment and leisure, especially concerning the impact of tourism. He is currently managing editor of the journal, *Leisure Studies*.

Jayne Stocks is a Senior Lecturer in Tourism at the University of Derby. She is a geographer by background and worked for a number of years in the airline industry. Her current active research interests centre on Ireland and Goa.

Jan van der Straaten is an environmental economist affiliated with the Department of Leisure Studies at Tilburg University. His research interests include tourism and the environment, and environmental policy. His publications include the Kluwer book *Tourism and the Environment*, with Helen Briassoulis.

Derek Hall is Head of the Leisure and Tourism Management Department of the Scottish Agricultural College (SAC). He has University of London degrees in geography and social anthropology (his 1978 PhD thesis was on aspects of urban community), and has written on tourism and other aspects of development and change in Central and Eastern Europe, on tourism and gender, and on transport and globalisation. He holds a personal chair in Regional Development.

Frank Howie is a Lecturer at Queen Margaret University College, Edinburgh, where he has specialised in tourism planning and sustainable development since 1989. Previously he worked in private consultancy and in the public sector. He holds degrees in the planning, tourism and environmental management fields from the University of Edinburgh and the University of British Columbia, Canada.

Guy Jackson is a former tourism industry management consultant who has lectured in Tourism Studies at Loughborough University since 1991. He was one of the co-authors of the European Commission DGXXIII project and publication *Sustainable Tourism Management: Principles and Practice* and has published more broadly on sustainable tourism development.

Theo Kamsma is a cultural anthropologist who graduated from the Vrije Universiteit Amsterdam, the Netherlands in 1991. His MA thesis deals with youth tourism in Amsterdam. He has been actively involved in the tourism industry in Amsterdam for many years. At present, he is a freelance journalist writing about tourism, leisure and Indonesia.

Joachim Kappert was born in 1960 in Düsseldorf, Germany. He studied Human Geography at the University in Düsseldorf, specialising in Tourism and World Transport. In Portugal, he intensified the social components of his studies by participating in courses on the Psychosociology of Organisations. He is currently writing a Masters thesis on Urban Tourism in the Faculty of Architecture at the University of Porto.

Elisabeth Kastenholz was born in 1964 in Cologne, Germany. She graduated in 1987 at the Institute of Public Administration in Bonn. After working for the German Foreign Ministry until September 1988, she studied 'Tourism Planning and Management' at the University of Aveiro, Portugal, where she has been a Lecturer and Researcher since 1994. She took an MBA at ISEE/University of Porto, and presented her MA thesis on rural tourism in Portugal in 1997. Since 1998 she has been working on her PhD thesis 'Destination images and their role in tourist behaviour and destination marketing'. She is a senior researcher in a project on the rural tourism market in North Portugal and is working on a development plan for the Archaeological Park of Foz Coa.

David J. Telfer is an Assistant Professor in the Department of Recreation and Leisure Studies at Brock University (Canada). Formerly he was a Lecturer in Tourism at the University of Luton (UK). His research interests are related to the local responses to tourism development, agritourism (in Canada and Indonesia), and the links between tourism and development theory.

Roger Vaughan is a Principal Lecturer in the School of Service Industries, Bournemouth University. His main current research interests are measuring the economic impacts of leisure activities, the adoption of information technology within the tourism sector and the analysis of people's perceptions of tourist destinations and leisure facilities.

Foreword

Tourism and Community Development was the theme of a Conference held by the European Association for Tourism and Leisure Education (ATLAS) in Viana do Castelo, Portugal in September 1997. Over 60 papers were presented at the conference on a wide range of subjects connected with tourism and the community. The papers selected for this volume are representative of the range of issues that were dealt with in the conference, and also reflect the breadth of debate on the relationship between tourism and community.

The conference would not have been possible without the support provided by DGXVI of the European Commission. It formed part of the EUROTEX Project, which aims to develop crafts tourism as a source of community development. The locations included in the EUROTEX project were integrated into the conference, and examples of local crafts and crafts tourism development were on display for its duration. Many of the papers presented therefore related to the problems of community development in rural regions such as the Alto Minho.

Our gratitude is due to the School of Management and Tourism at the Polytechnic Institute of Viana do Castelo, which hosted the conference. Particular thanks are extended to Carlos Fernandes, who formed the bridge between the global conference audience and the local community. The conference delegates were treated to a great deal of interaction with the local community in the Alto Minho Region, with conference sessions being held in cultural centres in Arcos de Valvadez, Soajo and Monçao.

This book could not have been produced without the help of a large number of people. In particular we would like to thank Leontine Onderwater, the ATLAS Project Manager for her help in collating the texts. We also extend our gratitude to the authors for their contributions, their positive responses to our suggestions and their ability to keep to deadlines (mostly!).

The final editing of the text also owes much to the ATLAS Winter University, an intensive course for an international group of students and staff in Kazimierz, Poland in January 1999. The medieval tranquillity of Kazimierz in winter provided the ideal environment to put the finishing touches to the text. Greg Richards extends his thanks to Bohdan Jung and Bozena Mierzejewska and their team from the Warsaw School of Economics for doing such a good job of the organisation and making it possible to spend time on the book.

Greg Richards and Derek Hall

1 The community: a sustainable concept in tourism development?

Greg Richards and Derek Hall

Introduction

As tourism becomes increasingly important to communities around the world, the need to develop tourism sustainably also becomes a primary concern. Human communities represent both a primary resource upon which tourism depends, and their existence in a particular place at a particular time may be used to justify the development of tourism itself. Communities are a basic reason for tourists to travel, to experience the way of life and material products of different communities. Communities also shape the 'natural' landscapes which many tourists consume. Communities are, of course, also the source of tourists; tourists are drawn from particular places and social contexts which in themselves will help shape the context of the tourist's experience in the host community.

Sustaining the community/particular communities has therefore become an essential element of sustainable tourism. The rationale of sustainable tourism development usually rests on the assurance of renewable economic, social and cultural benefits to the community and its environment. An holistic approach to sustainability requires that the continuing/improved social, cultural and economic well-being of human communities is an integral component of environmental renewal. This is equally applicable within notions of sustainable tourism; without community sustainability, tourism development cannot be expected to be sustainable. For this reason, as Taylor (1995: 487) argues, 'the concept of community involvement in tourism development has moved nearer to the centre of the sustainability debate'.

The concept of developing tourism sustainably for the community is not without its problems, however. While most models of sustainable development include the community as a cornerstone of the development process, the concept of community itself is not unproblematic. Whose community? How defined: in spatial/social/economic terms? Who in the community should benefit from tourism? How should the community be presented to the tourist? The nature of the community itself is also changing. Globalisation and localisation, increasing geographic and social mobility are questioning widely held beliefs about the composition and structure of 'community'. Who are the 'locals' in the local community? Where should one place the spatial or temporal boundaries of the 'local' community? The emergence of a 'global community' also problematises the concept of a local community. Further, aspatial communities, linked by bonds of common interest not place, exist within and across spatial communities.

The growing complexity of communities and the relationships between them pose significant challenges for the sustainable development of tourism. Local community structures can provide the source of both problems and potential solutions in the sphere of sustainable development. This is the major issue examined from different perspectives in this book.

This introductory chapter addresses the development of the two key concepts dealt with in the book: community and sustainability. The origins and applications of these concepts are analysed to illustrate how their application in tourism development and in the tourism literature has tended to converge in recent years. This is followed by a review of the tensions which have been revealed in local communities as a result of tourism development (or more often unsustainable tourism development), and of the growing literature focusing on the community impacts of tourism.

The changing concept of community

Even in the 1950s, dozens of different interpretations of 'community' could be identified (Hillery 1955). John Urry (1995) extended the Bell and Newby (1976) analysis of the concept to include four different uses of the term. First, the idea of community as belonging to a specific topographical location. Second, as defining a particular local social system. Third, in terms of a feeling of 'communitas' or togetherness; and fourth as an ideology, often hiding the power relations which inevitably underlie communities. Community as an ideology has certainly permeated the sustainability literature, and there are few sustainable tourism policies which do not refer to the importance of long-term benefits for the 'community'.

This renewed interest in the community as a basic unit of tourism development, management, planning and marketing can be traced to the changing meaning of the concept of community. Lash and Urry (1994) argue from a postmodernist perspective that having been initially threatened with extinction through modernist rationalisation and disembedding, through the increasing mobility of society and the 'end of geography' through global communications, the place-based notion of 'community' has actually re-emerged as a vehicle for rooting individuals and societies in a climate of economic restructuring and growing social, cultural and political uncertainty. As political, social and economic structures based on the nation state begin to be questioned, so local communities have come to be seen as essential building blocks in the 'new sociations' and political alliances of the emerging 'third sector'.

As well as providing the essential social 'glue' between locality and inhabitants, communities are increasingly being seen as providing the essential link between the local and the global. As Saskia Sassen (1991) points out, the global only becomes manifest where it is rooted in the local, because this is where the power relationships and integrations of globalisation are seen and felt, even though they may be formulated elsewhere – in the boardrooms of multinational corporations or at meetings of supranational unions. In this view, local communities are seen as the essential receivers and transmitters of the forces of globalisation. At the same time, however, local communities are the seat of resistance against the threatened homogenisation of globalisation.

As noted above, communities can be aspatial. The concept of 'community' has become explicitly disembedded from the local in its application to social, cultural or ethnic groups which may be spread throughout a nation or country, or even across the entire globe. This is a further effect of the detraditionalisation processes of modernity. Pre-modern societies were relatively sedentary, and often rigidly hierarchical, yet multiplex social relationships tended to be contained in a limited geographic area. When given political meaning, ethnic groups, as the accretions of spatialised communities, provided the building blocks for the development of a sense of nationhood. Nations arose often before the imposition of political boundaries, and particularly those of a colonial or imperial nature. With the rise of the nation state, however, out of the ashes of fragmented empires and alliances, there arose a need to reinforce a sense of nationhood which corresponded to the spatial boundaries of the state, essentially consolidating a wider feeling of community which extended beyond the physical boundaries of the local. This gave rise to what Andersson (1987) has termed 'imagined communities'. The nation state relied on its citizens being able to imagine themselves as members of a single nation, even though their social interaction with other citizens was limited. However, mobility and emigration, going back to at least the Jewish Diaspora, and later colonising activities of powerful states acted to spatially diffuse these imagined communities beyond those of the nation state itself. English, Scots, Irish, Italian, Indian and Chinese Diaspora 'communities' were established or transplanted in far-flung corners of the globe. The interaction between these communities and indigenous peoples, with local environments and with each other, evolved a global patchwork of communities which can be viewed as internalising the global–local dialectic.

Such communities thus created have often become prime tourist attractions – distinctive cultures spanning the gulf between the extended imagined community and the local. The Amish in North America or the Chinatowns found in many major cities are examples of such communities which have become tourist attractions in their own right. The mixing of different ethnic cultures which has occurred in the major metropolitan centres has also created a major source of tourist fascination, enabling tourists to encounter the smelting of the primitive, the modern and the postmodern (MacCannell 1993). Some communities are now beginning deliberately to exploit their multicultural nature for tourism, such as the 'world city Den Haag' campaign or the 'Citta del Duomo' promotion in Utrecht (Burgers 1992).

The fact that tourists may travel to experience such communities – if only at arm's length, interpreted through a guide or by dint of an evening's 'staged authenticity' – has placed renewed emphasis on the relationship between community and locality. Tourists travel partly to consume difference, to see how other societies live. They can also be repulsed from districts regarded as dangerous or hostile because of the nature of the 'community' within. There is, however, an assumption that differences can be experienced by travelling to specific locations which are associated with specific communities. The realisation that the community itself has become an object of tourism consumption has in turn encouraged some communities to reproduce themselves specifically for tourists. Through the process of site sacralisation, whole communities can begin to identify themselves with the way in which they are 'named' and 'framed' as tourist attractions. This in turn creates backstage and frontstage areas in the community, with the tourist gaze being carefully restricted to the 'staged' authenticity of the frontstage regions. This staging process is stimulated by tourists demanding 'authentic' local cultures which they may associate with a specific location. The whole community has to be reproduced to conform to the image that the tourist has of it. In the process, community relations themselves become commodified. Further, questions of land access and land rights in particular places relating to the residential pre/proscriptions of, for example, native Americans or Australian aboriginals, raise deeper questions concerning the extent to which tourism can actually assist equitable resource distribution and the upholding of human rights for communities of indigenous peoples who may have been marginalised by an explicitly different colonising society.

Communities are not simply victims of the globalisation process and commodification, however. Communities also become centres of resistance to the processes of modernisation. The homogenised global economic, social and cultural landscape emphasises local differences even more strongly, amid calls for devolution and political autonomy. Regions on the periphery of the global economy are asserting their identity as a means of preserving their cultural identity and developing their socio-economic potential (Ray 1998). This trend towards regionalism is strengthened by the tendency towards 'neotribalism' (Maffesoli 1996), which is related to the inherent need for group identity. The resurgence of local identities, perhaps reinforced by political structural change, creates the potential for tourism development, as evidenced by the presentation of Gaelic heritage in Scotland (MacDonald 1997).

The recognition that communities can have some influence over the development of tourism has created a growing stream of literature on community-based tourism and community development in tourism in recent years. Murphy's (1985) classic review of community tourism formed the basis for many later studies. As Telfer points out in Chapter 16 of this volume, Murphy emphasised the necessity for each community to relate tourism development to local needs. Building on this basic principle, later studies of community-based tourism have gradually broadened the scope of the term to include a wide range of issues, including ecological factors and local participation and democracy.

Although the concept of community has shifted in meaning and application in the tourism field over the years, the recent rediscovery of the 'local' and the growing importance of identity have placed 'community' at the forefront of discussions about tourism development. In particular, 'the local community' has become for many the appropriate context level for the development of sustainable tourism.

Sustainability

The detraditionalisation associated with modernity is also marked by a growing reflexivity both at individual and institutional levels. As Urry (1995) points out, one of the most important consequences of this reflexivity is an increased concern for the environment, and a growing awareness of the links between the local and the global environment. In the shift from an 'industrial' to a 'risk' society (Beck 1992), the need for development to be 'sustainable' becomes paramount. Local communities become not only important in terms of actions taken to preserve their own immediate environment, but also form part of wider alliances to preserve the environment globally (act local, think global). These involve the NGOs and pressure groups which, representing a membership of like-minded environmentally aware people, can themselves be viewed as communities of interest.

Sustainability is important because communities need to support themselves on the basis of available resources. As Jan van der Straaten points out in his study of sustainable tourism development in the Alpine region (Chapter 14), economic necessity is usually the driving force behind the growth of tourism. Without tourists, spatially marginal communities that find it increasingly hard to compete in other spheres with the major metropolitan centres may cease to exist. In this sense, environmental sustainability is inexorably bound up with concepts of economic, social, cultural and political sustainability. The 'principles of sustainable tourism management' (see Figure 1.1) outlined by Bramwell *et al.* (1998) indicate the need to involve local communities in the process of sustainable tourism management and development.

Place-based communities have become central to a holistic concept of sustainability, which embraces and integrates environmental, economic, political, cultural and social considerations. In this way there is an implicit recognition that to be truly sustainable, the preservation of the 'natural' environment must be grounded in the communities and societies which exploit and depend upon it. Most natural environments are culturally constructed (Richards 2000), and local communities and economic systems may hold the key to their survival or destruction.

1. The approach sees policy, planning and management as appropriate and, indeed, essential responses to the problems of natural and human resource misuse in tourism.
2. The approach is generally not anti-growth, but it emphasises that there are limitations to growth and that tourism must be managed within these limits.
3. Long-term rather than short-term thinking is necessary.
4. The concerns of sustainable tourism management are not just environmental, but are also economic, social, cultural, political and managerial.
5. The approach emphasises the importance of satisfying human needs and aspirations, which entails a prominent concern for equity and fairness.
6. All stakeholders need to be consulted and empowered in tourism decision-making, and they also need to be informed about sustainable development issues.
7. While sustainable development should be a goal for all policies and actions, putting the ideas of sustainable tourism into practice means recognising that in reality there are often limits to what will be achieved in the short and medium term.
8. An understanding of how market economies operate, of the cultures and management procedures of private-sector businesses and of public- and voluntary-sector organisations, and of the values and attitudes of the public is necessary in order to turn good intentions into practical measures.
9. There are frequently conflicts of interest over the use of resources, which means that in practice trade-offs and compromises may be necessary.
10. The balancing of costs and benefits in decisions on different courses of action must extend to considering how much different individuals and groups will gain or lose.

Source: Bramwell *et al.* (1998)

Figure 1.1 Principles behind the approach to sustainable tourism management.

This interdependence between the 'environment' in its widest sense, social communities and the tourists who visit them is fundamental to most models of sustainable tourism. The English Tourist Board's guidelines for sustainable tourism, for example, posit a triangular symbiotic relationship between the 'host community', 'place', or the environment and the 'visitor' or tourist.

The dependence of sustainability strategies upon local communities, however, raises questions about the nature, scope and function of the community itself. Community-led sustainable development requires an understanding not just of the relationship between local communities and their environment, but also of the political, economic and cultural tensions within communities. The relationships of 'local' and 'global' communities also needs to be better understood, particularly in the context of tourism.

Communities in conflict

The word 'community' implies a common interest, possession or enjoyment (*Collins Westminster Dictionary* 1966). However, the interests of those living in the local community do not always coincide. Not all local residents benefit equally from, or are equally happy with, tourism development. The literature on the bounding of 'community', conflict versus consensus and questions of inclusion and exclusion is, historically, considerable. The fact that elements of the same community can be in conflict over the aims or outcomes of tourism development (and indeed any other form of development) is often overlooked in the sustainable tourism literature, but is fundamental in raising questions over the validity and ideology of 'community'. The community is often treated as relatively homogeneous, with little internal conflict. Such assumptions of consensus themselves imbue 'community' with implicit ideological underpinnings.

As more studies have been made of the reactions of local communities to tourism, it has become clear that different groups or individuals may benefit or suffer disproportionately from tourism development. This leads to tensions within the community and sometimes open conflict (see, for example, Kamsma and Bras, Chapter 11 this volume). In some cases, however, there may be an 'altruistic surplus' effect, which leads individuals to recognise the communal good derived from tourism, and therefore lessens opposition to tourism development, even among those who may not benefit directly (Faulkner 1998).

In many cases, however, the development of tourism may only serve to highlight existing inequalities and differences in the community. The failure of existing institutions to address problems of inequality and deprivation are leading to the growth of the 'third sector', or grass-roots organisations, NGOs and other associations which operate outside existing formal structures. Such organisations are potentially able to operate in ways which circumvent existing structures, for example by directly linking local and global politics (Wilson 1996). The touchstone of these new alliances is 'empowerment', a concept which is based on the concept of generative, rather than distributive power. Most current power structures are distributive, in the sense that they presuppose a scarcity of resources which must be distributed. The various actors in the system are therefore forced to compete with each other for a share of the 'pie' in a zero-sum game. The generative, or positive-sum view of power, on the other hand, assumes that everyone has power, or skills and capabilities. The aim of individual and group empowerment, therefore, is to combine everyone's power in collective action for the common good. Such developments are essential if slogans such as 'think global, act local', or even 'think local, act global' (Swarbrooke 1996) are to be given some substance.

Although empowerment is a concept which is implicit in most versions of 'sustainable tourism' for example, most models of sustainable tourism assume a distributive form of empowerment to local communities from above, rather than generative empowerment from within. If these 'top-down' models are to be challenged, how is locally generated empowerment to take place? How are local communities to be linked with each other in order to create collaborative action at global level?

Structure of the text

The chapters presented in this book examine a number of different aspects of the concepts of community and sustainability, and the relationships between them, in relation to tourism development.

Part 1 of the book deals with issues of community participation and identity related to tourism. Understanding the way in which the community develops its identity in relationship to the development of tourism, and the stimulation of community participation are seen as being crucial issues which lay the basis for further parts of the book.

In Chapter 2 Bill Bramwell and Angela Sharman identify two analytical frameworks to assist researchers and policy-makers in understanding the approach to sustainable tourism found in the Hope Valley in Northern England. Evaluating both general approaches which a destination may take to tourism as a contribution to sustainable development, and approaches to community participation as one aspect of the sustainable tourism strategies found in destinations, the chapter then employs these frameworks to assess the approach taken to sustainable tourism development in the Hope Valley within the English Peak District National Park.

In Chapter 3 Tamara Rátz investigates the social and cultural impacts and consequences of tourism at Lake Balaton, one of continental Europe's major water bodies within a land-locked state. Through an evaluation of residents' perceptions, it is suggested that locals not only support the industry but are in favour of its expansion. While identifying as positive the employment opportunities afforded by tourism development, there is acknowledgment of the negative local effects of tourism-induced price inflation for goods and services.

Derek Hall's Chapter 4 evaluates the interaction between attempts at sustainable rural tourism development and local community participation within the context of the volatility that is contemporary Albania. The chapter draws attention to the wide range of contextual factors present in the Balkans influencing the nature and appropriateness of 'community' which cannot easily be accommodated within current tourism development models that present a Western-oriented perspective.

The second part of the book examines the relationship between sustainability and community. The chapters in this part highlight the complex interaction of the community with its natural and cultural environments, and the motivations of public and commercial sector actors that intervene in this process.

In Chapter 5 Brian Goodall and Mike Stabler argue that in order to develop tourism sustainably, communities need access to appropriate environmental indicators and performance measures. Current measurement techniques are piecemeal and haphazard, and lack community involvement. The authors examine various techniques for measuring environmental impacts, and argue that there is a need for both top-down and bottom-up approaches. They examine Environmental Performance Standards (EPSs) relating to the operation of individual enterprises and the Environmental Quality Standards (EQSs) which can only be determined at or above the destination level. They argue that bottom-up approaches alone cannot achieve sustainability, especially where impacts spread spatially beyond the destination itself. Control of factors such as air and water pollution will still need to be tackled by international, top-down command and control approaches. Progress with self-regulation has been slow.

The issue of measurement is also taken up in the case study of the Trossachs region of Scotland presented by Alison Caffyn in Chapter 6. Monitoring is an important part of the bottom-up approach adopted by the Tourism Management Programme established for the region. The monitoring was too ambitious, and lacked funding. Problems were encountered with defining the concept of sustainability, which means it is difficult to judge how sustainable the projects are. This limits progress towards sustainability. Caffyn stresses that it is important to incorporate the community into the process and to identify what they see as sustainable.

Frank Howie (Chapter 7) also analyses a bottom-up approach to sustainability in Scotland, this time in the urban setting of Edinburgh. He examines the way in which 'ordinary' areas of the city have been developed for tourism, as part of the search by 'new tourists' for 'authentic' experiences. In a series of case studies he demonstrates that the development of 'grey-area' tourism on the fringe of popular tourist areas may offer opportunities for local communities to become more involved in new forms of tourism development. However, the local community has very often been disenfranchised from the benefits of tourism, and often sees the economic impacts as accruing elsewhere.

Guy Jackson and Nigel Morpeth (Chapter 8) evaluate the effectiveness of the bottom-up approaches to sustainable tourism development envisaged under the Agenda 21 scheme. They conclude that local authorities have so far paid little attention to Agenda 21 in general, and to the need to develop sustainable tourism in particular. Long-term planning for sustainability is currently incompatible with the short-termism characteristic of most local authorities. Allied to lack of resources, the voluntary nature of Agenda 21 and the increasing consumer focus of much policy, has meant that implementation of Agenda 21 has been slow in the UK.

The third part of the book covers the relationship between enterprise and community in tourism development. The role of small entrepreneurs in creating tourism products which are deeply embedded in the social fabric of the community is examined in both urban and rural contexts. The formation of local networks linking the different actors in the tourism system is also considered.

The relationship between developed and developing economies is dealt with by Graeme Evans and Robert Cleverdon in Chapter 9. They argue that the development of 'fair trade' schemes is far more effective than aid in stimulating economic development and ensuring that the benefits of that development reach all sectors of the community. The success of such schemes depends on consumers from the developed world being willing to pay a fair price for the product, and usually depends on smaller, independent tour operators from developed countries developing niche markets for these products. Even if consumers are willing to pay more, however, the distribution of the economic benefits in the destination depends on creating appropriate networks and self-help initiatives to meet the demand for fair trade products.

Some of the obstacles to ensuring a fair distribution of the benefits of tourism development in the community are discussed in two Indonesian case studies. In Chapter 10 Heidi Dahles looks at the role of small entrepreneurs in community development. Growing demand for more flexible and individualised tourism products is causing a policy shift away from large-scale resort development towards stimulation of small-scale entrepreneurial activity. Dahles identifies two types of entrepreneurs: patrons, who control the means of production, and brokers, who act as intermediaries between the patrons and the tourists. Patrons, as owners of homestays and other small businesses, act largely within the formal sector of the economy, whereas the brokers (guides, street vendors, taxi drivers) operate more often in the informal sector. The brokers in particular enhance the flexibility of the tourism production system, and are often innovative. The position of these small entrepreneurs is precarious, however, because of the high level of competition and pressure from the authorities seeking to stamp out 'illegal' activities. Dahles argues that such entrepreneurs are vital to the vibrancy of the tourism product, and need to be utilised and supported rather than harassed by the state.

In Chapter 11 Karin Bras and Theo Kamsma examine the activities of small entrepreneurs on the Indonesian island of Gili Trawangan, where tourists are only just beginning to arrive in significant numbers. They describe how the local population initially responded to the growth of tourism by offering simple accommodation and other services suitable for the predominantly young visitors. The success of tourism has, however, attracted outside investors, who have begun to displace the locals. Again, tourism policies are not supportive of local small-scale entrepreneurs, but supportive of the development of 'quality' tourism based on luxury resorts. This case study indicates that national policy only pays lip service to the stimulation of small-scale entrepreneurship.

In a case study from the Netherlands, Janine Caalders (Chapter 12) argues that the development of networks can be a useful means of stimulating community development. Her study of local networks in rural Friesland shows that a combination of top-down and bottom-up approaches can work, providing there is sufficient integration of regional and local policies, and providing incentives are aimed at networks rather than individual entrepreneurs. Even so, there remain problems in involving the whole community in tourism development, because of differing levels of interest between individuals and groups in the community.

This is an issue also examined by Khoon Koh in Chapter 13, who raises the question why some communities are more successful than others in developing and supporting community entrepreneurship. Analysing two communities in Texas, he finds some empirical support for the hypothesis that entrepreneurship depends on the propensity of the local community to be entrepreneurial and the quality of the investment climate. He argues that local communities need to track the birth and death of tourism enterprises as a guide to policy.

The final part of the book looks specifically at the issues of community development in rural areas. Rural areas in developed and developing countries are facing increasing challenges from economic restructuring and the transformation of agriculture as the productive base. The chapters in this part analyse the way in which rural communities adjust to and resist the commodification of everyday life. The basic question being posed here is: how sustainable is the way of life of rural communities in the face of continued tourism development and marketing?

Jan van der Straaten's analysis of sustainable tourism in rural communities (Chapter 14) attempts to operationalise the concept of sustainable tourism and explain the limitations of current approaches to the issue. He argues that sustainable tourism is such a loosely defined concept that it allows the tourist industry to adopt it for marketing purposes. Rural regions are strongly affected by market forces that stimulate development. Community involvement is crucial to the success of sustainable tourism initiatives. However, marketing is the weak link for local communities. The combination of traditional agriculture and sustainable tourism, often posed as a 'solution' is not always successful, because traditional agriculture cannot support the infrastructure required for tourism.

Jayne Stocks (Chapter 15) examines the process of strategy development for tourism in the *Gaeltacht* – regions of Ireland where Irish is still spoken as a community language. She highlights the main action areas emanating from this process and considers a contemporary example of tourism development. From this, it is concluded that a community approach to tourism development appears to have been in action for the last thirty years and has been strengthened through grant aid which has enhanced the cultural heritage of the area, with economic benefits alleviating some of the problems of endemic out migration of young adults.

In Chapter 16, David Telfer first examines changes in development theory and considers how these changes have been reflected in tourism research and planning. He then goes on to address the question, through an Indonesian case study, of the appropriateness of agritourism as a path to community development. His conclusion is positive, although with reservations, and he points particularly to the need for a closer examination of evolving power relationships within the 'community'.

Joachim Kappert in Chapter 17 analyses community and regional development in northern Portugal and examines the extent to which accelerated growth of the tourism industry, guided by public and private policies, benefits the region. He addresses this question in terms of the kind and number of tourists atttracted and the nature and extent of the tourism infrastructure developed. The chapter also attempts to explain the reasons for shifts in tourism and supplier characteristics, to identify their implications for the region, and to suggest ways in which planning could better benefit rural communities.

In Chapter 18 Elisabeth Kastenholz evaluates the effective demand of summer tourists in rural north and central Portugal, based on a large-scale questionnaire survey. She finds that the market can be divided into four segments or clusters, and advocates policy orientation towards these. The chapter concludes by suggesting that the north and central Portuguese regions tend to attract different types of tourists, and that geographically related marketing policies would appear to be appropriate in targeting market segments.

In Chapter 19 Edwards, Fernandes, Fox and Vaughan explore the potential for the development of a brand image for the Alto Minho region of north-western Portugal through an exploration of visitors' perceptions. They argue that while a clear brand image emerges, the basis of such perceptions could be undermined by the very development of the brand if the underlying key determinant, the socio-economic structures of the local community, is undermined.

The concluding chapter of the volume summarises the issues raised in the book, looking in particular at the critical success factors identified in the case studies of community development. This chapter also examines conceptual frameworks, policy issues and future research agendas.

References

Andersson, A. (1987) *Culture, Creativity and Economic Development in a Regional Context*. Council of Europe Press: Strasbourg.

Beck, U. (1992) *Risk Society: Towards a New Modernity*. London: Sage.

Bell, C. and Newby, H. (1976) 'Communion, communialism, class and community action: the sources of new urban politics' in D. Herbert and R. Johnson (eds) *Social Areas in Cities*, vol. 2. Chichester: Wiley.

Bramwell, W., Henry, I., Jackson, G., Prat, A., Richards, G. and van der Straaten, J. (1998) *Sustainable Tourism Management: Principles and Practice*. Tilburg: Tilburg University Press (2nd edn).

Burgers, J. (1992) *De Uitstad*. Utrecht: Jan van Arkel.

Collins Westminster Dictionary (1966) London: Collins.

ETB (1991) *Tourism and the Environment: Maintaining the Balance*. London: English Tourist Board.

Faulkner, B. (1998) 'Tourism development options in Indonesia and the case of agro-tourism in central Java' in E. Laws, B. Faulkner and G. Moscado (eds)

Embracing and Managing Change in Tourism. London: Routledge, pp. 202–21.
Hillery, G. (1955) Definitions of community – areas of agreement, *Rural Sociology*, 20, 111–23.
Lash, C. and Urry, J. (1994) *Economies of Signs and Space*. London: Sage.
MacCannell, D. (1993) *Empty Meeting Grounds: The Tourist Papers*. London: Routledge.
MacDonald, S. (1997) 'A people's story: heritage, identity and authenticity' in C. Rojek and J. Urry (eds) *Touring Cultures: Transformations of Travel and Theory*. London: Routledge, pp. 155–75.
Maffesoli, M. (1996) *The Time of the Tribes*. London: Sage.
Murphy, P. (1985) *Tourism: a Community Approach*. London: Methuen.
Ray, C. (1998) Culture, intellectual property and territorial rural development. *Sociologia Ruralis*, 38, 3–20.
Richards, G. (2000) 'Cultural tourism' in J. van der Straaten and H. Briassoulis (eds) *Tourism and the Environment*. Dordrecht: Kluwer (2nd edn).
Sassen, S. (1991) *The Global City*. Princeton: Princeton University Press.
Swarbrooke, J. (1996) 'Towards the development of sustainable rural tourism in Eastern Europe' in G. Richards (ed.) *Tourism in Central and Eastern Europe: Educating for Quality*. Tilburg: Tilburg University Press, pp. 137–63.
Taylor, G. (1995) 'The community approach: does it really work?' *Tourism Management*, 17, 487–9.
Urry, J. (1995) *Consuming Places*. London: Routledge.
Wilson, P.A. (1996) 'Empowerment: Community Economic Development from the Inside Out'. *Urban Studies*, 33(4–5), 617–30.

Part 1

Community participation and identity

2 Approaches to sustainable tourism planning and community participation

The case of the Hope Valley

Bill Bramwell and Angela Sharman

Introduction

Today there is wide acceptance that sustainability is one of the most important issues faced by the tourism industry. This acceptance is reflected in the proliferation of publications attempting to define the principles and practice of 'sustainable tourism' and to relate them to the concerns of 'sustainable development'. The term 'sustainable tourism' usually denotes the application of the more general concept of sustainable development to tourism as a specific economic sector. The sustainable development concept was popularised in the late 1980s with the publication of *Our Common Future* by the World Commission on Environment and Development (WCED 1987). Given the importance and complexity of the issues, it is inevitable that there are now many different interpretations of what sustainable development is and what it means in the context of tourism.

This chapter does not add to the already substantial literature on what are regarded as particularly useful approaches to sustainability. It accepts that in practice there are many differing approaches to sustainable tourism and that different policies and practices may be appropriate in different circumstances. On that basis the chapter identifies two analytical frameworks to assist researchers and policy-makers to understand the approach to sustainable tourism found in a particular tourist destination. The first framework identifies some general approaches which a destination may take to tourism as a contribution to sustainable development, and the second examines one specific aspect of approaches to sustainable tourism. According to several commentators, sustainable tourism should enable members of the community living in a tourist destination to participate in decision-making about tourism activity which affects their lives. Hence, the second framework examines issues to consider in assessing approaches to community participation in tourism planning in a destination.

The two analytical frameworks are explained and then related to the circumstances of one tourist destination, the Hope Valley and Edale in Britain's Peak District National Park. The dramatic scenery of this destination consists of two valleys surrounded by high moorland. As a long-established tourist honeypot, it attracts numerous sightseers and outdoor recreationists, with the two valleys estimated to attract about 2.5 million visitors each year (PTP 1994a). The proportion of overnight visitors to day visitors is low, with crude estimates varying between 5 and 30 per cent depending on the time of year (PTP 1994b). This study examines only the three parishes of Castleton, Edale and Hope within the two valleys (Figure 2.1), which have a population of around 2,000. For simplicity the study area is called the Hope Valley. Tourism supports many local businesses and jobs, but there is growing concern among residents about tourism's impacts on the area's physical environment and ways of life. Between 1993 and 1995, thirty representatives of the local community and of outside stakeholder groups collaborated on the Hope Valley Visitor Management Plan Working Group to devise a visitor management plan for Castleton, Edale and Hope parishes, which applied sustainable tourism principles.

Numerous local documents and reports were consulted to develop the analysis. In addition, interviews were conducted with seventeen members of the Hope Valley Visitor Management Plan Working Group, who were members as representatives of: general community interests (7), environmental interests (4), local government (4), recreational interests (2), tourism interests (2) and other interests (2). Several working group members also represented other stakeholder groups, but these additional affiliations are identified above only for four members who mentioned another group in their interview. The interviewee sample represented a cross-section of members, with the interviews conducted between April and June 1995. Each interview was semi-structured, used non-directive questioning techniques, and was tape recorded.

The two analytical frameworks which are outlined are used to assess the approach to sustainable tourism contained in the visitor management plan for the Hope Valley and also the nature of the community participation involved in developing this plan. Issues related to the subsequent implementation of the plan are not evaluated.

Approaches to sustainable tourism

Issues affecting potential approaches

The report *Our Common Future* put forward the view that sustainable development involved not just gaining economic development but also issues of fairness between individuals and groups in today's society and between present and future generations (WCED 1987; Bramwell and Lane 1993). It proposed that fairness between generations involves ensuring that future generations are left with the natural and human-made resources required for them to meet all their likely needs. However, the concept of sustainable development has taken on diverse meanings, with these alternative interpretations reflecting different ethical positions and entailing varying policy objectives and management strategies. For example, Turner (1993)

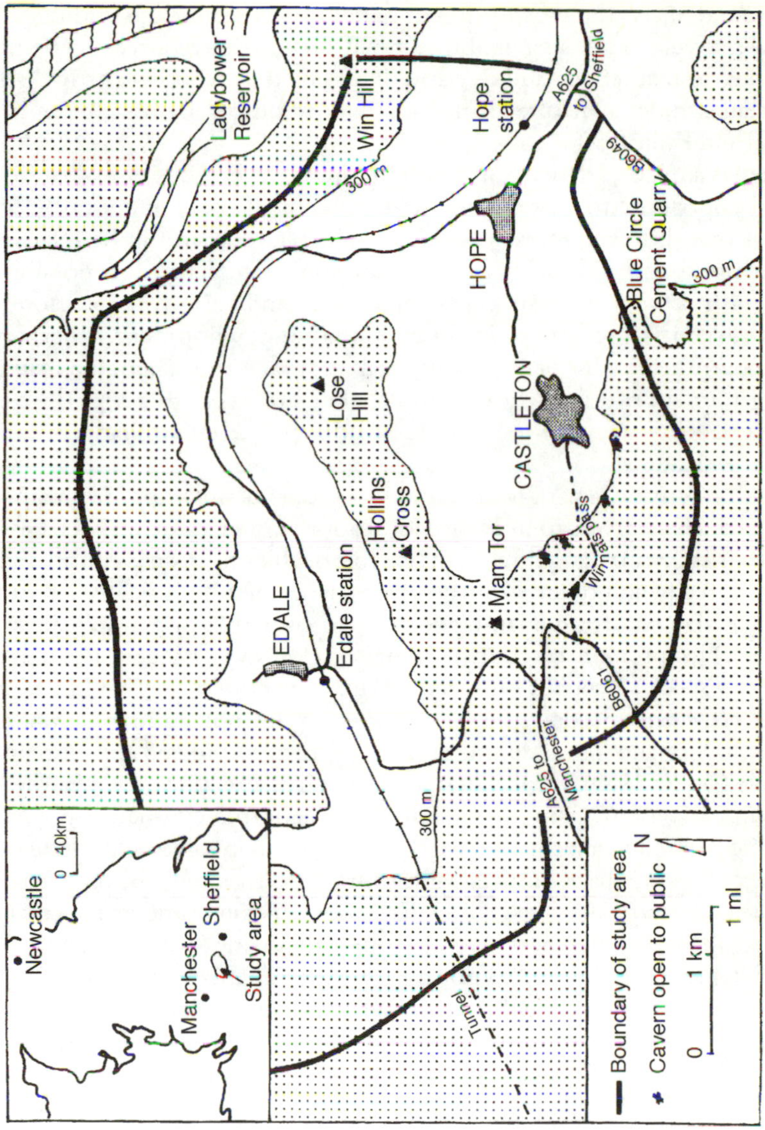

Figure 2.1 The Hope Valley parishes of Castleton, Edale and Hope.

describes four sustainability positions, these being very strong, strong, weak and very weak. A very strong sustainability position regards natural resources as having intrinsic or inherent value, holds that these resources must be conserved whether or not they are likely to provide particular benefit to society, and rejects the substitution of natural by human-made resources. By contrast, a very weak sustainability position allows resources to be used according to market demand, is strongly oriented to economic growth, and allows substitution of natural by human-made resources (Bramwell *et al.* 1996; Garrod and Fyall 1998).

There is similar variation and flexibility in interpretations of how tourism should contribute to sustainable development. Garrod and Fyall (1998) suggest that different approaches to sustainable tourism can be related to Turner's four positions on sustainable development. Hence, it is possible to distinguish between very strong, strong, weak and very weak positions on how tourism is used to assist in sustainable development objectives. Inevitably, there will not be a universal approach to sustainable tourism among tourism planners. For example, they are likely to adopt different approaches to sustainable tourism due to the unique circumstances in each tourist destination.

Mention is made here of four features of tourist destinations, from among others, which are likely to influence local approaches to sustainable tourism. First, tourist destinations differ in the extent to which tourism is considered to depend directly and overtly on the maintenance of a high-quality environment, whether natural, human-made or socio-cultural. For example, in some regions which promote ecotourism it may be clear to all stakeholders that the industry's success depends on careful management of attractive but highly fragile ecosystems. Second, the stage of development of a tourist destination is likely to influence the scale and types of tourism considered appropriate by planners (Butler 1980). Hence, in mass-tourism resorts it may be regarded as economically inefficient or socially unacceptable to abandon earlier investments in tourist facilities and to pull out of an involvement in tourism, and instead the emphasis of sustainable tourism policies could be on efficiency in waste and energy management by large hotels or on traffic management.

Third, there are differences between tourist destinations in the diversity and mix of stakeholders affected by tourism, which can affect the range of views and also the relative difficulty of developing appropriate mechanisms to incorporate divergent views (Williams and Gill 1994). The range of stakeholders may influence the ethical preferences and political processes which underpin the selection of a sustainable tourism strategy. Hence, a fourth and related issue is the political and institutional culture affecting the tourist destination, which may favour some stakeholders over others and may encourage or hinder wide participation in the planning process (Amin and Thrift 1995; Wahab 1997). What sustainable tourism strategy is implemented in a destination also typically meshes political preferences with the area's technical, financial and institutional realities (Bramwell *et al.* 1996).

There are many potential approaches to sustainable tourism which may be adopted by tourist destinations according to their unique circumstances. Hunter (1997: 864) describes 'at least four interpretations' or planning approaches to tourism which are intended to contribute to sustainable development. The typology of approaches has been adapted and summarised in Table 2.1. The four sustainable tourism planning approaches are labelled as tourism imperative, product-led, environment-led, and neotenous tourism. In the table each approach is related to the tourism development stage when it is likely to occur, the primary concern of the approach, the attitude to resource use on which it is based, and whether it represents a weak or strong interpretation of sustainable development. The typology in Table 2.1 will be related in the next section to the approach to sustainable tourism in the Hope Valley (Figure 2.1).

Approach in the Hope Valley

The Hope Valley Management Plan Working Group, with a membership of community residents and interested stakeholders from outside the area, developed the plan between 1993–95. What were the potential broad influences on the plan's approach to visitor management?

First, the plan puts a high value on maintaining the high-quality environment, including the vitality of the tourism industry. It also describes the area's proximity to major population centres, its established status as a popular tourist destination, and its many long-established tourist facilities. The plan contends that the 'combination of accessibility and attractiveness lies at the root of the problems that the area experiences' (PTP 1994b: 3). Several difficulties caused by the high volume of visitors are described. These include traffic congestion and parking problems in the villages, overcrowding at well-known sites, physical erosion of some walking routes, and conflicts caused by outdoor recreational activities such as mountain biking. It is suggested that many of the visitor pressures are concentrated in the summer months, weekends and public holidays. The above circumstances suggest, secondly, that the area has reached a mature or late stage of development in the destination life cycle.

The diversity of stakeholders among the local residents was another influence on the plan's approach. Among the local stakeholders are groups employed in tourism, in agriculture and at a large cement works, as well as many comparatively wealthy retired people and commuters to nearby cities. This economic and social heterogeneity seems to have increased social tensions around tourism issues. It is suggested in the plan that 'There is a widespread feeling that most local people don't benefit from visitors and that visitors are using the countryside for free', and also that 'there is a feeling from those not directly involved in tourism that those that are, reap handsome benefits largely at the expense of the rest of the community' (PTP 1994b: 11). A related issue is that there is 'a lack of understanding

Table 2.1 Four possible approaches to planning for tourism as a contribution to sustainable development

	Approach to tourism as a contribution to sustainable development	*Likely stage of tourism development*	*Primary concern*	*Attitudes to destination resources*	*Interpretation of sustainable development*
1.	Sustainable development through a *tourism imperative*	Early tourism development	Enhancing the development of tourism because it is comparatively more sustainable than other economic activities	Resource loss is acceptable as long as it is less than would otherwise occur and it does not affect the destination's attraction for tourists	Very weak
2.	Sustainable development through *product-led tourism*	Tourism is well developed	Developing new and maintaining existing tourism products because the well-being of local communities might be reduced unacceptably without the wealth generated by tourism	A wide range of concerns about destination resources but as a rule only when these very directly and clearly help to sustain the tourism products	Weak

3.	Sustainable development through *environment-led tourism*	• Perhaps most applicable to early tourism development • Conceivably applicable to developed tourist destinations seeking to reorientate tourism along more sustainable lines	Priority given to the destination's resources by: • promoting types of tourism which overtly rely on maintenance of high-quality environmental/ cultural resources • Establishing strong links between environmental quality and tourism success	• Perhaps a strong product focus but with a priority to conserving destination resources • Opportunities seized to provide support for other locally important economic activities • Opportunities seized to create touristic experiences which highlight resource conservation	Probably moderate/strong
4.	Sustainable development through *neotenous tourism*	Prior to significant tourism development	Active discouragement to tourism in order to protect destination resources	Minimising the use of destination resources	Very strong

Source: Adapted from Hunter (1997).

between tourism interests and the local community even though many enterprises are run by local people and many local people are involved in tourism' (PTP 1994b: 29).

Other broad influences on the plan relate to the political and institutional culture. One issue here is that the original initiative to involve the community in developing a visitor management plan for the Hope Valley came from the Peak Tourism Partnership, a public–private-sector partnership organisation established to promote sustainable tourism in the Peak District (PTP 1993a; PTP 1996). In turn, the Peak Tourism Partnership was established as a national project to demonstrate the application of approaches to sustainable tourism advocated in 1991 by the Government Task Force on Tourism and the Environment (English Tourist Board 1991). The Hope Valley plan discusses the visitor-management proposals in relation to the sustainable tourism principles outlined in the 1991 Task Force report.

A second issue is that the Hope Valley Visitor Management Plan Working Group, which was set up to devise the plan, included representatives from relevant organisations but it remained a separate organisation in its own right. For example, it was separate from the democratically accountable institutions of local government and from agencies such as the Peak Park Joint Planning Board. The plan recognises that getting the plan implemented 'will be almost entirely dependent on securing the interest and commitment of the various statutory bodies and other agencies . . . Without commitment from these agencies there will be no funding' (PTP 1994b: 32). While each practical measure in the plan is costed and related to likely sources of funding and possible lead organisations, it is recognised that getting the organisations to agree the plan even only in principle 'will need careful handling' (PTP 1994b: 33).

Two key observations underpin the approach taken in the visitor-management plan. These are that 'Although the area has long been a magnet for visitors there is increasing concern about the impact of tourism', and that 'there is also a recognition that visitors make an important contribution to the local economy and that people from outside the area have a legitimate right to come and enjoy its beauty' (PTP 1994b: 1). The plan's remit is to provide 'ways in which tourism might be better managed so that it can be accommodated and absorbed without overwhelming and destroying the area' (PTP 1994b: 1). The general aims for this improved management of visitors are 'to reduce the impact of visitors on the physical environment, improve the quality of life and benefits for local residents as well as providing a richer and better experience for the visitor' (PTP 1994b: 1). Related to these three general aims (of capacity management, improving community benefits from tourism, and enhancing the quality of the visitor experience) are the plan's specific objectives, shown in Table 2.2. The practical measures intended to address the three general aims are summarised in Table 2.3.

A key emphasis of the plan is visitor management within capacity constraints. Here the most relevant plan objectives are to prevent increases in visitor numbers on peak days, to reduce the number of cars on peak days,

Table 2.2 General aims and specific objectives of the Hope Valley visitor-management plan

Capacity management

- to prevent any increase in visitor numbers on peak days
- to reduce the number of cars on peak days
- to increase the proportion of visitors using public transport
- to reduce visitor pressure on the most sensitive locations.

Community benefits from tourism

- to maintain and if possible increase the economic return from tourism
- to improve the quality of life for the community and increase the benefits that tourism brings to the local community.

Quality of the visitor experience

- to enhance the visitor experience and increase appreciation of the area's special character.

Source: Adapted from PTP (1994b).

Table 2.3 General aims and related practical measures in the Hope Valley visitor-management plan

Capacity management

- Use car-parking capacity and pricing to control traffic and visitor volumes at peak periods.
- Introduce measures to discourage visitors from taking cars into the heart of the area.
- Reduce through traffic by changes to signing and road classification.
- Traffic calming and speed restrictions.
- Use car-parking revenues to subsidise capacity management and public transport measures.
- Promote access by rail and bus public transport.
- Improve management of visiting school groups which disturb local people.
- Upgrade certain footpaths as alternatives to more pressurised routes.
- Promote more robust recreational areas near cities to divert demand.

Community benefits from tourism

- Use capacity management to reduce tourism pressures on the community.
- Increase accommodation occupancy during the week and off-season.
- Work with the tourism industry to promote sustainable tourism.
- Open up dialogue between local residents and the tourism sector.
- Raise community awareness of the importance of tourism.
- Create mechanisms to allow visitors to make voluntary donations to support conservation work.

Quality of the visitor experience

- Provide information for visitors to increase their appreciation of the area's special qualities.
- Improve signing and information panels to orientate visitors and to put across key messages.
- Develop a new tourist information centre.

Source: Adapted from PTP (1994b).

and to reduce visitor pressure on the most sensitive locations. Among the related practical actions are steps to use car-parking capacity and pricing to control traffic and visitor volumes at peak periods, and to upgrade particular footpaths which can operate as alternatives to more pressured routes. These actions are intended to help reduce or prevent growth in tourist volumes on peak days and at sensitive sites. It could be argued that the plan does not examine issues surrounding the overall tourist-carrying capacity, and the emphasis on restricting tourism at peak times and in particularly sensitive places offers scope for tourism interests to increase tourism in the area at less busy times and in less sensitive places.

The approach to sustainable tourism in the visitor-management plan for the Hope Valley is closest to the third category in Table 2.1, that of sustainable development through environment-led tourism. It is clear that the Hope Valley is a developed tourist destination and that it is seeking to reorientate tourism along more sustainable lines. The plan seeks to strengthen the existing strong linkage between environmental quality and tourism success. The emphasis of the plan is on improving the management of existing types of visitors to the area, who are attracted by the existing tourist facilities. There is a relatively modest number of practical actions in the plan which relate to developing new tourism products, such as those based on increased visitor awareness of environmental and cultural issues. More generally, the plan does not make many proposals for actions to be taken by tourism businesses in the area. In this respect Hunter's (1997) description of the environment-led tourism category perhaps does not reflect the very strong emphasis in the plan on capacity management rather than on tourism product development.

Approaches to community participation in tourism planning

Issues affecting potential approaches

There are many potential benefits if the community living or working in a tourist destination is involved in tourism planning. Importantly, political legitimacy will be enhanced if this involvement means that community members have greater influence in decision-making which affects their lives (Benveniste 1989). Many commentators consider that sustainable tourism should involve taking account of people's views and choices on their present and future needs and welfare and on environmental, economic, social and cultural issues. As well as being politically more acceptable, community involvement in tourism planning will involve stakeholders with different interests and attitudes, and this diverse participation can lead to more consideration being given to tourism's varied economic, environmental and social impacts. Consideration of these many impacts may help to encourage sustainable development, although there is a potential dilemma that some communities may be relatively unconcerned about the long-term ecological or social sustainability of their decisions. Community participation in tourism planning can also build on the store of knowledge, insights and capabilities of the different stakeholders, and the sharing of ideas among these stakeholders can result in a richer understanding of issues and might lead to more innovative policies (Roberts and Bradley 1991).

While community participation in tourist destination planning can offer several advantages, it also presents some difficult challenges. For example, the resource allocations, policy ideas and institutional practices embedded within society may often limit how much influence some individuals and groups will have on the planning process. There are inequalities in the power of different stakeholders in the community and also in the power of local communities within the wider society. Hall (1994: 52) argues that 'power governs the interaction of individuals, organisations and agencies influencing, or trying to influence, the formulation of tourism policy and the manner in which it is implemented'.

Local circumstances will affect which of many potential approaches to community involvement in tourism planning is adopted in a specific tourist destination. Table 2.4 outlines an analytical framework of issues which affect the approach to community participation in tourism planning used in a particular destination. The framework focuses attention on whether and how relevant stakeholders living or working in the destination have a voice in the planning process, are involved in collective learning, and build consensual views across their differences. Within this framework consideration can be given to the extent to which power imbalances between stakeholders are reduced, if at all, within the community participation process. Three sets of issues are highlighted, these being the scope of the participation by the community, the intensity of their participation, and the degree to which consensus emerges among community participants. The framework addresses community participation issues in developing a tourism strategy; it does not consider issues arising from the subsequent implementation of the strategy.

One set of issues to consider relates to the scope of the participation by the community in tourism planning. A key question here is whether the range of participants from the community is representative of all relevant stakeholders. Relevant stakeholder groups include the whole community living in the destination as well as specific groups within it, such as owners of tourism businesses, retired residents and ethnic groups. A stakeholder is taken here to be 'any person, group, or organization that is affected by the causes or consequences of an issue' (Bryson and Crosby 1992: 65). Pertinent considerations include the balance among the participants in the planning process between stakeholders with power and those with little power and also between stakeholders who live in the destination and 'outsider' stakeholders who have an interest in the area but are less directly affected (Yuksel *et al.* 1999). A second key question relating to the scope of participation is the numbers who participate from among the relevant community stakeholders. One influence on the numbers involved will be the use of different

Table 2.4 Some issues affecting approaches to community participation in tourism planning

Set of issues	*Specific issues*
Scope of the participation by the community	• whether the range of participants from the community is representative of all relevant stakeholders • the numbers who participate from among the relevant community stakeholders.
Intensity of the participation by the community	• extent to which all community participants are involved in direct, open and respectful dialogue • how often the relevant community stakeholders are involved • extent to which all participants learn from each other.
Degree to which consensus emerges among community participants	• extent to which community participants reach a consensus about issues and policies • extent to which consensus emerges across the inequalities.

participation techniques, with questionnaires, for example, collecting the general opinions of many individuals and workshops revealing the more nuanced views of a small number of individuals (Ritchie 1985).

A second set of issues relate to the intensity of the participation by the community. Participation is more intense if it involves direct, open and respectful dialogue among the different stakeholders, and if the participants learn from each others' interests and attitudes (Healey 1997; Marien and Pizam 1997). A related consideration is how often the relevant community stakeholders are involved in the planning process (Gunn 1994). Without sustained attention being paid in the planning process to the interests and attitudes of all community participants, the participation may be seen as a token gesture and the views of the most powerful participants may prevail (Joppe 1996).

The third set of considerations concerns the degree to which consensus emerges among community participants in the planning process about the proposed policy direction. Healey (1997; 1998) contends that consensus-building among participants in the planning process should involve learning about each others' different points of view, reflecting on their own point of view, working together to establish a new discourse about the issues and the policy direction, and coming to value and respond to the new policy direction. Prentice (1993: 226) suggests that with community involvement in tourism planning 'there is no guarantee that differences in opinions can be resolved without dissension between beneficiaries and non-beneficiaries'. It may also be reasonable to expect only a 'partial consensus' to develop, with some divergence in views continuing between the participants. Policies resulting from consensus-building may also be rejected by some stakeholders and might reflect a continuing inequality between stakeholder groups (McArthur 1995; Goodwin 1998; Bramwell and Sharman 1999).

Each of the three sets of considerations in Table 2.4 is examined in the next section in relation to the approach to community participation used in developing the visitor-management plan for the Hope Valley.

Approach in the Hope Valley

The first set of issues examined here concerns the scope of the participation by the community in the planning process. More specifically, to what extent was the range of participants in the Hope Valley Visitor Management Plan Working Group representative of all relevant stakeholders living in the area? It was intended that the group of thirty members was representative but not so large it was unwieldy, with the members representing: general community interests (13), environmental interests (5), local government (5), recreational interests (3), tourism interests (3) and other interests (5). Some members had more than one affiliation, but this is identified only when mentioned in an interview. Each of the three parishes was represented by both a parish councillor and at least two residents. Over half of the thirty representatives lived and worked in the Hope Valley, with the others being stakeholders with an interest in tourism in the area but living elsewhere, such as representatives of national agencies or environmental groups. This balance of community and other representatives suggests there was a genuine attempt to involve residents. In the interviews conducted for this research, one parish representative commented that 'the parish council insisted that local representatives from the village were involved, not just elected parish councillors' (Interview 1995d). While two members principally represented tourism interests, and some others earned income from tourism, the working group was not dominated by local tourism industry representatives. However, there was perhaps a relatively small involvement by local shopkeepers or traders and a strong middle-class representation.

The scope of participation by the community was also influenced by the number of stakeholders resident in the Hope Valley who were involved in some way as part of the process to devise the plan. Many types of participation techniques were used and in combination these techniques involved large numbers of residents. For example, 60 people attended a community workshop which focused on 'developing a community agenda for tourism based on local needs rather than tourism demands' (PTP 1993b: 7). Following the workshop, the Peak Tourism Partnership established the working group which led development of the plan. In devising the plan, the working group took account of views contained in 'village appraisal' surveys of Castleton and Edale residents and also in interviews of relevant stakeholders, very many of whom lived in the area (PTP 1994b). Progress in developing the plan was also discussed in a workshop organised at the half-way stage (PTP 1994b).

The parish representatives also explained ideas being developed for the plan at parish council meetings and also at occasional public meetings held in the three parishes, with these representatives reporting back to the working group (Interviews 1995a, 1995b, 1995e, 1995f). Initial proposals for the plan were summarised in a newsletter sent to every local household, which also listed all community and parish council representatives and encouraged people to contact them concerning their views. However, while the consultations involved in devising the plan were quite wide-ranging, there was no use of questionnaire surveys giving all residents opportunities to explain their views on tourism and to respond to proposals. There was also scope for more extensive use of newsletters to report to local people on ideas being developed for the plan.

A second main set of issues here relates to the intensity of the participation by the community in developing the visitor-management plan. Among the seventeen working group members who were interviewed there was some division of opinion about the adequacy of the consultation with people living in the Hope Valley to develop the plan. Eight considered the consultation was generally adequate, six held some mixed views about its adequacy, and three considered it generally inadequate.

The respondents generally considered that dialogue in the working group meetings had been open and respectful. Several working group members commented on the vigorous but useful discussions in their regular meetings. One described how they 'had some heated debates', but that they were constructive and allowed people to express their views openly and discuss them at considerable length (Interview 1995c). Care had been taken to create conditions for open discussion in both the community workshop and working group meetings. For instance, the community workshop was held in a local school as a neutral venue which was central to community activities, and the working group met in a room in a local pub as it was an informal setting which was not intimidating or associated with any one particular organisation (PTP 1993b; Interview 1995c).

There were some difficulties related to how often the relevant community stakeholders were involved. Several respondents concluded that there should have been more consultation with the Castleton community prior to sending out the draft plan for comment and amendment. A tourism industry representative argued that 'people felt not consulted until the very late stages when the parish council representatives, for example, had a document in their hands and could call a village meeting with something concrete and say "What do you think about this?", by which time I think it was too late to start changing and taking people's views into it' (Interview 1995c). However, some working group members considered that such views arose because Castleton shopkeepers and traders objected strongly to certain proposals in the draft plan. One working group member living in Castleton observed that 'it was the traders of the village who were very much up in arms, saying they weren't consulted. And that wasn't true at all. Notices had been put around in our village, for example, and the interest was very small . . . local people are never interested in it until it hits them personally' (Interview 1995d).

Two respondents from the working group suggested that the participants in the meetings were not always prepared to learn from the views expressed by other people. One said that the views of parish representatives were not always taken into account, and decisions appeared to have been made prior to meetings (Interview 1995a). Another complained that the 'only people who were really being consulted were commercial interests', and that at meetings it seemed a 'ruling party caucus' had already decided what would be done (Interview 1995g). A factor here might be that one organisation, the Peak Tourism Partnership, had led the setting up of the working group, had established the framework to develop a visitor-management plan for sustainable tourism, and had provided continuing administrative support to the working group. Nevertheless, this organisation had intended to secure a 'close involvement of the community and the various agencies in the preparation of the plan' and had worked hard to include broad community representation on the working group (PTP 1994b: 1).

The third set of issues to be examined concerns the degree to which consensus emerged among working group participants about the policy direction of the visitor-management plan. It is claimed in the plan that its proposals 'have emerged and evolved as a result of extensive consultation and discussion and we believe that this process has helped build a strong consensus and support for the overall programme' (PTP 1994b: 17). In the interviews with working group members, about half were largely in favour of the plan proposals, with the others expressing some reservations. Typical more enthusiastic responses were that 'they came to a reasonable consensus . . . they have tried to strike a reasonable balance of opinions' (Interviews 1995b, 1995d). One respondent considered that the collaboration 'has been a good thing in my book as it has made people think beyond their narrow sectional interests and at overall management questions . . . now we've actually got a series of proposals I think we can work on, with a large measure of support' (Interview 1995i).

However, some respondents had reservations about the proposal in the plan for on-street car parking charges in Castleton village, which were intended to reduce problems of inconsiderate parking and traffic congestion (PTP 1994b). There were two reasons for their reservations. First, some considered it unfair that this scheme included proposals to charge Castleton residents for on-street parking permits, with some interpreting this charge as yet another cost of tourism to fall on local residents (Interviews 1995a, 1995c, 1995g). Second, some interviewees had regrets that the proposed on-street car parking charges had annoyed shopkeepers and traders in the village (because it might reduce the number of visitors) and had created local ill feeling (Interviews 1995b, 1995c). One respondent stated that 'once the traders suddenly felt that [it] . . . was going to restrict the number of visitors they were absolutely aghast and up-in-arms' (Interview 1995d).

The perceived threat to tourism businesses led to Castleton Chamber of Trade being formed, and this organisation protested strongly about the car parking proposals. As one respondent explained in relation to the working group: 'they had parish councillors – two or three of them on this committee – and I was at meetings where they certainly voiced their opinions very forcibly . . . but still there was a breakaway group from the Chamber of Trade from Castleton' (Interview 1995g).

Some concerns were expressed that the plan would benefit certain stakeholders more than others. One parish councillor commented that 'too much consideration is given to visitors as against residents, particularly in relation to parking', and another community representative suggested that the proposals 'are all in favour of the tourists, they do not seem to help the local population' (Interviews 1995a, 1995f, 1995g). In contrast, a tourism industry member argued that 'there are sections of the plan which, particularly from a tourism economy point of view, need very, very careful consideration, probably even re-consideration' (Interview 1995c). However, several respondents commented that the consultation process had successfully overcome many of their early suspicions, which often related to considerable distrust of one stakeholder organisation, the Peak Park Joint Planning Board, which is the local planning authority (Interviews 1995d, 1995e, 1995g, 1995h).

The spread of approval and reservations about the plan proposals is perhaps unsurprising as the consensus-building involved participants in discussion across significant differences of interest and outlook. Indeed, it might be claimed that it is common still to have tensions and even some dissension after such consensus-building. There was only a 'partial consensus' despite the ambiguity in the plan over the potentially divisive question of the overall tourist capacity of the area, which was discussed earlier. The objectives of reducing tourism pressure at peak times and in the most sensitive places had appeal to community members wanting to restrict tourism development. At the same time, these objectives left open the possibility of increasing tourism at less busy times and in less sensitive places, and this might have been attractive to local tourism interests. Nevertheless, the majority of interviewees did broadly support the visitor-management plan, even when it did not match their preferred outcome.

Conclusion

Bramwell *et al.* (1996) contend that sustainable tourism has seven dimensions, these being environmental, cultural, political, economic, social, managerial and governmental. This complexity and the value-laden nature of the concept mean that approaches to the development and marketing of sustainable tourism will vary between tourist destinations. This discussion also has suggested that the approach adopted will reflect such local circumstances as the stage of development of the tourist destination and its political and institutional culture, which in turn affect what is considered desirable, appropriate and feasible.

To date there has been relatively little examination in the literature of analytical frameworks to understand the sustainable tourism approaches adopted in different tourist areas. Two such frameworks were presented in this chapter and then used to assess the approach to sustainable tourism in the Hope Valley. The first framework identified general approaches which a destination may take to tourism as a contribution to sustainable development. It involved a typology of four general approaches and these are distinguished from each other by their primary concern, their attitude to resource use, the tourism development stage when they are likely to occur, and the relative strength or weakness of their interpretation of sustainable development. Even if the approach taken by individual destinations does not accord with one of the four simplified, 'ideal type' approaches, the variables behind the typology still provide a useful initial analytical tool. The second framework focuses on approaches to community participation as one aspect of the sustainable tourism strategies found in destinations. It considers three specific aspects of participation which are important to the overall participation approach. Further research is needed to evaluate in other tourist destinations these two analytical frameworks of sustainable tourism approaches as well as other relevant frameworks.

Note

Interviews were undertaken on the following dates:

1995a	19 April	1995f	14 June
1995b	5 May	1995g	16 June
1995c	15 May	1995h	26 June
1995d	7 June	1995i	28 June
1995c	9 June		

References

Amin, A. and Thrift, N. (1995) 'Globalisation, institutional "thickness" and the local economy' in P. Healey, S. Cameron, S. Davoudi, S. Graham and A. Madani-Pour (eds) *Managing Cities. The New Urban Context*, Chichester: Wiley, pp. 91–108.

Benveniste, G. (1989) *Mastering the Politics of Planning. Crafting Credible Plans and Policies That Make a Difference*, San Francisco: Jossey-Bass.

Bramwell, B. and Lane, B. (1993) 'Sustainable tourism: an evolving global approach', *Journal of Sustainable Tourism* 1, 1: 1–5.

Bramwell, B., Henry, I., Jackson, G. and van der Straaten, J. (1996) 'A framework for understanding sustainable tourism' in B. Bramwell, I. Henry, G. Jackson, A. G. Prat, G. Richards and J. van der Straaten (eds) *Sustainable Tourism Management: Principles and Practice*, Tilburg: Tilburg University Press, pp. 23–71.

Bramwell, B. and Sharman, A. (1999) 'Collaboration in local tourism policy-making', *Annals of Tourism Research* 26, 2: 392–415.

Bryson, J. M. and Crosby, B. C. (1992) *Leadership for the Common Good: Tackling Public Problems in a Shared-Power World*, San Francisco: Jossey-Bass.

Butler, R. W. (1980) 'The concept of a tourist area cycle of evolution', *Canadian Geographer* 24: 5–12.

English Tourist Board (1991) *Tourism and the Environment. Maintaining the Balance*, London: English Tourist Board.

Garrod, B. and Fyall, A. (1998) 'Beyond the rhetoric of sustainable tourism?' *Tourism Management* 19, 3: 199–212.

Goodwin, M. (1998) 'The governance of rural areas: some emerging research issues and agendas', *Journal of Rural Studies* 14, 1: 5–12.

Gunn, C. A. (1994) *Tourism Planning. Basics, Concepts and Cases*, Washington: Taylor and Francis.

Hall, C. M. (1994) *Tourism and Politics. Policy, Power and Place*, Chichester: Wiley.

Healey, P. (1997) *Collaborative Planning. Shaping Places in Fragmented Societies*, London: Macmillan.

Healey, P. (1998) 'Collaborative planning in a stakeholder society', *Town Planning Review* 69, 1: 1–21.

Hunter, C. (1997) 'Sustainable tourism as an adaptive paradigm', *Annals of Tourism Research* 24, 4: 850–67.

Joppe, M. (1996) 'Sustainable community tourism development revisited', *Tourism Management* 17, 7: 475–9.

McArthur, A. (1995) 'The active involvement of local residents in strategic community partnerships', *Policy and Politics* 23, 1: 61–71.

Marien, C. and Pizam, A. (1997) 'Implementing sustainable tourism development through citizen participation in the planning process' in S. Wahab and J. J. Pigram (eds) *Tourism, Development and Growth. The Challenge of Sustainability*, London: Routledge, pp. 164–78.

Prentice, R. (1993) 'Community-driven tourism planning and residents' preferences', *Tourism Management* 14, 3: 218–27.

PTP (Peak Tourism Partnership) (1993a) *Peak Tourism Partnership Newsletter 2, Winter 1993*, Hope: Peak Tourism Partnership.

PTP (Peak Tourism Partnership) (1993b) *Community Involvement in Tourism Management. A Pilot Scheme to Establish a Visitor Management Plan for Castleton, Hope and Edale. Consultancy Report by BDOR Limited*, Hope: Peak Tourism Partnership.

PTP (Peak Tourism Partnership) (1994a) *Castleton-Edale-Hope Local Interpretive Plan. Draft for Consultation*, Hope: Peak Tourism Partnership.

PTP (Peak Tourism Partnership) (1994b) *A Visitor Management Plan for Castleton, Hope and Edale. Consultancy Report by The Tourism Company, JMP Consultants Limited, and Transport for Leisure*, Hope: Peak Tourism Partnership.

PTP (Peak Tourism Partnership) (1996) *Peak Tourism Partnership Final Report, August 1992–October 1995*, Hope: Peak Tourism Partnership.

Ritchie, J. R. B. (1985) 'The nominal group technique: an approach to consensus policy formulation in tourism', *Tourism Management* 6, 2: 82–94.

Roberts, N. C. and Bradley, R. T. (1991) 'Stakeholder collaboration and innovation: a study of public policy initiation at the state level', *Journal of Applied Behavioral Science* 27, 2: 209–27.

Turner, R. K. (1993) 'Sustainability: principles and practice' in R. K. Turner (ed.) *Sustainable Environmental Economics and Management*, Chichester: Wiley, pp. 3–36.

Wahab, S. (1997) 'Sustainable tourism in the developing world' in S. Wahab and J. J. Pigram (eds) *Tourism, Development and Growth. The Challenge of Sustainability*, London: Routledge, pp. 129–46.

WCED (World Commission on Environment and Development) (1987) *Our Common Future*, Oxford: Oxford University Press.

Williams, P. W. and Gill, A. (1994) 'Tourism carrying capacity management issues' in W. F. Theobald (ed.) *Global Tourism. The Next Decade*, Oxford, Butterworth-Heinemann, pp. 174–87.

Yuksel, F., Bramwell, B. and Yuksel, A. (1999) 'Stakeholder interviews and tourism planning at Pamukkale, Turkey', *Tourism Management* 20, 351–60.

3 Residents' perceptions of the socio-cultural impacts of tourism at Lake Balaton, Hungary

Tamara Rátz

Introduction

Apart from obvious and visible effects on the economy and the physical environment, tourism can contribute to social and cultural changes in host societies, including changes in value systems, traditional lifestyles, family relationships, individual behaviour or community structure. In Hungary, most studies on the impacts of tourism have so far been restricted to economic analysis, and the subject of socio-cultural impacts of tourism has been under-researched.

The purpose of this study was to investigate the social and cultural impacts and consequences of tourism at Lake Balaton, in the Keszthely–Hévíz region. Tourism's social and cultural impacts are often difficult to measure, as, to a large extent, they are indirect. Consequently, the study is one of residents' perceptions of the effects of tourism upon their region rather than an attempt to measure the actual effects. To date, as reported in both the tourism research literature and the social impact assessment literature, the lack of accepted methodology prevents exact measurement.

Socio-cultural impacts of tourism

'Tourism is the temporary movement of people to destinations outside their normal places of work and residence, the activities undertaken during their stay in those destinations, and the facilities created to cater to their needs' (Mathieson and Wall 1982: 1). During their stay in the destination, tourists interact with local residents and the outcome of their relationship is changes in the host individuals' and host community's quality of life, value systems, labour division, family relationships, attitudes, behavioural patterns, ceremonies and creative expressions (Fox 1977; Cohen 1984; Pizam and Milman 1984). The larger the cultural and economic difference between tourists and local residents, the more obvious and more significant these changes are (Mathieson and Wall 1982). Changes in the host community's quality of life are influenced by two major factors: the tourist–host relationship and the development of the industry itself.

Tourist–host encounters occur in three main contexts: where the tourist is buying some good or service from the host, where they are in the same place at the same time, and when they meet and share ideas and information (de Kadt 1979). As the last type of encounter is far less common than the first two, tourism often fails in promoting mutual understanding among different nations and stereotypes prevail (Nettekoven 1979; Krippendorf 1987; O'Grady 1990).

The tourist–host relationship is characterised by a number of features: (a) it is transitory, unequal and unbalanced, lacks spontaneity and is limited by spatial and temporal constraints (UNESCO 1976); (b) the tourist usually stays in the destination for a short time, so there is no opportunity to develop the superficial relationship into a more meaningful one (Sutton 1967); (c) the traditional spontaneous hospitality turns into commercial activity (de Kadt 1979; Jafari 1989); (d) tourists are on holiday, served by locals, which results in different attitudes and behaviour (Sutton 1967); and (e) the obvious relative wealth of the tourists often leads to exploitative behaviour on the hosts' side (Nettekoven 1979).

The main impacts of the tourist–host relationship are the demonstration effect, when the hosts' behaviour is modified in order to imitate tourists (Duffield and Long 1981; Crandall 1987; Pearce 1989; Tsartas 1992); the change in language usage in the destination (White 1974; Brougham and Butler 1977; Jeffs and Travis 1989; Wallace 1997); the growth of alcoholism, crime, prostitution and gambling (Young 1973; Graburn 1983; O'Grady 1990) and the transformation (revitalisation or commoditisation) of the material and non-material forms of local culture (UNESCO 1976; Mill 1990; Evans 1994).

Besides the physical presence of tourists and their encounters with local residents, the development of the tourism industry also contributes to changes in the quality of life, social structure and social organisation of local residents. Rapid and intensive tourism development results in different and usually less favourable impacts than organic and small-scale development (de Kadt 1979; Krippendorf 1987; Pearce 1989; Peck and Lepie 1989).

The development of the tourism industry is often credited for generating new employment in the destination (Crandall 1987; Pearce 1989). However, much of this employment is seasonal, unskilled and low-paid (Vaughan and Long 1982; Allcock 1986), and the community's traditional work patterns might be seriously affected, resulting in the abandonment of agricultural occupations (Verbole 1995; Crick 1996). Other significant impacts of tourism development are changes in the size and the demographic characteristics of the host population (Crandall 1987; Jeffs and Travis 1989); alteration of community structure (Duffield and Long 1981; Haukeland 1984); increased mobility of women and young adults (Mason 1990; Kousis 1996); infrastructural development in the destination, increased supply of services, and, consequently, improved quality of life for local residents (Garland 1984; Milman and Pizam 1988; Coccossis 1996).

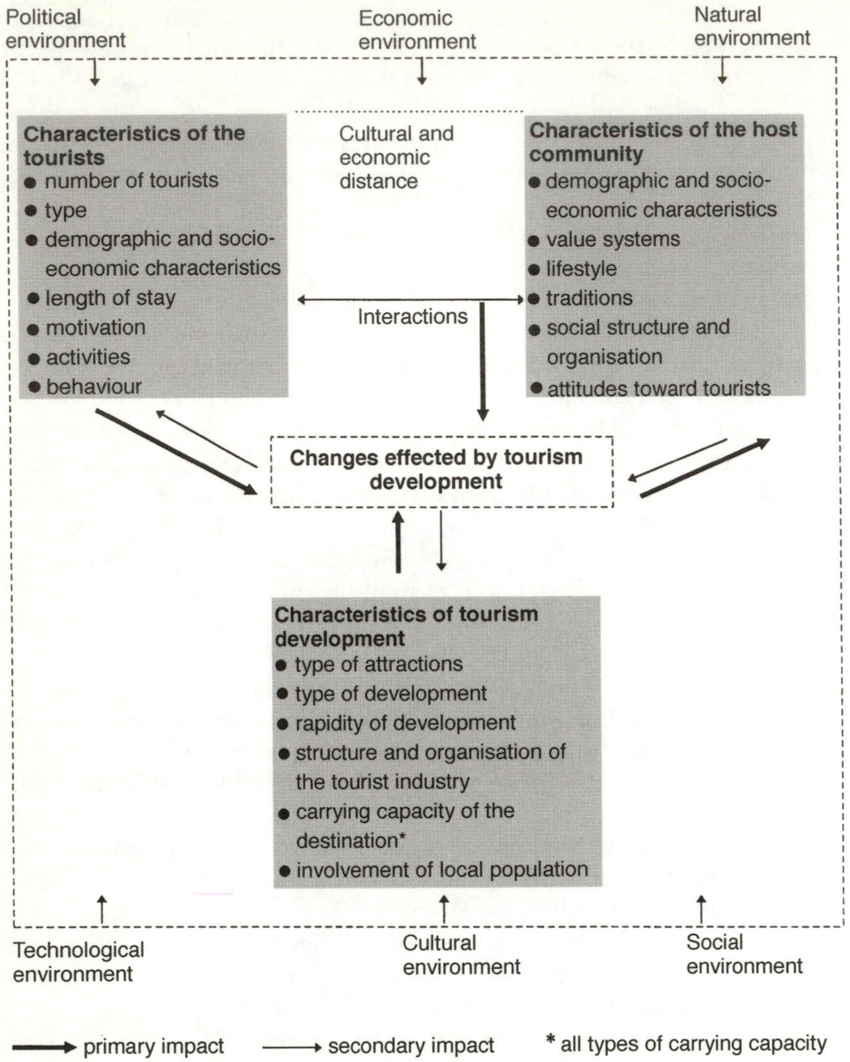

Figure 3.1 Tourism's socio-cultural impacts within the framework of wider social change.

The social and cultural characteristics of the host community are, of course, also continuously influenced by the political, economic, technological, social, cultural and natural aspects of their wider environment (Figure 3.1). The problem of separating tourism's impacts from these influences is unsolved yet (Crick 1996), so this research deals only with impacts perceived by residents as the impacts of tourism development. The main objective of socio-cultural impact analysis is to provide developers, local authorities and all other parties concerned with information on the host communities' perceptions of and attitudes to tourism development in their destination, so that perceived positive impacts could be reinforced and perceived negative impacts could be minimised.

Study area

Lake Balaton, the largest freshwater lake in Central Europe is located in the Western part of Hungary (Figure 3.2). The area is the second most important tourist destination of the country (after Budapest). Tourism is the largest industry at Lake Balaton. In 1994, close to 2.5 million tourists visited the lake. These visitors purchased goods and services worth 60–80 million Ft ($240,000–$320,000) (estimates vary due to the presence of the grey economy). The Keszthely–Hévíz region is located in the western part of Lake Balaton, consisting of five settlements: Keszthely, Hévíz, Gyenesdiás, Vonyarcvashegy and Cserszegtomaj. The total number of inhabitants of the region is approximately 32,300, the number of domestic and international tourists – according to the official statistics – was approximately 154,900 in 1994.

Concerning the importance and potential effects of tourism, all the five settlements of the region are different. Keszthely is a historical university town being less dependent on tourism (the tourist–resident ratio of the town is 2.54 as opposed to 17.9 in Hévíz), Hévíz is a spa resort with little seasonal fluctuation in tourist numbers, Gyenesdiás and Vonyarcvashegy are typical lakeside resorts where most of the residents depend on tourism – and the six-week-long tourist season in July and August – as their major income source. Cserszegtomaj is a hillside village with a slowly developing tourism industry. The area is host to seasonal mass tourism. Although the annual number of arrivals is not high due to the short season, the destination is overcrowded during July and August. The characteristics of tourist inflow meet the criteria of mass tourism defined by Smith (1989). The average length of stay in the area is relatively short (7–8 days), and average expenditure is low (approximately $18/day) (Lengyel 1995).

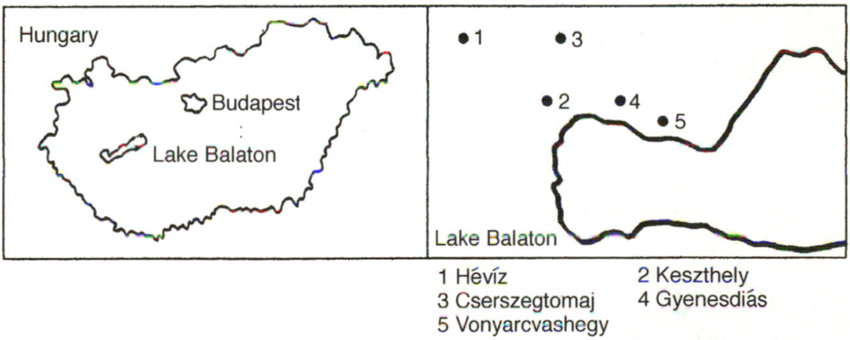

Figure 3.2 The study area.

Methodology

The basis of this study was a survey of residents in the five settlements of the Keszthely–Hévíz region. In order to gather information on residents' perceptions, structured and unstructured personal interviews, both with the inhabitants and the local representatives of the tourist industry, were carried out in the region over a six-week period in summer 1995. The questionnaire in the survey was conducted partly by the author and partly by a student of the Budapest University of Economic Sciences. The unstructured personal interviews with tourist experts and with the representatives of the industry were conducted by the author.

Sample size was set at 220 for the region. The sample size, though relatively small by social science standards, represents a pragmatic compromise between level of precision and cost of data collection. In setting sample sizes for each settlement, both the number of inhabitants and the number of tourists were taken into consideration. The demographic profile of the sample represents the population of the region as reported by local authorities of the settlements. The majority of the respondents (about 77 per cent) have lived in the Keszthely–Hévíz region over ten years.

Overall attitude toward tourism

In general, the residents of the Keszthely–Hévíz region had a positive attitude toward tourism. About 75 per cent of the respondents favoured the presence of tourists in the region (Table 3.1). Since the success of tourism depends very much on the human factor, i.e. the attitudes and behaviour of the residents of a destination towards tourists, this seems to be an encouraging result for the future of tourism development. But we should also take into consideration that almost one-fifth of the respondents oppose the presence of tourists, which is a relatively high proportion. There is no significant difference between residents living in the different settlements ($\chi^2 = 8.22, p = 0.41$).

A majority (about 88 per cent) also felt that the image of the Keszthely–Hévíz region had improved somewhat or significantly as a result of tourism activities (Table 3.2). There is a significant difference between the settlements ($\chi^2 = 27.99, p < 0.05$), the most positive impact being perceived in Vonyarcvashegy and Gyenesdiás, where the tourism industry is developed and economic dependence on tourism is high.

Table 3.1 Feelings about the presence of tourists in the Keszthely–Hévíz region

Support for tourist presence	*n*	%
Favour tourist presence	165	75.0
Neither oppose nor favour	15	6.8
Oppose the presence of tourists	40	18.2

Table 3.2 The impact of tourism on the image of the Keszthely–Hévíz region

Impact	*n*	*%*
1. Significantly worsen	2	1
2. Worsen somewhat	20	2
3. Not make any difference	5	9
4. Improve somewhat	95	43
5. Significantly improve	98	45

Mean: 4.22

Relationship of tourists and residents

Even if the region is overcrowded during the few weeks of the tourist season, 55 per cent of respondents would prefer the number of tourist arrivals to increase to a certain extent, as opposed to only 6.3 per cent who would prefer tourist numbers to decrease (Table 3.3). Compared with the feelings about the presence of the tourists, we can see a contradiction: economic dependence on tourism is so significant in the region that not even those respondents who oppose the presence of tourists would ask for a decrease in numbers.

Tourists in the Keszthely–Hévíz region are mainly Hungarians or from Germany or Austria (i.e. from the same or a similar European culture, with common history and traditional relationships), so the cultural distance between tourists and local residents is not supposed to be large (at least compared to international tourism in Third World destinations). Findings of the survey are in contradiction with this belief: the majority of respondents

Table 3.3 Residents' opinions on the number of tourists coming to their settlement

Opinion	*n*	*%*
1. The number of tourists should increase significantly	49	22.3
2. The number of tourists should increase somewhat	73	33.2
3. The number of tourists is appropriate	84	38.2
4. The number of tourists should decrease somewhat	10	4.5
5. The number of tourists should decrease significantly	4	1.8

Table 3.4 Differences perceived by residents between tourists and themselves

	No. of respondents	*%*	χ^2
Significant difference	76	34.5	
Some difference	96	43.6	
No difference	48	21.8	
$\sum$	220	100	15.45*

Note
* $p < 0.05$

Table 3.5 Types of difference perceived by residents

	Significant difference		*Some difference*		$\sum$		χ^2
	no.	%	no.	%	no.	%	
General behaviour	38	17.3	43	19.5	81	36.8	8.17
Financial situation	49	22.3	55	25.0	104	47.3	11.00
Leisure activities	28	12.7	44	20.0	72	32.7	7.89
Attitude towards the environment	12	5.5	17	7.7	29	13.2	8.30
Other	9	4.1	5	2.3	14	6.4	13.3

Note
220 = 100%

(78 per cent) perceived significant or some difference between themselves and tourists (Table 3.4). Their perceptions are probably partly the consequence of the different social roles of hosts and guests: hosts are stationary, at home, living their everyday life, catering for the needs of visitors, while guests are flexible, mobile, enjoying their leisure time, away from home (Mathieson and Wall 1982). The economic characteristics of tourists are also different from those of the hosts, especially in the case of Western tourists (even if the manifested difference in the destination is not equal to the actual difference).

Responses differ significantly between settlements, the largest differences being perceived in Vonyarcvashegy and Keszthely. Concerning the types of difference, the most significant one occurred in the financial situation of tourists and local residents, which is a natural consequence of the different characteristics of being a tourist on a holiday and being a host at home (Table 3.5). Further differences were perceived in general behaviour and also in the types of leisure activities and in the amount of leisure time spent by various activities. Environmental attitudes do not really differ according to the responses, though similar research carried out in the same region on the environmental impacts of tourism had slightly different conclusions.

Impacts of tourism development

Residents were also asked to express their opinions on the impact of the tourism industry on a variety of areas. Forty-five areas (variables) were evaluated with regard to the impact of tourism (Table 3.6 and Table 3.7).

As can be seen from Table 3.6, employment opportunities, income and standard of living, general infrastructure, quality of life, quality of restaurants, cultural facilities, opportunity for meeting interesting people, opportunity for learning about other nations and language skills were perceived to improve as a result of tourism development in the Keszthely–Hévíz region, and availability of real estate and morality were perceived to somewhat deteriorate following tourism.

Table 3.6 Residents' mean response to tourism's positive effects on the region†

Variable	*Mean*	*Std.dev.*	*S.E.*	χ^2
Employment opportunities	4.61	0.60	0.04	42.58***
Language skills	4.56	0.67	0.05	12.54
Income and standard of living	4.46	0.67	0.04	45.39***
Opportunity for learning more about other nations	4.45	0.56	0.04	5.46
General infrastructure	4.13	0.80	0.05	44.48***
Quality of restaurants	4.12	0.96	0.06	16.07
Opportunity for meeting interesting people	4.04	0.89	0.06	23.91
Quality of life	4.01	0.82	0.05	28.06*
Cultural facilities (theatres, cinemas, museums, etc.)	3.96	0.75	0.05	49.42***
Opportunity for shopping	3.89	1.07	0.07	44.04**
Leisure facilities	3.83	0.94	0.06	37.59**
Tolerance toward difference	3.69	0.98	0.07	16.61
Attitude toward work	3.62	0.97	0.06	28.10
Sports facilities	3.59	0.76	0.05	25.07*
Conservation of old buildings	3.57	1.14	0.07	64.62***
Cultural identity	3.23	1.03	0.07	33.68*
Relationship between generations	3.06	0.87	0.06	18.38
Religion	3.05	0.87	0.06	42.06**
Housing conditions	3.04	1.12	0.07	45.33***
Public security	2.44	1.26	0.08	97.64***
Morality	2.40	0.94	0.06	48.53***
Availability of real estate	2.17	1.02	0.07	21.26
Traffic conditions	1.86	1.12	0.08	65.59***

Notes
† Response range was 1 to 5
1 = Significantly worsen
2 = Worsen somewhat
3 = Not make any difference
4 = Improve somewhat
5 = Significantly improve
* $p < 0.05$ ** $p < 0.01$ *** $p < 0.001$

According to Table 3.7, costs of living, costs of land and housing, general prices for goods and services, a settlement's overall tax revenue, residents' pride in their settlement, hospitality and courtesy toward strangers, residents' concern for material gain, prostitution, gambling, organised crime, individual crime, noise and congestion were perceived to increase due to tourism development in the region, and unemployment and to a lesser extent mutual confidence among people were perceived to decrease.

As to the rest of the variables, their mean (not above 4.0 and not below 2.0) may indicate that the current level of tourism has had relatively less impact on them, although in certain cases (where the standard deviation is relatively high) positive and negative effects cancel each other out. For

Table 3.7 Residents' mean response to tourism's negative effects on the region†

Variable	*Mean*	*Std.dev.*	*S.E.*	χ^2
General prices for goods and services	4.71	0.67	0.04	63.84***
Cost of land and real estate	4.57	0.77	0.05	29.82*
Residents' concern for material gain	4.46	0.69	0.05	16.18
Congestion	4.40	0.82	0.05	87.60***
Settlement's overall tax revenue	4.36	0.89	0.06	61.91***
Noise	4.22	0.75	0.05	80.13***
Organised crime	4.21	0.90	0.06	52.07***
Individual crime	4.15	0.92	0.06	67.30***
Hospitality and courtesy toward strangers	4.12	0.76	0.05	18.42
Residents' pride in their settlement	3.97	0.91	0.06	23.12
Prostitution	3.95	1.03	0.07	101.54***
Gambling	3.83	1.04	0.07	70.73***
Littering	3.83	1.01	0.07	64.78***
Vandalism	3.77	0.82	0.05	55.34***
Politeness and good manners	3.66	0.85	0.06	20.46
Drug abuse	3.56	1.28	0.08	64.41***
Alcoholism	3.53	0.89	0.06	34.03*
Sexual permissiveness	3.51	1.07	0.07	36.55*
Honesty	2.73	1.01	0.07	17.35
Mutual confidence among people	2.73	1.09	0.07	16.36
Unemployment	1.73	0.82	0.05	39.98**

Notes
† Response range was 1 to 5
1 = Significantly decrease
2 = Decrease somewhat
3 = Not make any difference
4 = Increase somewhat
5 = Significantly increase
* $p < 0.05$ ** $p < 0.01$ *** $p < 0.001$

example, in the case of housing conditions, respondents perceived an improvement in interior design in order to meet the requirements of the tourists, and in the quality of new buildings due to increased financial resources, but they also considered that during the main season many families move out of their house to garages or cellars in order to accommodate tourists.

The impact of tourism on the conservation of old buildings is also both positive and negative. Tourists look for and appreciate local architecture and authentic traditional style, so a few buildings have been renovated or conserved as mainly tourist attractions. But, on the other hand, old buildings did not have enough capacity to provide accommodation for the growing number of tourists, so they were demolished to make room for new and large guest houses or family houses with appropriate capacity and supply of tourist facilities.

Table 3.8 Hosts' attitudinal/behavioural responses to tourist activities in the region (*n*/% of population)

	Active	*Passive*
Positive	127/57.7	38/17.4
Negative	10/4.5	30/13.6

Reactions of residents to tourist activities

Table 3.8 shows a matrix representing the attitudes and behaviour of local residents of the region to tourism. The matrix is based on the framework developed by Bjorklund and Philbrick (1972) and applied to tourism (to the tourist–host relationship) by Butler (1980). The two dimensions of analysis are the attitudes of local residents towards tourism (positive–negative) and their behavioural responses (active–passive). Attitudes and reactions of individuals change in time, according to the process of tourism development in an area.

In the Keszthely–Hévíz region, the majority of residents (57 per cent) actively support and promote tourist activities and tourism development in the region, mainly by running their own tourist businesses, but also by other means, such as learning languages in order to be able to communicate with tourists, or singing in a choir which gives concerts to tourists in the summer season. Only 5 per cent actively oppose further tourism development in the region, mainly for environmental reasons. The remaining 6.8 per cent of the population have neither a positive nor negative attitude toward tourism development.

Conclusion

The results of the study show that support for the tourism industry is strong among the local residents of the Keszthely–Hévíz region. Furthermore, residents not only support the current size of the industry, but are also in favour of its expansion. Despite the overall positive attitude toward tourism, local residents also perceived negative changes as consequences of the impacts of the tourism industry on the region. The most strongly perceived positive impact was the improvement of employment opportunities; the most strongly perceived negative one being a general increase in the prices of goods and services.

Although the results of this study are mainly confirmed by similar international studies, to make them applicable to the whole area of Lake Balaton and to other destinations, it would be necessary to carry out a more exhaustive and comprehensive investigation and analysis over a longer period.

References

Allcock, J .B. (1986) 'Yugoslavia's Tourist Trade: Pot of Gold or Pig in a Poke ?' *Annals of Tourism Research* 13, 4: 565–88.

Bjorklund, E. M. and Philbrick, A. K. (1972) 'Spatial Configuration of Mental Process', unpublished paper, Department of Geography, University of Western Ontario.

Brougham, J. E. and Butler, R. W. (1977) *The Social and Cultural Impact of Tourism: A Case Study of Sleat, Isle of Skye*, Edinburgh: Scottish Tourist Board.

Butler, R. W. (1980) 'The Concept of a Tourist Area Cycle of Evolution: Implications for Management of Resources', *Canadian Geographer*, 24, 1: 5–12.

Coccossis, H. (1996) 'Tourism and Sustainability: Perspectives and Implications' in G. K. Priestley, J. A. Edwards and H. Coccossis (eds) *Sustainable Tourism? European Experiences*, Oxon: CAB International, pp. 1–21.

Cohen, E. (1984) 'The Sociology of Tourism: Approaches, Issues and Findings', *Annual Review of Sociology*, 10: 373–92.

Crandall, L. (1987) 'The Social Impact of Tourism on Developing Regions and Its Measurement' in B. J. R. Ritchie and C. R. Goeldner (eds) *Travel, Tourism and Hospitality Research*, New York: John Wiley & Sons, pp. 413–23.

Crick, M. (1996) 'Representations of International Tourism in the Social Sciences: Sun, Sex, Sights, Savings, and Servility' in Y. Apostopoulos *et al.* (eds) *The Sociology of Tourism*, London: Routledge, pp. 15–50.

de Kadt, E. (1979) *Tourism: Passport to Development? Perspectives on the Social and Cultural Effects of Tourism in Developing Countries*, New York: Oxford University Press.

Duffield, B. S. and Long, J. (1981) 'Tourism in the Highlands and Islands of Scotland: Rewards and Conflicts', *Annals of Tourism Research*, 8, 3: 403–31.

Evans, G. (1994) 'Fair Trade: Cultural Tourism and Craft Production in the Third World' in A. V. Seaton (ed.) *Tourism. The State of the Art*, Chichester: John Wiley & Sons, pp. 783–91.

Fox, M. (1977) 'The Social Impact of Tourism: A Challenge to Researchers and Planners' in B. R. Finney and A. Watson (eds) *A New Kind of Sugar: Tourism in the Pacific*, Santa Cruz: Center for South Pacific Studies, University of California, pp. 27–48.

Garland, B. R. (1984) *New Zealand Hosts and Guests. A Study on the Social Impact of Tourism,* Palmerston North: Market Research Centre, Massey University.

Graburn, N. H. H. (1983) 'Tourism and Prostitution', *Annals of Tourism Research*, 10, 3: 437–43.

Haukeland, J. V. (1984) 'Sociocultural Impacts of Tourism in Scandinavia', *Tourism Management*, 5, 3: 207–14.

Jafari, J. (1989) 'Sociocultural Dimensions of Tourism. An English Language Literature Review' in J. Bystrzanowski (ed.) *Tourism as a Factor of Change*, Vienna: The Vienna Centre, pp. 17–60.

Jeffs, S. and Travis, A. (1989) 'Social, Cultural and Linguistic Impact of Tourism in and upon Wales' in J. Bystrzanowski (ed.) *Tourism as a Factor of Change. National Case Studies*, Vienna: The Vienna Centre, pp. 90–114.

Kousis, M. (1996) 'Tourism and the Family in a Rural Cretan Community' in Y. Apostopoulos *et al.* (eds) *The Sociology of Tourism*, London: Routledge, pp. 219–32.

Krippendorf, J. (1987) *The Holiday Makers. Understanding the Impact of Leisure and Travel*, Oxford: Heinemann.

Lengyel, M. (1995) *A Balatoni Turizmus Fejlesztési Koncepciója*, Budapest: KIT Képzõmûvészeti Kiadó.

Mason, P. (1990) *Tourism. Environment and Development Perspectives*, Godalming: World Wide Fund For Nature.

Mathieson, A. and Wall, G. (1982) *Tourism: Economic, Physical and Social Impacts*, New York: John Wiley & Sons.

Mill, R. C. (1990) *Tourism: The International Business*, Englewood Cliffs, NJ: Prentice Hall.

Milman, A. and Pizam, A. (1988) Social Impacts of Tourism on Central Florida, *Annals of Tourism Research*, 15, 2: 191–204.

Nettekoven, L. (1979) 'Mechanisms of Intercultural Interaction' in E. de Kadt *Tourism – Passport to Development? Perspectives on the Social and Cultural Effects of Tourism in Developing Countries*, New York: Oxford University Press, pp. 135–45.

O'Grady, R. (ed.) (1990) *The Challenge of Tourism*, Bangkok: Ecumenical Coalition on Third World Tourism.

Pearce, D. G. (1989) *Tourist Development*, 2nd edn, Harlow: Longman.

Peck, J. G. and Lepie, A. S. (1989) 'Tourism and Development in Three North Carolina Coastal Towns' in V. L. Smith (ed.) *Hosts and Guests: the Anthropology of Tourism*, Philadelphia: University of Pennsylvania Press, 2nd edn, pp. 203–22.

Pizam, A. and Milman, A. (1984) 'The Social Impacts of Tourism', *UNEP Industry and Environment*, 7, 1: 11–14.

Smith, V. L. (ed.) (1989) *Hosts and Guests: The Anthropology of Tourism*, 2nd edn, Philadelphia: University of Pennsylvania Press.

Sutton, W. A. (1967) 'Travel and Understanding: Notes on the Social Structure of Touring', *International Journal of Comparative Sociology*, 8, 2: 218–23.

Tsartas, P. (1992) 'Socioeconomic Impacts of Tourism on Two Greek Islands', *Annals of Tourism Research*, 19, 3: 516–33.

UNESCO (1976) 'The Effects of Tourism on Socio-Cultural Values', *Annals of Tourism Research*, 4, 2: 74–105.

Vaughan, R. and Long, J. (1982) 'Tourism as a Generator of Employment: A Preliminary Appraisal of the Position in Great Britain', *Journal of Travel Research*, 21, 2: 27–31.

Verbole, A. (1995) 'Pros and Cons of Rural Tourism Development: A Discussion on Tourism Impacts and Sustainability, with a Case from Slovenia', Paper presented at the XVIth Congress of the European Society for Rural Sociology, Prague.

Wallace, J. M. T. (1997) *Putting Culture into Sustainable Tourism: Negotiating Tourism at Lake Balaton, Hungary*, Department of Sociology and Anthropology, North Carolina State University.

White, P. E. (1974) *The Social Impact of Tourism on Host Communities: A Study of Language Change in Switzerland*, Oxford: School of Geography, University of Oxford.

Young, G. (1973) *Tourism: Blessing or Blight*, Harmondsworth: Penguin.

4 Identity, community and sustainability

Prospects for rural tourism in Albania

Derek Hall

Introduction

Most existing models of community-based tourism and notions of sustainability are based on the experience of relatively stable societies. This chapter is concerned with approaches to tourism development processes in a country, Albania, which has been experiencing short- to medium-term domestic instability embedded within longer-term Balkan regional instability (Hall and Danta 1996). It assesses the context and prospects for community-based tourism in Albania in the face of the likely mass tourism development which will take place when internal stability is restored.

The chapter first briefly discusses the concept of 'community' in local development processes. It next examines the post-communist context for tourism development in Albania, and explores hypotheses surrounding notions of redundant human capital and tourism being a refuge for those 'dislocated' by systemic change. Within this context the chapter focuses on the role of rural tourism and specifically of community-based approaches to its development. Finally the chapter draws some conclusions both for our understanding of local development processes within conditions of change and instability, and for the wider European context within which a country such as Albania is placed.

Community and development?

Following the United Nations Conference on Environment and Development (UNCED), there emerged global consensus that sustainable development should be based on local-level solutions derived from community initiatives. 'Co-management' – an appropriate sharing of responsibilities for natural resource management between national and local governments, civic organisations and local communities – has become an accepted policy approach of international development agencies. This approach contrasts favourably with earlier concepts of 'development' being driven by state agendas and resource control (Holland 1998).

As a widely accepted criterion of sustainable tourism (e.g. Pigram 1994), 'community participation' has been widely promoted and debated for several reasons. First, local involvement in development processes is likely to assist the formulation of more appropriate decisions and to generate an increase in local motivation. Second, support for environmental conservation and protection measures is likely to be greater (Tourism Concern 1992). Third, as a service industry, tourism requires the goodwill and co-operation of host communities (Simmons 1994). And finally, visitor satisfaction is likely to be greater where 'hosts' support and take pride in their tourism (Cole 1996).

Such an emphasis on local, bottom-up approaches, would appear to derive its legitimacy from an implicit assumption of the cohesion of local 'communities'. 'Community' can embrace notions of spatial contiguity, social interaction, reflexivity, notions of shared aspirations and values. It tends to be employed implicitly in a consensual sense, even when potential conflict and competition, particularly in relation to the power relations involved in decision-making processes, are the reality of place- (and non-place) based social relations (Joppe 1996; Pearce *et al.* 1996). 'Communities' may not be geographically bounded or homogeneous entities, but can be socially differentiated and diverse. Considerations of gender, income, age, class/caste, origins, and other aspects of social identity can be reflected in the holding of conflicting values and resource priorities rather than the predominance of shared interests and aspirations (Leach *et al.* 1997). The consensual model, therefore, as revealed in the case of Albania, may not be appropriate or indeed desirable. Notions of 'participation', especially when linked to 'community', may embrace widely different levels and qualities of involvement at the local level (Pretty 1995).

Transforming Albania

When political change finally came to Albania in 1991, its impact was relatively more profound than in any other country of Central and Eastern Europe (CEE). After being hermetically 'protected' from alien influences for almost half a century, Albanians' realisation of the impoverishment and inadequacies of their country in comparison to its neighbours, motivated large numbers to leave, or attempt to leave, the country. By 1993, up to 400,000 people – about 10 per cent of Albania's total population and some 15 per cent of the country's labour force – had left, many to seek work within the EU (IMF 1994). This experience has been unique in Central and Eastern Europe's post-communist 'transition'. Significantly, however, these émigrés have despatched remittances back to the country estimated to be worth between $300 million and $1 billion per annum, representing Albania's major source of external income after aid, being equivalent to perhaps a third of GDP, and acting as a major economic lifeline for many local areas and families.

Several Albanian border villages and southern towns have been substantially depopulated as a result of both temporary and permanent emigration, the latter particularly undertaken by ethnic Greeks in the south. At the other end of the country, however, an in-migration of Kosovar Albanians, more than 300,000 by 1994 (Anon. 1994), reacting to Serb pressures, has continued to exacerbate rural conditions in those parts of the country, adjacent to Kosovo, which are some of the most impoverished in Albania. This has accelerated a knock-on process of rural to urban and inter-regional migration (Hall 1996). It also contributed to the growth of grey economies involved in circumventing United Nations sanctions against Serbia, ostensibly for the benefit of ethnic Albanians in Kosovo and Montenegro. The availability of such potentially lucrative activities may have also acted to deflect local effort away from food production, and, alongside the growth of organised crime, into marijuana and coca cultivation, helping to nurture local mafia development in the process (Gumbel 1997a, 1997b; Willan 1997).

Some observers had argued that the Albanian post-communist economic and structural transition should have been easier than elsewhere in Central and Eastern Europe because of the poverty of the economy and small scale of the country (Åslund and Sjöberg 1992). A notable upturn in the country's agricultural fortunes, sustained up to the end of 1996, came to be viewed as a reflection of the success of privatising the country's most extensive employment sector (Mancellari *et al.* 1996). Albania began to be seen as a model for comprehensive structural adjustment programmes (Christensen 1994), and despite an uncertain domestic and external Balkan environment, by the end of 1996 the country's economic development was being highlighted by some observers as one of the success stories of the 'transitional' economies, returning apparent growth rates of 11–15 per cent, some of the world's highest (EIU 1996).

However, such indices were based on an economy developing from virtually year zero in 1991–2, and were dependent to a great extent on foreign aid and émigré worker remittances, with foreign direct investment (FDI) remaining relatively small: by March 1996, cumulative FDI had reached only $218 million. Most new employment was being generated by the 'kiosk' economy: the buying and selling of consumer goods and household items usually imported from neighbouring countries and fuelled by expatriate remittances. A growth in services such as restaurants, taxis, repairs and construction was largely the result of small-scale entrepreneurial endeavour. Across south-eastern Europe such activities have been more effective and profitable in urban areas, where most of the beneficiaries of the PHARE-sponsored SME Foundation have operated (Dana 1996). The villages continued to be poorly provisioned, and the status of rural work remained low.

Further, many domestic savings were being channelled into the country's pyramid investment schemes which collapsed during the first months of 1997, bringing domestic chaos in their wake, and emphasising some of the problems of attempting to introduce concepts of sustainability within a context of collective economic and social deprivation. An estimated 70 per cent of all Albanian families had been willing to submit their savings to such schemes for financial returns of up to 50 per cent per month. Such savings had come from several sources, including émigré remittances. But many families had received their flats or houses virtually for nothing following the privatisation of state housing in 1992. This distortion of the perceived value of housing led many to sell their easily gained homes in order to acquire investment cash for the pyramid schemes. The pyramid failures thus left many homeless, placing greater strains on welfare provision and exacerbating the growth of apparently rootless sub-cultures within the country, which in their turn have provided breeding grounds for organised crime.

Loss of the savings has curtailed much potential SME growth in the short to medium term. On the other hand, reflecting the fact that some have clearly benefited from the pyramid scams, the role of organised crime and grey economies has been strengthened, particularly in the south of Albania, by the civil anarchy which followed the pyramids' collapse.

The renewed social, economic and political instability revealed a vacuum in the processes and framework for sustaining civil society whereby alternative social networks, including the resurgence of clan-based loyalties in the north of the country, emerged. Formal financial aid programmes, notably from the World Bank and the European Bank, were suspended as a result of such upheaval, even though estimates from such sources suggested an infusion of $20 million was urgently needed to sustain the country's basic infrastructures and distribution systems which were gradually being improved (Republic of Albania 1996). A vicious circle of instability, lack of investment and a lack of confidence in the future has thus resulted.

Tourism: any port in a storm?

In the early 1990s tourism was identified as a key Albanian development sector, being viewed as (potentially) able to meet requirements of environmental sustainability, employment, investment and foreign exchange generation, infrastructural enhancement and diffusion of economic benefits (Touche Ross and EuroPrincipals Limited 1992). Even today tourism is viewed by many Albanians as a means of attaining the country's salvation. The apparent relative wealth generated by tourism in Greece, Italy and Croatia created a demonstration effect which has conditioned thinking, as evidenced during the parliamentary process to approve the establishment of five tourism priority zones for investment concentration within the country (Ministry of Tourism and EBRD 1993). Political representatives saw the designation of such zones as a short-cut to economic prosperity and many lobbied for their own constituencies to be included in the priority list. The result, as indicated by the then Tourism Minister (Spaho 1993b), was virtually nonsensical, as almost all likely recreational areas within the country were listed as priority investment areas, including those with vulnerable natural environments which the government had earlier acknowledged as needing priority protection (Hall 1994).

The previous half-century experience of relative isolation from external innovation, an emphasis on developing heavy industry, and an education system severely constrained by ideological dogma, produced in Albania an extreme example of the post-communist phenomenon whereby the human capital generated by the old system proved largely inappropriate for the new environment (Vecernik 1992). Such a redundancy of human capital (Szivas and Riley 1997), reflecting the 'dislocation' of people separated from their accumulated education and experience, was explicitly manifested in Albania through mass emigration and continued attempted emigration, domestic upheaval and the growth of organised crime. This dislocation has taken place at a societal level, and has entailed profound psychological as well as political and economic change.

Evidence from Hungary has suggested to Szivas and Riley (1997) that tourism can be a 'refuge' industry under such circumstances in the sense that expanding service industries with perceived low skill requirements become the recipients of those dislocated from other industries that are contracting or have collapsed, even though tourism may be otherwise unpopular as an employment destination because it has low status (Saunders 1981) and apparently low income. In addition to the demonstration effect noted above, however, tourism employment in Albania appears to have been viewed positively for a number of interrelated factors: (a) domestic upheaval has speeded up the acceptance of change and has encouraged opportunism; (b) tourism has a strong association with the informal economy, and (c) the absolute poverty of so many Albanians has profoundly influenced the opportunity costs of new economic activities.

Service industries such as tourism encourage the formation of small businesses (Goffee and Scase 1983) not least because they tend to be labour intensive, large numbers of small-scale enterprises makes it difficult to enforce statutory requirements, and their widespread use of cash money is associated with the operation of informal economic mechanisms. The existence of grey economies may be essential to economic restructuring by providing a training ground for future legitimate enterprise; tourism entrepreneurship can be part of this process. The position of some tourism enterprises in relation to the informal economy might suggest that actual earnings and benefits from (at least some forms of) tourism employment may be significantly higher than official earnings data indicate (Szivas and Riley 1997).

Thus two linked hypotheses suggest: (a) redundancy of human capital in Albania has been experienced on a mass scale with individuals finding themselves irrevocably separated from their education and previous experience; and (b) tourism employment might provide a role of refuge sector, where such a workforce, finding that its human capital is devalued or even redundant, turns to an industry which is perceived to have growth potential and skills that are relatively easy to learn (Szivas and Riley 1997).

Rural tourism: a sustainable, community-based strategy?

As a country of great diversity within an area no larger than Wales or Belgium, Albania possesses a wealth of natural and cultural heritage. In addition to long stretches of sandy beach, there are mountain ranges, lakes, forests, varied and indigenous wildlife, Classical, Byzantine and Ottoman archaeology and architecture, folklore, customs and crafts. All suggest opportunities for developing small-scale niche specialist tourism opportunities in the interior of the country. But tourism infrastructure is severely limited (e.g. see Albania 1998); substantial employment training is required and Albanian culinary arts are in particular need of attention.

Recognition of the vulnerability of the country's natural and cultural heritage resources (Hall 1991, 1992, 1993) was expressed in an early post-communist government commitment to sustainable forms of tourism development (Atkinson and Fisher 1992; Spaho 1992, 1993a). But there have been substantial pressures on government from commercial sources, emanating from Malta, Italy, Greece and Kuwait, amongst others, to respond to open up the Albanian coast to large-scale resort complex development.

Acknowledging that pressures for such coastal developments are likely to be irresistible, and that 'community'-based projects are vital in rural areas to reduce the chances of villages becoming impoverished in their backwash, the sponsorship of a number of small-scale 'sustainable' tourism projects has been attempted by UK (Holland and Plymen 1997), Dutch (Pelgröm 1997) and other organisations in the Albanian countryside, notably in some depopulated southern hillside villages such as Qeparo and in the north of the country near Shkoder. The tourism component in such projects has included plans for abandoned houses to be refurbished in vernacular style using indigenous materials to help stimulate and/or retain local craft skills. Local ownership has been encouraged for the accommodation of small numbers of implicitly high-value, low-impact rural tourists (Fisher *et al.* 1994; Fisher 1996). Local social and economic sustainability has thus been a major priority of such schemes.

The southern project, for example, supported by EU PHARE-LIEN, WWF (Worldwide Fund for Nature, UK) and Voluntary Service Overseas' East European Programme, aimed to help fill the vacuum of initiative which the communists had left, by empowering local people through participatory community development and transferring key economic skills in agriculture, horticulture, tourism, finance and marketing. By helping to create employment opportunities for the resident population, the project also hoped to attract emigrant workers back to the region. In so doing, it sought to ensure that local leaders and NGOs had skills in public participation and that the communities had a focus for information, advice and resources which would provide a solid base for putting in place mechanisms to help develop long-term practical partnerships within the region, the country and Europe.

Encouragement of community involvement and ownership of development schemes, now being more widely advocated for Albania's planning framework (Nientied 1998), has been constrained, however, by the legacy of almost half a century of centralised, top-down civil administration, which afforded local people little real opportunity to experience bottom-up development or genuine opportunities to participate in meaningful local decision-making. Past experience also debased notions of co-operation and communal collaboration: any collective action is likely to be viewed sceptically as a reversion to communism and, in an economy of scarcity, self-interest has tended to dominate thinking. In the Qeparo area it had been hoped that the pursuit of a participatory rural appraisal (PRA) would help to develop village action plans, as PRAs have done in development projects in parts of India and Latin America. But any sense of 'community' and collectivism was found by the project co-ordinators to be critically much less in Albania, with local residents expressing through their individualism basic aims of needing to accumulate as rapidly as possible.

Resistance to co-operation was particularly characteristic of one village where distrust of outsiders and between villagers was a major impediment. This was expressed when collective economic activity, such as setting tourism accommodation standards and pricing, was discussed, which was often viewed by villagers in terms of gains to be had for one part of the community at the expense of others. That unified standards and pricing would benefit the village's long-term tourism development through collective proven reliability and quality was not easily understood. The desire for short-term gain inhibited such a longer-term and collective view (Holland 1998).

In such circumstances, Sztompka (1993) has distinguished between 'hard' institutional and organisational frameworks such as economy, law and administration, and 'soft' intangibles and imponderables such as interpersonal bonds, loyalties, values and networks. Without the latter, accumulation of trust is inhibited and economic development is more difficult. Where trust is weak, as in Albania, people fail to realise the productivity and other gains which can be achieved by effective co-operation.

Sensitive social integration is a vital, if obvious, prerequisite for the sustainability of rural tourism initiatives. Albanians have fallen back on traditional precepts for social relations embodied in the Canon of Lek Dukajin, a compendium of customs concerned with family and clan honour handed down from the Middle Ages and employed by clan elders to regulate the mountains and to control the blood feud which has re-emerged as a major force and constraint on development. Conflict over land and property rights following a less than perfect privatisation process has contributed to the existence of some 2,000 blood feuds, involving as many as 60,000 of the country's three million people (Jolis 1996; Konviser 1997; Holland 1998).

Distrust both of neighbours and of any authority has provided a vacuum into which the mafia has been able to interpose, adding an institutional layer of 'authority' between state and society. Its strength of social structure, stressing small group inclusiveness, avoidance of co-operation outside the group, and intolerance of competition has neatly fitted the dislocated nature of post-communist Albanian society. For community development to attempt to empower local actors when organised crime has superimposed itself on parts of the country, raises questions of identity, participation, consensus and power taking on a local configuration which 'Western' models cannot easily accommodate.

Conclusion

Despite the growth of niche tourism marketing and the attempted 'greening' of many aspects of the tourism industry, global mass coastal tourism continues to grow in absolute terms, and Albania has yet to play a full part in the Mediterranean sun-sand-surf syndrome. The redundant human capital hypotheses might suggest that tourism is a 'natural' employment sector for Albanians. The country has the potential mass appeal of being situated a shorter flying time from the major European tourist generating markets than destinations such as Tunisia, Greece, Turkey and the Canary Islands. However, a mass market role and the capital investment which precedes it, is likely to be allayed until at least stability and the rule of law have firmly re-established themselves within the country and until relations with neighbouring states are improved.

The postmodern tourist quest to find the unspoilt and the authentic gained momentum just as political and economic restructuring took hold in Central and Eastern Europe. Albania offers a potentially valuable addition to the global niche tourism portfolio. Ironically, the country's reputation and representation (Hall 1999) as amongst the poorest in Europe could actually act in its favour, offering a rustic tourist escapist experience. Further, the Albanian diaspora, notably its North American components, represents a potentially lucrative market for those expatriate Albanians wishing to retrace their origins.

But as a means of reducing the negative backwash effects of mass tourism development and as a more appropriate model for the scale and nature of Albania, the likely sustainability of non-mass tourism development within the country would seem currently handicapped by a number of major constraints. Most obviously, a meaningful sense of 'local community' has been all but destroyed through a combination of a half century of abuse of notions of communal co-operation, and the post-communist rise of exclusive, introspective and particularist refuges of clan-based loyalties and mafia association that many of the country's 'dislocated' have resorted to during a period of considerable upheaval.

Second, rural attractions will need to act as a basic resource for tourism organised and sustained through locally owned small enterprises, and to act as a vehicle for integrated rural development. However, post-communist structural adjustment has seen a significant rural to urban population shift

(Hall 1996), and rural living and rural economic pursuits now tend to be viewed as inferior, not least because rural populations were compelled, by administrative and other methods, to remain living in rural areas (Sjöberg 1992, 1994), and there has been a subsequent natural reaction against that imposition. Further, most images of modernisation transmitted to the region have tended to be explicitly urban based.

Third, while commercial promoters may point to the lack of local infrastructure in Albanian villages as representing a pristine, uncommercialised environment having a strong attraction for niche rural tourists, local residents involved in small-scale rural tourism development schemes are likely to have supported them on the understanding that tourism will be the vehicle specifically to help bring change and improvement to their local, and particularly their own, infrastructure. This will include aspirations for better transport access, improvement in the quality and quantity of water supply, electricity, telephone and consumer goods provision. Such 'host' perceptions may not be compatible with, for example, ecotourism marketing.

Fourth, any implicit objective in such development schemes of the empowerment of local women faces a double burden following the half century of general subservience under state socialism, and the underlying male-oriented nature of society. The latter can place gender-based structural obstacles in the way of genuine participation, such that (mass) tourism-induced migration away from rural areas may, ironically, bring new opportunities for individual migrants, particularly young women, which traditional social norms in the home village may stifle. In theory, involvement of women in rural tourism development processes can help to shift the balance of economic power within households (Kinnaird and Hall 1996). But the reality of the consequent domestic implications may well be short-term local tension and conflict rather than the successful addressing of questions of long-term structural inequality. Nonetheless, attempts to rectify this situation are being pursued through a rural development strategy which recognises the need for transitional (1998–2000) and consolidation (2000–2004) phases of development (Ministry of Agriculture 1998).

Our models of 'community'-based processes within sustainable tourism development processes appear weak within contexts of instability, as faced in Albania. Although this chapter has drawn upon concepts of redundant human capital, dislocation and refuge, there appears the contradiction in Albania that those very structural conditions which might encourage indigenous participation in tourism development processes – a dislocated labour force, significant grey economies, orientation towards small-scale, family-based entrepreneurial activity – have emerged within a context of instability which has undermined most attempts to kick-start a tourism industry, whether small-scale rural niche or coastal mass.

References

Albania (1998) *Tourism in Albania*, http://www.albania.co.uk/main/tourism.html

Anon. (1994) 'Keeping out', *The Economist*, 21 May.

Åslund, A. and Sjöberg, Ö. (1992) 'Privatisation and transition to a market economy in Albania', *Communist Economies and Economic Transformation*, 4, 1: 135–50.

Atkinson, R. and Fisher, D. (1992) *Tourism investment in Central and Eastern Europe: structure, trends and environmental implications*, London: East West Environment Ltd.

Christensen, G. (1994) 'When structural adjustment proceeds as prescribed – agricultural sector reforms in Albania', *Food Policy*, 19, 6: 557–60.

Cole, S. (1996), 'Cultural heritage tourism: the villagers' perspective. A case study from Ngada Flores' in Proceedings of the International Conference on Tourism and Heritage Management, *Towards a sustainable future: balancing conservation and development*, Yogyakarta: Gadjah Mada University Press.

Dana, L. P. (1996) 'Albania in the twilight zone – the perseritje model and its impact on small business', *Journal of Small Business Management*, 34, 1: 64–70.

EIU (Economist Intelligence Unit) (1996) *Bulgaria, Albania: country report 4th quarter 1996*, London, EIU.

Fisher, D. (1996) 'Sustainable tourism in southern Albania', *Albanian Life*, 59: 27–9.

Fisher, D., Mati, I. and Whyles, G. (1994) *Ecotourism development in Albania*, St Albans: Ecotourism Ltd/Aulona Sub Tour/Worldwide Fund for Nature UK.

Goffee, R. and Scase, R. (1983) 'Class, entrepreneurship and the service sector: towards a conceptual clarification', *Service Industries Journal*, 3: 146–60.

Gumbel, A. (1997a) 'Albania's export boom in vice and drugs', *The Independent*, 2 December.

Gumbel, A. (1997b) 'The gangster regime we fund', *The Independent*, 14 February.

Hall, D. R. (1991) 'Albania', in D. R. Hall (ed.) *Tourism and economic development in Eastern Europe and the Soviet Union*, London: Belhaven, pp. 259–71.

Hall, D. R. (1992) 'Albania's changing tourism environment', *Journal of Cultural Geography*, 12, 2: 33–41.

Hall, D. R. (1993) 'Albania' in F. W. Carter and D. Turnock (eds) *Environmental Problems in Eastern Europe*, London: Routledge, pp. 7–37.

Hall, D. R. (1994) *Albania and the Albanians*, London: Frances Pinter.

Hall, D. R. (1996) 'Albania: rural development, migration and uncertainty', *GeoJournal*, 38, 2: 185–9.

Hall, D. R. (1999) 'Representation of place: Albania', *Geographical Journal*, 165.

Hall, D. and Danta, D. (eds) (1996) *Reconstructing the Balkans*, Chichester and New York: John Wiley & Sons.

Holland, J. (1998) *Sustainable tourism development: the case of Albania*, Paper, Rural Tourism Management: Sustainable Options, international conference, Ayr: SAC Auchincruive, 9–12 September.

Holland, J. and Plymen, J. (1997) *Report of the PRA Training Workshop and Fieldwork in Himara and Borsh Communes, Albania – Community Based Planning for Sustainable Economic Development*, Godalming: World Wide Fund for Nature UK.

IMF (International Monetary Fund) (1994) *IMF economic reviews no. 5: Albania*, Washington DC: IMF.

Jolis, B. (1996) 'Honour killing makes a comeback', *The Guardian*, 14 August.

Joppe, M. (1996) 'Sustainable community tourism development revisited', *Tourism Management*, 17, 7: 475–9.

Kinnaird, V. and Hall, D. (1996) 'Understanding tourism processes: a gender-aware framework', *Tourism Management*, 19, 2: 95–102.

Konviser, B. (1997) 'Where revenge is a way of life', *The Financial Times*, 8 November.

Leach, M., Mearns, R. and Scoones, I. (eds) (1997) 'Community-based sustainable development: consensus or conflict?', *IDS Bulletin*, 28, 4.

Mancellari, A., Papapanagos, H. and Sanfey, P. (1996) 'Job creation and temporary emigration: the Albanian experience', *Economics of Transition*, 4, 2: 471–90.

Ministry of Agriculture (1998) *The strategy of agricultural development in Albania*, Tirana: Albanian Ministry of Agriculture.

Ministry of Tourism and EBRD (European Bank for Reconstruction and Development) (1993) *Albania: investing in tourism*, Tirana, Albanian Ministry of Tourism.

Nientied, P. (1998) 'The question of town and regional planning in Albania', *Habitat International*, 22, 1: 41–7.

Pearce, P., Moscardo, G. and Ross, G. (1996) *Tourism community relationships*, Oxford: Pergamon.

Pelgröm, H. (1997), *SNV: at home in the South*, The Hague: SNV Netherlands Development Organisation.

Pigram, J. J. (1994) 'Alternative tourism: tourism and sustainable resource management', in V. L. Smith and W. R. Eadington (eds) *Tourism alternatives*, Chichester and New York: John Wiley & Sons, pp. 76–87.

Pretty, J. (1995) 'The many interpretations of participation', *In Focus*, 16: 4–5.

Republic of Albania (1996) *Public investment programme 1996–1998*, Tirana: Council of Ministers Department of Economic Development and Aid Coordination.

Saunders, K. C. (1981) *Social stigma of occupations*, Farnborough: Gower.

Simmons, D. (1994) 'Community participation in tourism planning', *Tourism Management*, 15, 2: 98–108.

Sjöberg, Ö. (1992) 'Underurbanisation and the zero growth hypothesis: diverted migration in Albania', *Geografiska Annaler*, 74B, 1: 3–19.

Sjöberg, Ö. (1994) 'Rural retention in Albania: administrative restrictions on urban-bound migration', *East European Quarterly*, 28, 2: 205–33.

Spaho, E. (1992) 'Tourism: promising contracts', *Albanian Economic Tribune*, 6, 10: 17–19.

Spaho, E. (1993a) Personal interview, Tirana: Deputy Minister of Tourism, 26 March.

Spaho, E. (1993b) Personal interview, Tirana: Minister of Tourism, 10 December.

Szivas, E. and Riley, M. (1997) 'Tourism employment and entrepreneurship in the changing economy of Hungary', in V. Edwards (ed.) *Proceedings of the Third Annual Conference on Central and Eastern Europe in a Global Context*, Chalfont St. Giles: Buckinghamshire College.

Sztompka, P. (1993) 'Civilisational incompetence: the trap of post-communist societies', *Zeitschrift für Soziologie*, 22, 2: 85–95.

Touche Ross and EuroPrincipals Limited (1992), *Albania tourism guidelines*, Tirana and London: The Ministry of Tourism, Government of Albania and the European Bank for Reconstruction and Development.

Tourism Concern (1992) *Beyond the green horizon*, Godalming: Worldwide Fund for Nature UK.

Vecernik, J. (1992) 'Labour force attitudes on the transition to the market: the Czechoslovakia case', *Journal of Public Policy*, 12, 2: 177–94.

Willan, P. (1997) 'Mafia linked to Albania's collapsed pyramids', *The European*, 13 February.

Part 2

Sustainable tourism and the community

5 Environmental standards and performance measurement in tourism destination development

Brian Goodall and Mike Stabler

Introduction

From merely uttering statements of intent, the acceptance of sustainability principles and urging businesses and tourists to adopt good environmental practices, there is a growing recognition that the attainment of sustainable tourism must move forward to the setting up of the requisite organisations and management structures, derivation and implementation of appropriate instruments and introduction of monitoring devices and procedures. However, while there is evidence of a number of initiatives by tourism bodies, trade associations and businesses (World Travel and Tourism Council 1991; Hotel Catering and Institutional Management Association 1991; Dingle 1995), the process of putting in place effective mechanisms and systems has hardly begun, the impression being that action is piecemeal and almost haphazard (Stabler and Goodall 1997). This is unsurprising given the absence of coherent and co-ordinated direction by international bodies, governments and environmental agencies. Furthermore, although the environmental consequences of tourism development for destinations have been recognised for several decades and there has been support for involving host communities in such development decisions, there is little evidence that community participation has become widespread practice or been effective in influencing the nature and scale of development in most tourist destinations.

This chapter, following on from two papers by the authors stemming from the Agenda 21 action programmes concerning, respectively, the awareness of sustainability issues and action by the hospitality sector (Stabler and Goodall 1997) and the principles determining environmental standards (Goodall and Stabler 1997), considers the difficulties of establishing standards and measuring the environmental performance of tourism businesses arising from the problems of specifying appropriate indicators. The issue can be considered at two levels. First, there is the lack of information at the individual business or establishment level as to the target standards which should be set and how, therefore, environmental performance may be measured. Second, at the spatial or aggregate level, especially the destination, there is acute uncertainty as to the impact of tourism activity and development. For example, in the hospitality sector, the installation of state-of-the-art energy-saving technology and waste-management procedures by an hotel can almost immediately have an impact on costs and may increase revenue if the organisation advertises its 'green credentials' and so attracts more eco-conscious clients. However, this may well lead to an increase in the level of aggregate demand and consequently an overall rise in total energy use and waste generation. Furthermore, the larger number of visitors puts unintended and increased pressure on the local infrastructure, services and environment, the effect of which is uncertain, especially with regard to the additional financial burden which is likely to fall not only on the public sector but the private (business) and personal sectors (perhaps receiving no direct benefit from tourism), because of the increased costs of mitigating the adverse effects of tourism, particularly dealing with waste.

Therefore, issues arise, such as what environmental standards or targets should be set, particularly relating to the thresholds of unacceptable levels of environmental damage, what should be quantified to indicate the impact of tourism development and how activity should be monitored? These issues are discussed with appropriate examples to illustrate the problems and to suggest the way forward to introduce enforceable procedures and systems, including guidelines for environmental performance measures at business and destination levels.

The chapter considers first the environmental impact of tourism businesses and establishments in destination areas by tracing the sources of materials and energy inputs and outputs which are respectively used in, and arise from, the delivery of the tourism product. The ramifications of this local activity for the wider environmental context are identified, particularly the consequences for the biosphere of waste emissions and discharges. Then the alternative, but not necessarily exclusive methods for implementing environmental principles are examined, namely 'top-down' and 'bottom-up', in order to link the achievement of improvements in environmental quality and performance by attaining defined standards in each.

Subsequently, attention is turned to establishing what are environmental quality standards or objectives, with some reference being made to tourism industry codes of conduct and practical advice in the form of environmental manuals. How environmental performance can be measured is explained by considering its input, internal management and output dimensions. The problems of evaluating changes in environmental performance are discussed, using an hotel as an illustrative example, emphasis being placed on the difficulties of operationalising quality and performance standards. A number of examples of quantitative and qualitative measures of the responses are given which underline how the evaluation of performance could be effected.

In the concluding section the possible directions of the process of moving from principles to measurement and operationalisation and ongoing periodic monitoring are identified. The feasibility of bottom-up self-regulation by the tourism industry is assessed. The limitations of this are exposed in the light of the divergence between the tourism industry's objectives and the self-perception, identities, and needs of the local community, in conjunction with greater empowerment emerging from the restructuring of local government, and the recognition that a degree of top-down regulation is inevitable.

The context of tourism environmental impact

While it is accepted that much environmental impact of tourism activity is local, for instance noise, visual intrusion, congestion, solid and effluent-based discharges and atmospheric emissions, the wider ramifications of the demand for and supply of the product in destinations is inescapable. It is not only the supply of materials and energy resources, most of which are likely to emanate from outside the immediate area, or even region, but also the distribution of services and the transit of tourists to and from the destination which must be acknowledged. In effect the impact on the biosphere of inputs and outputs in origin, transit and destination areas must be taken into account.

Economics has long recognised the interrelationships between economies, industries, markets and consumers (Richardson 1972) in the concept of input–output analysis which traces the origins and final destinations of productive activities involving the extractive, manufacturing and distributive industries and the use of inputs such as land, capital, raw materials and energy. In the environmental economics field this has been extended to include the contribution of the natural environment as an input and as a recipient of outputs in the productive process (Turner *et al.* 1994), hitherto largely ignored in the traditional approach. In Figure 5.1, the idea of input–output analysis has been simplified and adapted, in the form of a flowchart to emphasise the environmental impact of the delivery of the tourism product, by which is meant the multi-component composition of travel, accommodation, facilities, services and the natural and human-made resource base. The figure shows, working from top to bottom, the resources drawn from the natural environment which are required to produce the tourism goods and services which are ultimately supplied in the destination. These are not restricted to those consumed by the tourists, but also include items such as boilers, refrigerators, telephones, vehicles, etc. which form the capital equipment of suppliers as well as consumables, for instance food and drink, brochures, stationery, electricity, gas and oil. As indicated, a broad distinction is made between materials and energy and the various stages of production are identified, largely reflecting the three industrial sectors – extractive or primary, manufacturing or secondary and distribution/service or tertiary. At each stage discharges and emissions into the biosphere occur, essentially affecting the three main elements, air, land and water.

At the destination level, Figure 5.1 depicts not only inputs and outputs flowing into and out of the supply side but also directly to and out from tourists, for instance, the use of a private car to get from the point of origin

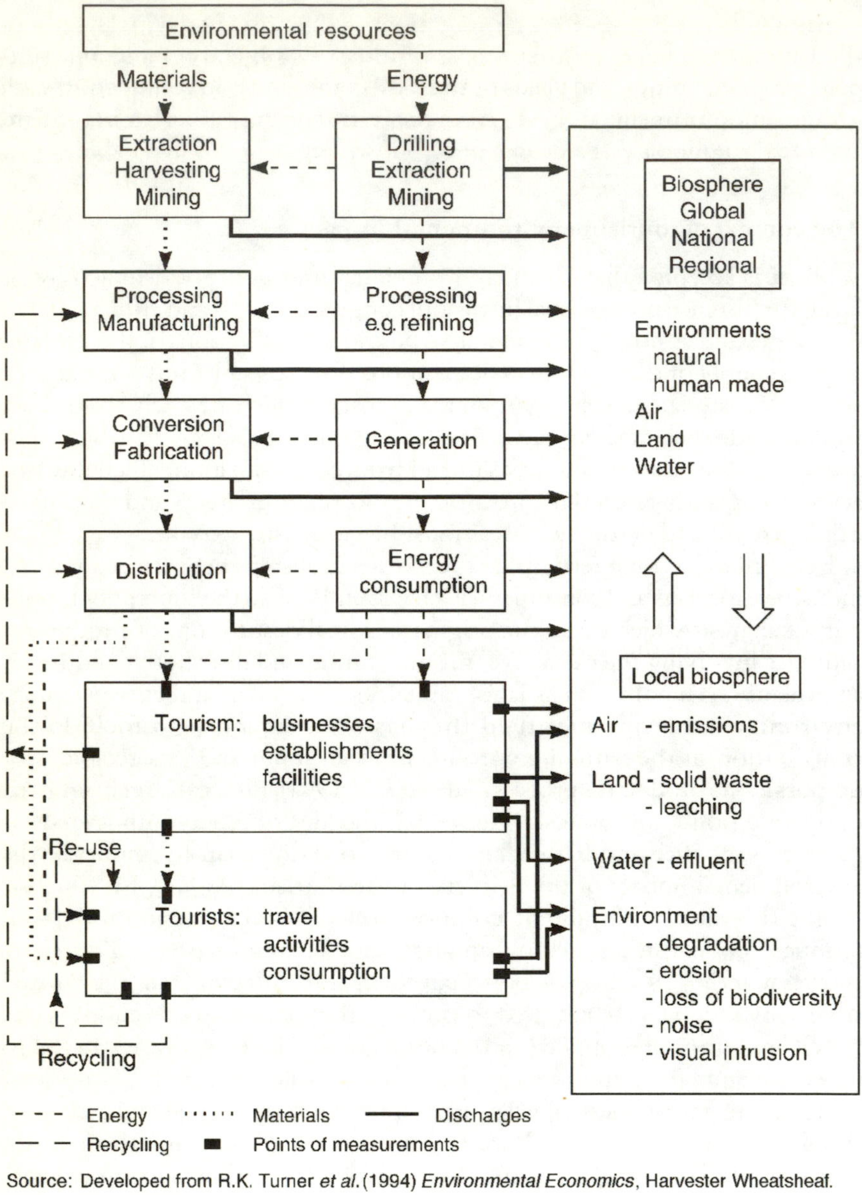

Figure 5.1 Tracing the environmental impact of tourism in destinations.

to the destination. Also included are possible re-use and recycling of resources, the former being distinguished from the latter in not involving transformation or reprocessing. The possible points of measurement of these flows are indicated for the destination only. It should also be noted that there is a linkage between the local biosphere and the wider regional, national or even global one. Thus, for example, CO_2 (carbon dioxide) emissions at the local level contribute to total global warming whereas solid waste disposal might only become a wider problem if it is necessary to ship it out of a destination, for example, an island with no further landfill sites. Others, such as those listed under environment at the bottom right of the figure are local problems, except perhaps for biodiversity whereby loss of rare fauna and flora species may contribute to its overall global diminution. What the figure does not do of course is show the benefits and costs of tourism development or expansion and therefore its impact in monetary terms. To express such effects would necessitate an economic appraisal applying cost–benefit analysis (CBA) to establish the net benefit or cost. This issue is not pursued further here; it has been considered elsewhere by Stabler (1998).

In sum, therefore, Figure 5.1 demonstrates the interrelatedness of destinations with both origin and transit zones and supply industries within a destination itself or elsewhere in the global economy. That is to state that there is not only a horizontal (spatial) linkage but a vertical (industrial sector) one. It is not a one-way relationship for, as suggested, there is a feedback, certainly biospherically, from the destination into the wider environment.

Implications of the nature of environmental inputs and outputs for the implementation of sustainable tourism

Given the pervasive nature of the environmental impacts of tourism activity and the linkages outlined in the previous section, in implementing policies there is a need for 'top-down' and 'bottom-up' approaches. Clearly, for instance, to deal with atmospheric pollution, environmental quality targets for greenhouse gas reduction might have to be set in order to meet often internationally agreed standards for air quality. Nevertheless, this does not preclude initiatives at a local level to set higher standards and targets, such as the improvement of air quality. For example, a distinction can be drawn between rural and urban tourism destinations because the concentration of certain air pollutants, e.g. low-level ozone and lead, is likely to be more severe in densely populated and congested urban areas than in more dispersed rural ones. Thus, a destination municipal authority may take legal powers to enforce compliance with set standards and also encourage action, through informational programmes, and self-regulation by the tourism industry and tourists. It may also decide to constrain tourism reflecting the wishes of the community, a significant proportion of which may be opposed to tourism because of the detrimental effects it suffers.

However, while there is such a connection between top-down and bottom-up methods of implementation, as shown in Figure 5.2, two distinctions need to be drawn. The first relates to Environmental Quality Standards (EQSs) and Environmental Performance Standards (EPSs) and the second is the difference between the destination as a whole and tourism businesses within it. Taking both these matters together, in essence it is only at the destination

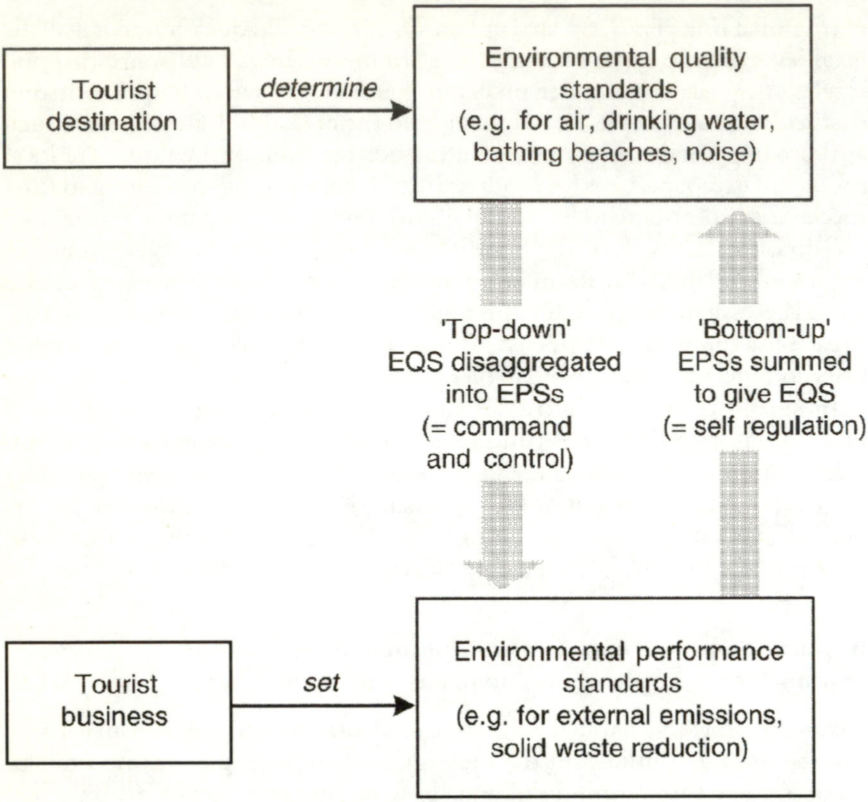

Figure 5.2 Environmental performance and quality – alternative links between a tourist destination and its tourism businesses.

level, or above, that EQSs can be determined. For example, as depicted in the figure, standards derived from environmental scientific research for air, drinking and bathing water and beach quality can be set by international directives, such as in the European Union, or by national and local government as already suggested above. The pro-active destination authority can therefore influence environmental quality from the viewpoint of safeguarding host community interests. On the other hand EPSs are more concerned with the specific or lower level of businesses in terms of the environmental effect of simply operating the enterprise. Thus, a particular business can exercise some control over inputs, for example by purchasing as far as possible materials coming from renewable resources, local sources, or which can be recycled. Similarly, it can endeavour to reduce outputs, such as air emissions, effluent discharges and the generation of solid waste. It can evaluate its performance by measuring the change in the volume of these outputs into the biosphere. Effectively businesses can set their own targets and monitor their attainment. In this sense they are exercising a self-regulatory bottom-up approach to improving environmental quality which can contribute to improving that in the destination and wider afield. This suggests that the top-down and bottom-up methods can be applied and complement each other at the local destination level as well as down from the international and national level and up from the destinations. Issues surrounding these notions are examined in more detail below.

From principles to practice in the tourism industry

Increasing awareness of tourism's interaction with the environment should lead tourist destinations and tourism businesses to behave environmentally responsibly as they recognise the consequences of poor environmental performance for, respectively, their place and corporate images and their commercial success. Environmentally responsible behaviour goes well beyond simple compliance with the law. The tourism industry has a major role to play in providing solutions to the environmental problems stemming from its activities.

Agenda 21 and the tourism industry

Agenda 21, the sustainable development action programme agreed at the 1992 Earth Summit, argues for environmental self-regulation by industry as a priority action for achieving its goals (UNCED 1992). It contains thirty-two provisions targeted at transnational corporations and these provide appropriate benchmarks against which environmental commitment and progress towards environmental improvement can be evaluated.

How does the tourism industry measure up to these provisions? At a general level, comparing the World Travel and Tourism Council's *Environmental Principles* (1991) against the Agenda 21 provisions reveals:

- on corporate environmental responsibility and environmental management systems partial conformity on four out of eight provisions
- on environmentally sound production and consumption patterns partial conformity on three out of eight provisions
- on environmental risk and hazard minimisation partial conformity on two out of ten provisions
- on full cost environmental accounting no mention of the four provisions
- on international environmental support activities no mention of the two provisions (UNCTAD 1996).

Certain provisions, for example those covering risk and hazard minimisation, are most relevant to industries, such as chemicals, which are most destructive of the environment. However, this comparison does suggest, in the case of the tourism industry, that there is an opportunity to review and strengthen the guidance emanating from the 'top' of the industry. Additionally, within the context of developing local Agenda 21 initiatives, major industries and operators, on a 'bottom-up' basis, should be involved in partnership arrangements with local authorities. If businesses are particularly innovative, such initiatives can give the opportunity for destination communities to voice their opinions and participate in the process of agreeing local environmental objectives and targets, systems and procedures.

Codes of conduct in the tourism industry

Translating recognition of the industry's environmental responsibility into action in the case of tourism has produced a multiplicity of codes of conduct (Mason and Mowforth 1995; UNEP/IE 1995). These codes emanate not only from the industry and its sectoral trade associations but also from governments, at national, regional and local levels, and from a variety of non-governmental environmental organisations. They emphasise environmental commitment and delimitation of responsibility, consideration of environmental factors in the planning and development of tourism facilities, and the development of environmentally sound products, operations and processes (Goodall 1997). Codes may be targeted not only at tourism businesses but may also be directed at tourists and destination communities. Such codes are frequently 'statements of ideal' (Mason and Mowforth 1995, 53) or principles, but fail to give practical advice, so badly needed by many tourism businesses and destinations, on how to implement best environmental practice at business unit or site scale.

Environmental manuals in the tourism industry

Lack of practical advice is just beginning to be addressed on a number of fronts. Technical advisory manuals are becoming increasingly available through trade associations, e.g. Hotel Catering and Institutional Management Association (1991), International Association of Antarctica Tour Operators (1993), Canadian Restaurant and Foodservices Association (Wight 1994a), British Holiday and Home Parks Association (Mason and Mowforth 1995), International Hotels Environment Initiative (1993), as well as from public-sector tourism organisations which include destination authorities, in the form of general advice, e.g. Scottish Tourist Board (1993), RDC/ETB/CC (1995), or user-friendly do-it-yourself green audit kits, such as the South Devon Green Tourism Initiative (Dingle 1995) now to be made available nationally (Beioley 1995).

Key aspects of environmental management in business are dealt with in these manuals, such as energy and water conservation, water and air quality, waste minimisation and disposal, noise abatement, environmental considerations in product purchase, land management and local external environment, and use of transport and community issues. The success of individual tourist destinations and tourism businesses in implementing a particular aspect of environmental management is often illustrated by summary case studies. The widespread availability of manuals demonstrates that the knowledge to improve environmental performance exists but it should be stressed that the take-up and implementation within the tourism industry is voluntary. Not surprisingly adoption of environmentally responsible behaviour and, therefore, environmental performance varies markedly between tourism businesses (Goodall 1995; Goodall 1996; Stabler and Goodall 1997).

However, where guidance is given, albeit infrequently, on how to assess current environmental performance and calculate the feasibility of making improvements little or, usually, no indication is given of what constitutes appropriate environmental quality standards (EQS) for destinations and environmental performance standards (EPS) for businesses. Moreover, the problems of measuring environmental performance and the interpretation of the results are almost certainly overlooked.

Operationalising environmental quality and environmental performance standards for tourism

Before the environmental impacts of tourism can be managed effectively they must be measured accurately. The data to effect such measurement – which relates to existing (not proposed) tourism activity – are frequently lacking at the level of both the tourist destination and the individual tourism business. The difficulties of obtaining data should not be underestimated but a clear understanding is required of how tourist destination and tourism business environmental performance should be measured and judged.

In any tourist destination there is a relationship between the overall level of tourist activity, the quality of the environment and the sum of the environmental performances of its individual tourism businesses. Principles underpinning environmentally responsible behaviour have to be expressed as workable environmental objectives or standards against which actual levels of environmental quality and performance can be compared, i.e. definable, preferably quantifiable, targets. This presupposes agreement on a set of appropriate environmental issues for which measures can be established. Environmental quality standards defining some desired state of the destination environment need to be agreed, for example the quality of bathing waters and beach amenity required at a seaside resort. Such EQSs prescribe the level of pollution or nuisance not to be exceeded and should allow for the capacity of the destination environment to assimilate and disperse pollutants and recover from damage. In practice, the determination of these critical loads or threshold limit values which identify safe minimum standards are often problematic because, in many instances, of limited scientific understanding of environmental processes and/or an absence of data (Goodall and Stabler 1997).

Implied in the above reasoning is an absolute EQS but to what extent is human-induced environmental change allowable? All tourism activity has an

Table 5.1 What environmental standards can be set for tourism?

At the destination level	*At the tourism business unit level*
Air quality	Emissions to the atmosphere
Drinking water quality	Water quality
Bathing water quality	Sewage and liquid effluent discharges
Waste arisings and disposal	Wastes generated
Noise levels	Noise levels
Soil quality	–
Derelict land	–
Landscape protection	Visual intrusion
Heritage protection	–
Land covered by urban development	–
Environmentally managed land	Environmental management of grounds
Biodiversity	–
Habitat protection	–
Water use	Water consumption
–	Energy consumption
Local resource use	Other resource use intensity

impact on the environment but change is not necessarily damage. Wight (1994b) poses the question as to what are the 'limits of acceptable change'? Since a value judgement is involved, differing interpretations of this concept exist: ranging from the 'no observed effect level' (NOEL), through the 'no observable adverse effect level' (NOEAL), and the 'lowest observable adverse effect level' (LOEAL), to 'as low as reasonably achievable' (ALARA) (UNEP/GESAMP 1986). Furthermore, it might also be expected that destination EQSs acknowledge concepts of sustainable development/yield and best practicable environmental option (BPEO). Agreeing EQSs at the destination level is therefore not straightforward.

What environmental standards for quality at the destination level and performance at the tourism business unit level can be introduced? Table 5.1 suggests an illustrative range of environmental characteristics for which standards might be set. The environmental standards for the destination must cover as full a range as is feasible of environmental media which, taken together, will influence the destination's environmental capacity for tourism. For the tourism business, the environmental standards relate to all its activities and products, from its acquisition of both materials and energy resource inputs, through its production and service processes, to its generation of wastes and therefore cover 'footprint' impacts (Rees 1992) occurring beyond the tourist destination (see Figure 5.1).

Measurement of environmental performance

The setting and monitoring of environmental standards, at both destination and tourism business unit levels, requires environmental performance to be measured. Such performance measurement must reflect tourism–environment interactions in a systematic way in order to determine the extent to which improvement in environmental performance matches up to the targets or standards set. Moreover, it should be noted that where environmental degradation is severe staged targets may be set *en route* to achieving the ultimate EQS. In addition, such performance measures should also provide early warning of impending environmental problems, for example where performance is lagging and targets are not being met. Also, they should assist identification of areas where environmental targets/standards should be strengthened, for example because scientific knowledge has advanced understanding of the pollution concentration damaging to the environment. Measurement of environmental performance is, however, not only a question of what should be measured but also of how best to measure it.

Much recent research on environmental measurement has focused on so-called 'environmental indicators' targeted at monitoring the state of the environment especially at the national scale in the context of sustainable development policies (MacGillivray 1995; OECD 1996; Department of the Environment 1996). These 'indicators' seek to measure how environmental conditions are changing over time. They are effectively the environmental equivalent of inflation and unemployment rates as measures of an economy's changing state/performance.

The research is mostly based on the pressure–state–response concept (OECD 1993). Thus, tourism exerts pressures on the environment, changing the latter's state in terms of its quality and its stocks of natural resources. Tourist destinations, through policy, and tourism businesses (and tourists), through behavioural actions, respond to these environmental changes thus reducing the pressures caused by tourist activities. Measurement could target the pressures, for example the consumption of resources or output of pollutants; the state of the environment, for example the quality of the environment or the stock and quality of resources; or the responses, for example actions to reduce pollution or resource consumption. A typical case is that of bathing water quality at a seaside resort where pressures could be measured by the volume of untreated sewage effluent released into the sea per unit of time; the state of the environment by the concentration of a pollutant, such as faecal coliforms, in the bathing waters; and the response by investment in actions to treat raw sewage before disposal.

There is considerable debate about whether a particular measure identifies pressure or state of the environment or both (OECD 1996) and whether measures of pressure and response can be separated (Department of the Environment 1996). With respect to tourist destinations it is suggested that pressure and state of the environment measures are particularly important because the quality of that environment (a vital factor in attracting tourists) may be put at risk by certain levels and types of tourist activity. For tourism businesses it may be argued that response measures, and perhaps also pressure ones, are most relevant for assessing the effectiveness of their actions to improve their environmental performance. It is such actions at the level of the individual tourism business which, in aggregate, generate the improvements in environmental performance which result in destination EQSs being achieved.

Concentrating on the tourism business, using an hotel as a particular example, environmental performance measures must cover the full range of its activities, including emissions and wastes, energy and other natural resource consumption, product and process design, supply chain linkages and effectiveness of its environmental management systems. Furthermore, an integral part of this measurement process is the establishment of an environmental baseline to which the hotel's performance can be indexed. In terms of what should be measured certain key performance areas can be distinguished:

- The input dimension is concerned with material inputs (e.g. energy consumption), product purchases (e.g. food and drink), and supply chains (e.g. need for transport) and where, in addition to amounts, questions of renewability, recyclability and sourcing from local suppliers need to be considered.
- The internal dimension is concerned with the processes, procedures and operations in the hotel (e.g. cleaning routines) and the design of its product (e.g. availability of car collection/delivery service to local airport), and including effectiveness of any environmental management systems.
- The output dimension is concerned with the direct environmental impact arising from the hotel's emissions to the atmosphere and water, noise from its operations and wastes requiring disposal.

These key areas may well be interrelated since a decision on the heating, ventilation and air-conditioning system to be used in an hotel has implications for energy consumption and emissions generated. The best measures of environmental performance are quantitative ones but where the environmental issues are not directly quantifiable, qualitative measures can be used. Table 5.2 gives examples for the key performance areas identified above of the quantitative and qualitative measures that might be used in the case of an hotel.

It is no easy matter deciding how best to prescribe an environmental measure. Take energy consumption or efficiency of an hotel as a case in point. In its routine operations the hotel is likely to use a combination of fuels, say electricity for lighting, ventilation and air conditioning, natural gas for cooking and central heating, and diesel/petrol for its motor vehicles. Given the possibilities for substitution, such as gas or electricity for cooking, gas or oil for central heating, fuel composition is as important from an environmental point of view as overall energy consumption because different fuels have different environmental impacts. The hotel's decision to switch

Table 5.2 Quantitative and qualitative measures for evaluating an hotel's environmental performance

Quantitative	*Qualitative*
Input dimension	
Extent of source reduction of inputs	Trading partner/supplier assessment
Phasing out of environmentally damaging inputs	
Energy efficiency	
Use of renewable resources	
Water consumption	
Local sourcing of inputs	
Internal dimension	
Level of environmental investment	Effectiveness of environmental management systems
Internal air quality	Use of range of environmental audits
Internal water quality	Disclosure of environmental performance information
	Training of employees on environmental issues
Output dimension	
Emissions reduction	Support for environmental organisations/ projects
Liquid effluent reduction	
Waste reduction	
Recovery and recycling ratios	
Noise abatement	
Legal compliance: number of breaches of regulations, prosecutions, fines	

between fuel supplies for certain purposes will be influenced by their relative prices and the overall financial costs. The decision does not necessarily reflect all the environmental costs of using the specific fuel concerned, e.g. gas produces less damaging emissions than coal or oil.

Providing the amounts of the different fuels consumed in the hotel in any given period are known, conversion factors exist which allow all forms of energy consumption to be standardised to kilowatt hours (kWh). The hotel's total (annual) energy consumption should then be related to some function of its level of activity. Should this be the size of the hotel? That is, calculate kWh/m^2 of floor space/year. It should be noted that certain energy will be consumed irrespective of its activity level and use of the hotel, e.g. by electronic security systems and keeping unoccupied rooms above frost-stat temperatures in winter. Or, should it be related to occupancy, e.g. by calculating kWh/guest night? Calculation of the latter is further complicated since occupancy rates may be quoted gross, i.e. number of potential bed-days available, or net, i.e. effective number of bed-days actually available.

Alternatively, energy consumption could be related to a measure of business activity, e.g. kWh/unit of revenue (or profit). If comparisons are to be made between hotels it is probably desirable to use a standardised occupancy-related energy consumption figure for agreed categories of hotel size because larger hotels are likely to have more sophisticated and extensive facilities, such as indoor swimming pools and gymnasia, which increase energy consumption. Similar considerations affect other measures of environmental performance, especially the quantifiable ones, and apply equally at destination or business unit level.

Evaluating improvements in environmental performance in tourism

Notwithstanding that the measurement of environmental performance has also provided a starting environmental database for the destination or tourism business there are, nevertheless, problems in interpreting the nature and extent of the improvement in environmental performance over time.

Taking the comparison of the current with the immediate past environmental performance of a destination or tourism business, how obvious is it, in comparing two figures, that environmental performance has improved? Consider the hotel case again where it can be demonstrated that total energy consumption has decreased from one year to the next. The simple explanation of such a decrease could be a fall in the annual occupancy rate since fewer guests require less energy to be consumed. Even where energy consumption is standardised for occupancy, the explanation could be found in seasonal occupancy rate fluctuations, for example if in an English hotel summer months' occupancy increased and winter months' occupancy decreased leading overall to less energy being consumed. Furthermore, energy efficiency would appear to have improved where kWh/guest-night for year two is lower than for year one but, in any instance where the percentage increase in guest-nights is greater than the percentage decrease in kWh/guest-night, total energy consumption will have increased. Attention therefore has to be paid to total as well as 'standardised' energy consumption.

Complications also arise because of variations in weather conditions. Even where there is no change in occupancy rates, total energy consumption in UK hotels is lower for years with mild winters. When comparing an hotel's energy consumption between two years such differences can be allowed for because energy for heating buildings is closely related to 'degree-days', which measure the extent to which mean daily temperatures fall below a base temperature, conventionally set at 15.5°C. Whilst the lower energy consumption arising from such circumstances is beneficial from an environmental viewpoint, it does not represent an improvement in environmental performance stemming from hotel management actions. Isolating those improvements in environmental performance, in this case hotel energy efficiency, following changes to operating practices, such as consolidation of food storage in refrigerators or regular cleaning of light fixtures and investment in capital equipment, for example energy-efficient lamps, roof insulation, or automatic controls of heating and lighting to respond to actual loads, is not straightforward. It is, however, the objective of the evaluation exercise.

Is it sufficient for a destination or an hotel to compare their current performance simply with their past performance? The comparison ought to be extended beyond the particular destination and the individual hotel. For example, even where an hotel has improved its environmental performance over the previous year, that performance may still be inferior to that of a competitor of similar size, location, facilities and occupancy. Best practice within the hospitality industry can be identified via competitive benchmarking (Coker 1996). This would allow an hotel's environmental performance to be measured against its best competitors or leading hotels in the environmental field and indicate feasible environmental targets to set itself.

Should the comparison, in the hotel case, be confined to the hospitality/tourism industry? If superior environmental performance is known to exist in other service industries, for example retailing, then generic benchmarking (Coker 1996) would be able to identify lessons/practices transferable to the hospitality/tourism industry. However, it may be possible to improve on best practice elsewhere. An alternative approach would ask how the hotel's current environmental performance measures up against what is known to be the best possible technical and economically feasible performance. This latter approach follows from accepted pollution control policy and is fully consistent with the 'polluter pays' principle. It is based on best available technology (BAT), but commonly incorporated into environmental protection legislation as best available technology not entailing excessive costs (BATNEEC) (Department of the Environment 1990), and allows for the up-grading of required environmental performance standards from time to time in line with technological advance. Normally BAT applies, although the presumption can be modified where it can be clearly demonstrated that the costs of applying it would be excessive compared with the environmental improvement/protection achieved. Certainly it can be argued that tourism businesses should be aware of environmental achievements and possibilities beyond their immediate industry.

Similar problems in evaluating changes in environmental performance could also be detailed at the destination level. Consider the case of bathing water quality at a seaside resort if, after several years failing to meet an EQS, such as that set under the European Commission's *Bathing Water Quality Directive*, the resort's bathing waters now comply with the EQS, does this represent a real improvement in environmental quality? Where there has been investment in capital schemes for sewage treatment, the considerable cost often at least partially being borne by the local community, the quality of the waters for bathers will certainly have improved. In the absence of such investment, the change is more likely to be a function of the sampling frequency and number, which vary between EU member states and may not always be complied with (National Rivers Authority 1995), and the inherent variability of data collected over the bathing season. Compliance with the EQS is assessed as 95 per cent of samples meeting the mandatory coliform standards but there are other parameters, for example, physico-chemical conditions and presence of heavy metals, which are also significant for human health. Indeed, it is the perceived risk to human health that dominates the quality assessment in this case rather than wider considerations of environmental quality affecting marine ecosystems.

Conclusion

The foregoing discussion suggests that to deliver environmental objectives, at destination or tourism business level, effective, committed policy-makers and managers are needed. This is emphasized by Welford and Gouldson (1993, 76–7) in the case of businesses in general, but can equally be applied to the public and voluntary sectors locally and at higher levels. 'Companies and managers who take the environment seriously change not only their processes and products but also their organisation. The ability to do this effectively, profitably and in an environmentally friendly way depends on the qualities of management itself and the effectiveness of systems in place.' Environmental performance measures therefore need to be established which are:

- representative of a destination's and/or tourism business's environmental concerns and consistent with their environmental policy objectives
- scientifically valid and transparent, focused on something that is measurable, preferably (but not exclusively) in quantitative terms
- understandable and appropriate, resonant and meaningful, i.e. straightforward to interpret and measuring an aspect of destination activity and/or tourism business behaviour thought to be significant
- available with only a short time-lag and capable of regular up-dating in order to show trends over time, give early warning of negative signs, and sensitive to destination development and/or tourism business changes
- based on readily available data of known quality (or data obtainable at reasonable cost)
- allow comparability by providing a target or guideline against which actual performance can be measured (based on Department of the Environment 1996).

Notwithstanding the impression given by the presence in major resorts of large international chains, especially in the accommodation, fast food, transport and, to an extent, retail sector, the tourism industry in destinations is still largely a fragmented one with many small independent businesses making decisions and operating on the basis of their own perceptions of trading conditions and prospects. Despite the setting up of local tourism authorities and associations and some initiatives on environmental action referred to above, there is not necessarily any cohesive and commonly agreed strategy on tourism development or protection of the natural and human-made resource base. Thus, the likelihood of effective and concerted action is limited, as the authors have demonstrated in previous research (Goodall 1995; Goodall and Stabler 1997 and Stabler and Goodall 1997). Therefore, the prognosis for bottom-up self-regulation by the industry itself is a gloomy one. However, in making the recommendations above one must be mindful that tourism businesses do not exist in isolation from the broader economic, political, social and cultural structure of the destination. Communities in destinations have their own identities, objectives and needs which do not necessarily coincide with those of the tourism industry, despite, very often, a heavy reliance on it for the local economy to remain viable in terms of income and employment. Accordingly, the potential exists for the community to counter the possibly detrimental impacts of what can be widely regarded as unwanted external pressures and to take control of their own destiny. This is a particularly important issue where tourism development or expansion is contemplated or in its early stages. In this respect bottom-up approaches to reduce the impact on and protect the environment, defined in its widest sense to embrace social and cultural aspects, as well as the natural environment, are not confined to tourism industry actions. Indeed, even with the more direct impacts, such as waste generation, emissions and discharges, on which attention has been concentrated in this chapter, local communities may have their own agendas and thus set standards and targets which accord with what is acceptable to them. For example, at seaside resorts, political action might be taken to promote land-based sewage treatment plants rather than continue to allow untreated discharges into the sea. Via participation in the land-use planning system local residents might influence building design, traffic-control schemes and noise-abatement schemes, to meet their needs rather than those of the tourism industry.

There is growing evidence of such community action in these fields and, in recent years, in the natural resources field also where there has been a perceptible movement towards local community control. For example, with regard to the natural environment, Shackley (1996) and Sinclair and Stabler (1997) offer a number of cases of wildlife programmes where local communities have shared in or assumed responsibility. Similarly in a more general tourism context, Wilson (1997), Cole (1997) and Hamzah (1997) cautiously indicate how local inhabitants can influence sustainability objectives. Community involvement with and control of natural resources is demonstrated in the literature which increasingly recognises its value, for instance the advent of a new journal entitled *Human Dimensions of Natural Resources*. Furthermore, as in the European Union, local devolvement of powers is acknowledged through the confirmation of subsidiarity. This move, which primarily concerns economic and political matters, should necessarily encompass environmental issues as well.

Notwithstanding these trends, the extent of bottom-up approaches to attaining sustainable tourism cannot alone achieve it, whether the initiatives arise from the tourism industry or the local community or both, acting in a concerted way. For certain environmental actions, especially those where the spatial impact is displaced or overspilled beyond the tourist destination, the approach must be a holistic top-down one insofar as international bodies and national governments and agencies determine the framework within which bottom-up methods operate. Thus, for instance, the decision as to who is responsible for mitigating adverse environmental effects (polluter pays principle), setting of safe minimum standards and targets, enactment of the requisite instruments and the determination of systems and procedures for pervasive environmental impacts, such as air emissions exacerbating global warming, the discharge of heavy metals and radioactive substances and degradation of internationally important resources, must rest at the highest level. Moreover, it is only at this level that control can be exercised over the aggregate impact.

This interpretation of the current position suggests that while the flexibility of a bottom-up approach might be possible, to promote sustainability at the local level for certain forms of environmental impacts, for example noise, visual intrusion, congestion, land-use changes, ultimately top-down, command-and-control methods must be exercised. Certainly, it is becoming increasingly apparent that the tourism industry will be required to comply with more stringently defined environmental quality and performance standards. Progress on a voluntary basis has been too slow and uncoordinated to be effective. While compliance with the environmental standards and performance discussed in this chapter will not be any more painful for the tourism industry than for any other economic activity, such as manufacturing, the emerging issue of its free use of the natural and human-made environment, which constitutes its resource base, is much more contentious. If the industry is compelled to pay for its use of such resources the ramifications for it and tourists will be dramatic giving rise to far-reaching effects on holiday costs and prices.

References

Beioley, S. (1995) 'Green tourism – soft or sustainable?', *Insights* 7, B75–89.

Coker, C. (1996) 'Benchmarking and beyond', *Insights*, A139–44. London: English Tourist Board.

Cole, S. (1997) 'Anthropology, local communities and sustainable tourism development' in M.J. Stabler (ed.) *Tourism and Sustainability: Principles to Practice*. Wallingford: CAB International.

Department of the Environment (1990) *A Guide to the Environmental Protection Act 1990*. London: HMSO.

Department of the Environment (1996) *Indicators of Sustainable Development for the United Kingdom*. London: Government Statistical Service, HMSO.

Dingle, P. A. J. M. (1995) 'Practical green business', *Insights*, C35–45.
Goodall, B. (1995) 'Environmental awareness and management response in the hospitality industry', *Revista Portuguesa de Gestao*, II/III, 35–45.
Goodall, B. (1996) 'The limitations of environmental self-regulation'. Paper presented at the *International Conference on Integrating Economic and Environmental Planning in Islands and Small States*, Valletta, Malta, 14–16 March 1996.
Goodall, B. (1997) 'The role of environmental self-regulation within the tourism industry in promoting sustainable development' in W. Hein (ed.) *Tourism and Sustainable Development*. Hamburg: German Overseas Institute, 271–93.
Goodall, B. and Stabler, M. J. (1997) 'Principles influencing the determination of environmental standards for sustainable tourism' in M. J. Stabler (ed.) *Tourism and Sustainability: Principles to Practice*. Wallingford: CAB International.
Hamzah, A. (1997) 'Evolution of small-scale tourism in Malaysia: Problems, opportunities, and implications for sustainability' in M. J. Stabler (ed.) *Tourism and Sustainability: Principles to Practice*. Wallingford: CAB International.
Hotel Catering and Institutional Management Association (1991) 'Environmental issues', *Technical Brief No. 13*. London: HCIMA.
International Association of Antarctic Tour Operators (1993) *Guidelines of conduct for Antarctica tour operators as of November 1993*. Kent, WA: IAATO.
International Hotels Environment Initiative (1993) *Environmental Management for Hotels: The Industry Guide to Best Practice*. Oxford: Butterworth-Heinemann. (2nd edn 1996).
MacGillivray, A. (ed.) (1995) *Environmental Measures: Indicators for the UK Environment*. London: New Economics Foundation, World Wide Fund for Nature-UK, and Royal Society for the Protection of Birds.
Mason, P. and Mowforth, M. (1995) 'Codes of conduct in tourism' *Occasional Papers in Geography*, 1. Plymouth: Department of Geographical Sciences, University of Plymouth.
National Rivers Authority (1995) *Bathing Water Quality in England and Wales in 1994*. London: HMSO.
OECD (Organisation for Economic Co-operation and Development) (1993) *Core Set of Indicators for Environmental Performance Reviews: a synthesis report by the Group on the State of the Environment*, Environment Monographs No. 83. Paris: OECD.
OECD (Organisation for Economic Co-operation and Development) (1996) *Environmental Performance in OECD Countries: Progress in the 1990s*. Paris: OECD.
RDC/ETB/CC (Rural Development Commission/English Tourist Board/Countryside Commission) (1995) *The Green Light: A Guide to Sustainable Tourism*. London: ETB.
Rees, W. (1992) 'Ecological footprints and appropriated carrying capacity: what urban economics leaves out', *Environment and Urbanization* 4, 2: 121–30.
Richardson, H. W. (1972) *Input–Output and Regional Economics*. London: Weidenfeld and Nicolson.
Scottish Tourist Board (1993) *Going Green: Guidelines for the Scottish Tourism Industry*. Edinburgh: Scottish Tourist Board.
Shackley, M. (1996) *Wildlife Tourism*. London: Thomson International Business Press.
Sinclair, M. T. and Stabler, M. J. (1997) *Tourism Economics*. London: Routledge.
Stabler, M. J. (1998) 'Environmental aspects of tourism: Applications of cost-

benefit analysis' in T. Baum and R. Mudambi (eds) *Economic and Management Methods for Tourism and Hospitality Research*. Chichester: John Wiley.

Stabler, M. J. and Goodall, B. (1997) 'Environmental awareness, action and performance in the Guernsey hospitality sector', *Tourism Management* 18, 1: 19–33.

Turner, R. K., Pearce, D. W. and Bateman, I. (1994) *Environmental Economics: An Elementary Introduction*. London: Harvester Wheatsheaf.

UNCED (United Nations Conference on Environment and Development) (1992) *Agenda 21: A Guide to the United Nations Conference on Environment and Development*. Geneva: UN Publications Service.

UNCTAD (United Nations Conference on Trade and Development (1996) *Self-Regulation of Environmental Management*. New York and Geneva: United Nations Publications.

UNEP/GESAMP (United Nations Environment Programme/United Nations Joint Group of Experts on the Scientific Aspects of Marine Pollution) (1986) *Environmental Capacity: An Approach to Marine Pollution Prevention*. UNEP Regional Seas Reports and Studies No. 86. Geneva: UNEP.

UNEP/IE (United Nations Environment Programme: Industry and Environment) (1995) *Environmental Codes of Conduct for Tourism*. Paris: UNEP.

Welford, R. and Gouldson, A. (1993) *Environmental Management and Business Strategy*. London: Pitman.

Wight, P. (1994a) 'The greening of the hospitality industry: economic and environmental good sense' in A. V. Seaton (gen. ed.) *Tourism: The State of the Art*. Chichester: John Wiley, 665–74.

Wight, P. (1994b) 'Limits of acceptable change: a recreational-tourism tool in cumulative effects assessment'. Paper presented at *Cumulative Effects in Canada: from Concept to Practice*, National Conference of the Alberta Society of Professional Biologists and the Canadian Society of Environmental Biologists, Calgary, Canada, 13–14 April 1994.

Wilson, D. (1997) 'Strategies for sustainability: Lessons from Goa and the Seychelles' in M. J. Stabler (ed.) *Tourism and Sustainability: Principles to Practice*. Wallingford: CAB International, 173–98.

World Travel and Tourism Council (1991) *WTTC Policy: Environmental Principles*. Brussels: WTTC.

6 Developing sustainable tourism in the Trossachs, Scotland

Alison Caffyn

Introduction

There has been much discussion of sustainability in the academic literature over the last ten years. Over the same period a number of practical initiatives addressing tourism and the environment have been established, trying to put some of the theory into practice. This chapter uses the case study of Scotland's Tourism and the Environment Task Force and one of the local Tourism Management Programmes (TMPs) to try to evaluate what progress such initiatives are making towards their stated aim of developing sustainable tourism.

The chapter investigates what can be achieved through the mechanism of a tourism partnership to involve the local community in developing sustainable tourism in a sensitive environment. It uses the example of the Trossachs to focus on three key elements: community participation in tourism partnerships; the monitoring of tourism development and the partnership's impacts; and the progress towards achieving sustainable forms of tourism development. The problems involved in all these areas are debated and the strengths and weaknesses of the approach taken in the Trossachs are evaluated.

The chapter first gives a brief background to Scotland's Tourism and the Environment Task Force and then summarises the nature, objectives and activities of the Trossachs Trail Tourism Management Programme (TTTMP). It then evaluates the degree to which the local community has been involved and the implications of this. An unusual feature of the initiative in the Trossachs is the emphasis given to monitoring. The mechanisms introduced are evaluated in relation to the TMP's overall aim of developing sustainable tourism. The final sections briefly consider the future mechanisms for managing tourism in the Trossachs and elsewhere and then try to draw the three main themes together, blending the theoretical issues with the practical experience from the case study. The conclusions also identify lessons for other areas about the relationships between tourism partnerships, local communities and sustainable tourism development.

Background

Tourism is an increasingly important industry in Scotland. It attracts £2.2 billion spending and employs over 8 per cent of the workforce. Scotland receives 1.8 million overseas and 9 million UK visitors each year and has a significant leisure day trips market with its population of 5 million, mostly concentrated in the central belt.

The Tourism and the Environment Task Force (the Task Force or TETF) was established to take forward the proposals in the report *Tourism and the Scottish Environment: a sustainable partnership* (STCG 1991).[1] Early action included publishing *Going Green* a guide for tourism businesses (TETF 1993a) and developing a number of Scotland-wide initiatives on issues such as footpaths and caravan sites. A more targeted approach was also developed by establishing a series of area-based TMPs. The work of the Task Force now dovetails with the overall *Scottish Tourism Strategic Plan* (STB 1994) alongside other initiatives on arts, activity holidays, training, seasonality and visitor attractions.

The Task Force's mission is: 'to promote sustainable use of Scotland's world class natural and built environment in order to optimise the wealth of opportunities for the Scottish tourism industry'. Four specific aims are listed:

1. to promote awareness and understanding of the interactions between tourism and the natural and built environment
2. to develop a planned approach to tourism development which addresses visitor management and other tourism and the environment issues in an integrated way
3. to market Scotland as a tourism destination based on the sustainable use of our natural and built environment
4. to promote adoption of environmentally sensitive practices (TETF 1996b: 2).

Twelve TMPs have been developed to date, formulated and implemented by locally based bodies, with the support of national agencies. The Task Force published initial guidelines setting out the broad process of developing a TMP, the key elements of which were to involve an integrated local approach aimed at increasing visitors' enjoyment of an area, protecting and improving its environmental quality, and encouraging responsible interactions between tourism and the environment (TETF 1993b: 4). The guidelines have been recently revised and are now much more detailed, looking at the whole process of partnership based on experience from existing TMPs (TETF 1997a).

All the TMPs are concerned with the management of the environment in a popular tourist destination but vary significantly in their scale, focus, structure and levels of funding. Examples include area-based programmes such as the Great Glen or Pitlochry; those focused on specific sites such as

Callanish; those in an urban setting such as St Andrews and others concentrating on a particular topic such as the Skye and Lochalsh footpaths project.

The research focused on the Trossachs Trail Tourism Management Programme (TTTMP) in central Scotland. It was chosen due to the author's involvement in establishing the TMP in the early 1990s and it should not be regarded as representative of other TMPs, which vary widely. The research involved a desk-based study of documentation from the Task Force, the TTTMP and related academic literature. In addition a series of semi-structured interviews were carried out with the 'lead officer' from each partner organisation and the TMP Development Officer. The research has therefore largely addressed the issues from a public-sector viewpoint.

Trossachs Trail Tourism Management Programme

The Trossachs is one of Scotland's outstanding scenic areas. It has been described as 'Scotland in miniature'. The area lies in central Scotland across the geological fault line where the Highlands meet the Lowlands (Figure 6.1).

The spectacular scenery of mountains, forests and lochs and its associations with Rob Roy MacGregor have made it a popular destination for tourists and day visitors for over 150 years. The Trossachs area was originally popularised by Sir Walter Scott's novels in the nineteenth century (Gold and Gold 1995). It is within easy reach of Glasgow and the whole central belt of Scotland which has a population of approximately three million.

The Trail, which has been promoted as a car touring route since the late 1980s, follows a 50 km circuit linking the villages of Callander, Brig O'Turk, Aberfoyle, Thornhill and Doune (Figure 6.2). The original aim of the trail was to give the area a more cohesive identity by linking all the villages, visitor attractions and sites of interest and thereby enhancing the quality of the visitor's experience, encourage people to stay in the area longer and increase the subsequent economic benefits.

The Trossachs Trail Tourism Management Programme (TTTMP) was established in 1992. It covers an area of 260 km^2 and includes the natural attractions of Ben Ledi, Ben A'an, Loch Katrine, the Lake of Menteith and the Queen Elizabeth Forest Park.

The members of the TTTMP are:

- Argyll, the Isles, Loch Lomond, Stirling and Trossachs Tourist Board
- Forest Enterprise
- Forth Valley Enterprise
- Rural Stirling Partnership
- Scottish Natural Heritage
- Scottish Tourist Board
- Stirling Council.

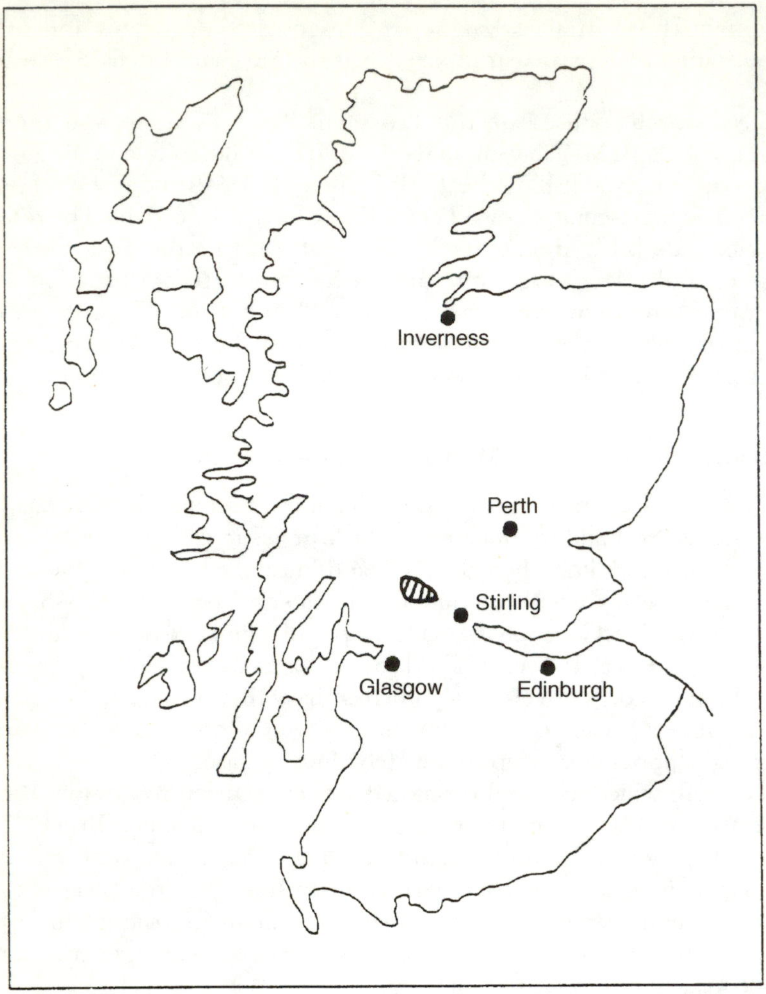

Figure 6.1 Location of the Trossachs area.

One of the early actions was to commission a strategy document to guide the work of the TMP. This set out a strategic framework for the initiative, analysed the market, assessed the local environment characteristics, recreational opportunities and visitor facilities, drew up aims and objectives and proposed an action programme with a list of over thirty potential projects. It also made detailed recommendations about implementation and monitoring.

The overall aim of the TTTMP as set out in the strategy is: 'to manage and develop tourism and day visit opportunities in a manner which brings maximum benefits to local communities and the local economy, while sustaining and enhancing the quality of the Trossachs environment which comprises the area's prime attraction' (Scott 1993: 8). The main objectives are listed below:

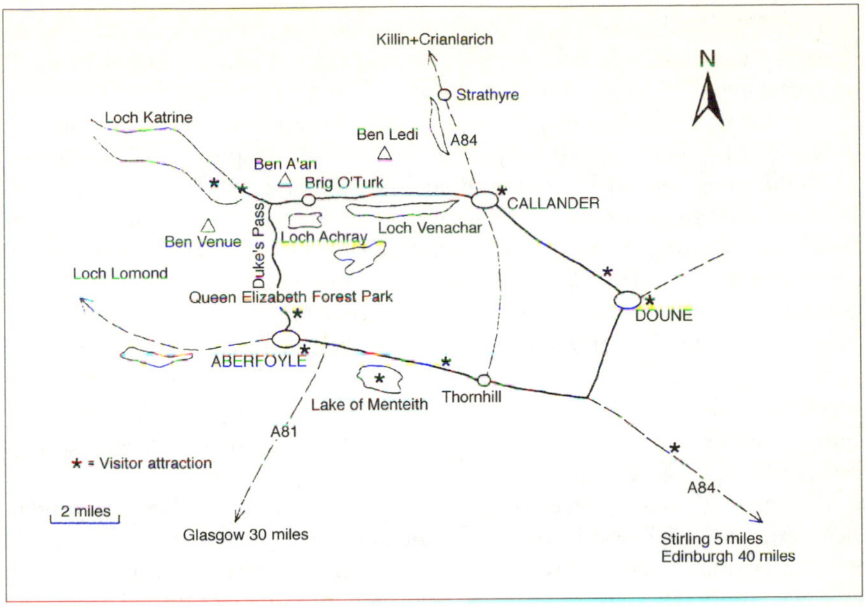

Figure 6.2 The Trossachs Trail.

- to safeguard and enhance the key environmental assets of the Trossachs and establish effective mechanisms for their longer-term management
- to strengthen the viability of the local tourist industry and improve the quality of the visitor experience in the Trossachs, by enhancing the range and quality of visitor attractions, accommodation, activities
- to increase the contribution which visitors make to the local community, economy and environment of the Trossachs
- to increase visitors' understanding and respect for the Trossachs environment and for the interests of the local communities and land managers
- to monitor levels of visits, visitor activities and the environmental impact of visitors, throughout the Trossachs and at specific key locations, and to establish mechanisms to manage visitor flows and control their impacts on the natural environment, on road and path systems and on local communities.

During 1993–4, while the strategy was being developed, an early action programme with a budget of £657,000, contributed by partner agencies, was implemented tackling urgent improvement projects. The TMP then used a software package to help partners prioritise the projects set out in the strategy. They then agreed to concentrate largely on the top ten ranked projects over the next two years, moving on to work on other desirable, but less urgent or significant, projects at a later stage. A TMP Development Officer was appointed in June 1994 to help co-ordinate and implement projects.

The TMP budget brought together funding commitments from partner organisations and allocated it to priority projects on an annual basis. It included existing funds for related but independent projects, for example the spending by Forest Enterprise on upgrading certain sites. In addition new funding was specifically allocated by some partners, including £150,000 from Scottish Tourist Board in year one, whilst other agencies tailored existing budgets to the new priorities and tried to identify extra funding. No additional money has been made available from central government for TMPs. When existing funding and new commitments were combined, the budget for years two and three totalled over £1.5 million per year. This packaging of a wide range of projects enabled partners to secure significant new funds for a high-profile initiative and achieve high levels of leverage. It also enabled funding to be found for a number of relatively small but important projects such as signposting which on their own would have been unlikely to attract funding.

Hughes (1995) has emphasised this constructive relationship between capital-rich agencies (such as local enterprise companies) and revenue spenders (usually local authorities). The Trail was a 'package' not just to attract visitors but also to attract commitments from partner agencies. A wide range of planned projects in the area were pulled together into a coherent programme, which acquired national backing from the Task Force and became a much more attractive proposition for the local and national agencies to invest in.

Hughes saw the primary purpose of the Trail as 'spreading the load' of tourism around a wider area (1995). However, no references were made to this in the strategy or during interviews. In fact the 'load' was not perceived as the main problem in itself but the poor visitor experience on offer throughout the area. Facilities had deteriorated due to lack of investment and maintenance, many key attractions had become dated and there were gaps in provision (for example signing and interpretation). Some deterioration will have been due to pressure of numbers, e.g. on footpaths, but others such as broken fencing, crumbling litter bins or the 1950s decor at Loch Katrine simply needed upgrading. The aim was to improve visitors' experiences and introduce better visitor management for the future. The Trail did aim to link-in the lesser visited villages of Doune and Thornhill to create a logical circular drive with a range of attractions spread throughout its length and encourage greater visitor spending in quieter areas. Thus a greater spread of attractions and benefits were objectives but there was no perceived need to spread visitors further afield simply because of pressure of numbers. Examples of TTTMP projects are given below:

Access projects

- major repair work to eroded paths on mountains
- upgrading of car parks in villages and at viewpoints
- creation of new sections of the 'Central Highland Way' cycleway
- creation of new forest trails and short walks around villages.

Visitor attractions and facilities

- refurbishment of the Forest Visitor Centre, including disabled access
- establishment of the Trossachs Trundler bus service around the Trail (1953 Bedford bus)
- new farm attraction near Thornhill (private-sector farm diversification)
- major redevelopment of facilities at Loch Katrine
- new exhibition at the Rob Roy and Trossachs Visitor Centre, Callander
- development of a Trossachs Discovery Centre, Aberfoyle.

Environmental improvements

- improvement works to streets, car parks and toilets in villages
- creation of links between villages and walks/cycleways
- renewal of riverside walk and new children's play area
- removal of unnecessary litter bins and repainting of Victorian mile-posts.

Information and interpretation

- road signing for the Trail
- improved signing of attractions, footpath waymarking and village welcome signs
- Trossachs Trail promotional leaflets (in five different languages)
- new information and interpretation points
- new walking guides.

Thus, the TMP has achieved many improvements on the ground as well as a more strategic approach to tourism in the area. In 1994 it was awarded a Scottish Tourist Board 'Thistle' Award for best Area Tourism Initiative in Scotland.

A variety of cuts in public-sector funding have led to reduced budgets in the years since 1995, however the wider region was designated as a European Union Objective 5b area and some projects have been able to apply for additional funding from this source.

The TMP partners have been very aware of the process they are involved in, and their obligations to monitor progress and re-examine priorities at regular intervals. The comprehensive 1993 strategy is regularly consulted, the Development Officer recently stated 'For us, the original strategy is still very much a working document. You see copies that are all dog-eared with use' (TETF 1997b: 4). In 1996 a 'taking stock workshop' was held to reflect on achievements to date, changes in resources, remaining objectives and overall progress towards sustainable tourism development.

Three aspects of the TTTMP are now examined in more detail: community and business participation, the monitoring programme and the issue of sustainability. These are areas identified by the Task Force as 'emerging weaknesses' amongst many TMPs and have also been criticised by other commentators (MacLellan 1996 and Hughes 1995; 1996). The TTTMP itself highlighted the need for more work on all three issues in its two-year report (TTTMP 1995).

Community participation

There has been a relatively slow realisation of the importance of community participation in the TTTMP. As its two-year report states: 'Much of the programme's early work has been through public sector support for a wide range of projects. However, we hope this emphasis will shift to local businesses and communities as they will have a vital role in future success' (TTTMP 1995: 18).

Whilst a community seminar was held in the early stages as the strategy was being drawn up and the consultants talked to local residents and businesses, community participation did not emerge as a particularly strong element in the strategy itself. The references to community and voluntary involvement were directed towards harnessing people's efforts for particular projects such as organising events or enhancing local open spaces (Scott 1993). Little consideration was given to involving community and business representatives in the TMP partnership itself. As the partnership achieved some of the more obvious projects and widened its perspective it has become clear that local involvement is important in developing further ideas and in implementation. The problem is that having not involved the community from the start it becomes much harder to bring them on board at a later stage. The community is known to harbour some resentment at not having been part of the process from the start.

More recently, action has been taken to involve the community to a greater extent. A Community Forum was established in 1996 which now meets on a twice yearly basis.[2] The Forum discusses the TMP work programme and priorities and is able to influence projects and raise issues of concern. The main concerns so far have included litter problems, maintenance, signposting, lay-bys, public transport provision, the advantages and disadvantages of extending the season and general perceptions of tourism amongst the local community. The level of participation when related to Arnstein's model (1971) is relatively low. The Forum has initially simply been responding to the TMP agenda, but in time it may develop its own agenda, priorities and aspirations for the area. Recently the Forum has begun to raise new issues and one or two joint projects have been initiated. Examples include the installation of new benches in a town square and the replacement of broken barriers along a scenic road.

Amongst the partners there was a general recognition of the TMP's failings in this area. Most were very supportive of the principle of participation. However, others saw difficulties in attracting interested parties and in achieving a fair representation from the scattered towns and villages and the fragmented business community. There is also recognition that the TMP has an image problem. Local perceptions of the Trossachs Trail and of the TMP are hazy (System 3 1997).

However, the TMP has avoided significant local opposition. Hughes compared the level of community support favourably in contrast to the St Andrews TMP which ran into opposition to some of its early projects (Hughes 1995). This may be partly due to luck rather than to a specific strategy to get the community on board. However, the partners and locally based Development Officer had a good appreciation of community concerns and priorities through regular contacts with community councils, the Rural Stirling Partnership, the Tourist Association, etc. In addition, most of the initial work involved uncontroversial infrastructural improvements which were unlikely to attract much local criticism.

The Task Force highlighted community participation as the TTTMP's main weakness stating 'there is little private sector involvement to date. There is a degree of scepticism within some local communities and the partnership will need to continue to rally support from the trade and the public' (TETF 1996a: 18). It highlighted amongst the emerging weaknesses of all TMPs the 'lack of human resource development, particularly where community participation is involved' (p. 12) and it identifies community involvement as an important success factor 'to ensure grass roots participation and ultimately self-motivation' (p. 11). The Task Force's new guidelines for TMPs discuss levels of participation explicitly and suggest that 'functional and interactive participation offer the best opportunities for long term success' (TETF 1997a: 3) whereby local people take a considerable degree of control.

There are a number of further issues regarding community participation which there is not space to explore in detail. (See Caffyn 1998 for a wider discussion of community participation and power relationships in partnerships.) These include:

- The equality of representation from businesses and communities.
- The mixture of views within the community about the local environment and tourism.
- The question of whether encouragement of participation by the community is simply tokenistic or even a way of placating local people.
- The development and empowerment of the community as a major objective of a partnership in itself. As Hughes states 'If tourism strategies are to be sustainable . . . they must be developed, not simply in conjunction with the public, or through public participation, but as forms of community development' (1995: 59).

Whilst the difficulties of achieving successful participation over a large area with fragmented communities must be acknowledged, significant efforts were made only when the TMP was moving into its later stages. The motivations of the public-sector partners included a belief in the principle of community participation but may also have been influenced by the trend towards participation as the new orthodoxy and the need to placate local concerns. In the case of the Trossachs the verdict regarding community participation must be 'too little, too late'.

Monitoring

Monitoring has been a key element of the TTTMP from the start. The TMP identified that:

> 'We must:
> - be able to look back and assess our achievements and failures and look forward by plotting trends
> - understand environmental changes and visitor characteristics. Visitors have an impact on the environment and we must understand this relationship
> - find out the views of visitors
> - monitor economic changes' (TTTMP 1995: 18).

Monitoring needs to be a compulsory element of a sustainable tourism strategy. The Trossachs strategy made detailed recommendations for visitor monitoring over a ten-year period. It also recommended that expert advice be sought on developing a programme of environmental monitoring. The partners acted on both these recommendations establishing a range of visitor monitoring mechanisms and commissioning a study into the most appropriate techniques for monitoring visitor impacts on the environment.

The visitor monitoring programme incorporates measures to assess traffic flows and economic impacts. The aim was to monitor a range of indicators to enable a broad picture of relevant factors to be assembled. The importance of collecting baseline data early on in the initiative was stressed and traffic counters, people counters and a major visitor survey were organised in the first year of the TMP. The complete programme is summarised below.

- Major visitor survey 1994, repeated 1997
  – supplemented by self-completion surveys in intervening years
- Visitor counts at attractions and car parks
- 'Tourism Economic Activity Monitor'
- Traffic counters at five points around trail
- Car park surveys
- Counters installed at key points on footpaths and cycleways
- In-depth survey of individual businesses.

One-off surveys of new attractions/initiatives:
- Rural retailing (business survey) 1992
- Trossachs Trundler (passenger survey) 1993
- Tigh Mor Trossachs (Holiday Property Bond customers) 1994

- Loch Katrine (feedback on new facilities) 1995
- Rob Roy and Trossachs Visitor Centre (feedback on new exhibition) 1996
- Rob Roy/Braveheart films (impact on motivations/perceptions) 1996.

In interviews the partners all stressed the importance of the monitoring programme. The data collected so far had proved useful in revealing visitor motivations, spending and views and it was found that visitor numbers on the ground had been grossly underestimated in places. Partners felt the information helped them make the case for funding and helped them target spending where it was most needed.

However, progress on the environmental monitoring was slower. The report commissioned to develop a monitoring programme to determine the environmental impacts of tourism was completed in 1994 (Dargie *et al.*). It reviewed existing methodological concepts such as carrying capacity and limits of acceptable change and concluded that a new monitoring framework should be developed drawing on elements of these approaches. A site-based framework was proposed to monitor impacts and changes to a variety of site types, incorporating variables such as levels of use, scale of damage and management regime. They identified twelve sites including mountain paths, low-level paths, loch shores and a scenic drive to be monitored using detailed mapping techniques, measurements of path widths and other damage, plus photographic evidence.

The report was welcomed by partners but there were obvious and substantial funding implications. Scottish Natural Heritage, which had led the project, were not able to commit funds for the additional staff and equipment that would be required. There was also concern that the proposals were over-elaborate and technical – one commented 'it would have been using a sledgehammer to crack a nut'. After a considerable delay a much scaled down programme has been adopted including the installation of more people counters at sensitive sites, some photographic monitoring and also the better harnessing of existing information from countryside rangers. There had been some frustration amongst partners at the delays in this fundamental element of the monitoring and concerns that the later the monitoring was implemented the less useful the information would be to measure changes over the period of the TMP. However, there was recognition that the monitoring process needed to be simple and easy to sustain on an ongoing basis.

MacLellan hints that the Dargie report (1994) was not implemented for political as well as economic reasons and because it might be restrictive of further tourism development (1996: 22). Little evidence of this was found. Whilst some partners saw it as more important than others they were all keen to see environmental monitoring procedures set up quickly. Perhaps if additional funding had been available via the Task Force the full proposals might have been implemented as a pilot project. But, as MacLellan emphasises, no additional funding has been provided from the Scottish Office for any Task Force project.

Current priorities for the TMP are firstly, to develop ways of monitoring local people's attitudes to the TMP and tourism in general. This aspect of monitoring the social impacts of tourism in the area has been the weakest element to date. The second priority is to develop an effective methodology for monitoring the performance and views of local businesses. There are acknowledged weaknesses in the economic monitoring so far and the Tourism Economic Activity Monitor model is notoriously limited in detail. Recent research is developing better business monitoring and trying to tie it in with the environmental objectives of the TMP (System 3 1997). This begins to address the Task Force's criticism that one of the TMPs' emerging weaknesses in general is the 'failure to make the economic links between the environment and the tourism industry' (TETF 1996a: 12).

MacLellan also raises concerns that due to economic pressures monitoring of TMPs might be reduced to 'measuring "the feel good factor", of the business community or *ad hoc* visitor surveys, notoriously fickle and unreliable short term indicators of tourism performance' (1996: 22). However, whilst the Trossachs monitoring programme is by no means perfect it has been developed to give a wide range of information within an overall framework. It tries to produce the most useful data for management and monitoring purposes using reasonably pragmatic methods, and feedback is specifically sought on projects carried out by the TMP. The main challenge for the TTTMP is now to co-ordinate, synthesise and interpret the results of these monitoring mechanisms and put the findings to good use. It must also distinguish between monitoring the trends and impacts of tourism and the impact of the TMP itself.

The Task Force action plan points out that monitoring techniques should be tested over long periods of time and that no TMP has yet been running long enough to be accepted as best practice. They offer guidance to TMPs on monitoring, recommending that baselines are identified and that 'Measurement should be simple, cost effective and repeatable and able to detect change. However, the framework needs to be sophisticated enough to cope with an area's complexity and rigorous enough to provide reliable results that can be used as a sound base for policy and management' (TETF 1997a: 5). The fundamental question remains however – do these systems actually monitor sustainability, the main aim of the TMPs? So far the answer would have to be – only partially. The Task Force proposes that more fundamental sustainability monitoring be introduced in time and is leading work on devising meaningful sustainability indicators (see below). Until that time coherent evaluation of sustainability will remain elusive.

Sustainability

The research found much ambivalence amongst partners about whether their aim of sustainability was being achieved. As the two-year TTTMP report states: 'We have to address the fundamental question at the heart of the initiative: are we moving towards sustainable tourism in the Trossachs? While it is easy enough to ask the question, it is not so easy to answer' (1995: 18).

The Task Force notes that one of all TMPs' general weaknesses has been the 'lack of attempt to define sustainability and demonstrate its application' (TETF 1996a: 12). Most partners agreed with this but, as one interviewee said 'if the experts can't define sustainability, how can we?'.

A number of interviewees felt there was confusion about the term and that the partners all interpreted the concept differently: 'sustainability can mean all things to all people'. Nevertheless there appeared to be genuine efforts by all partners to develop sustainable projects and to work towards sustainability for the area as a whole. However, many interviewees felt real sustainability is unattainable: 'We're pale green not dark green'. There was evidence of a more pragmatic approach: 'we need to reach a balance of economic benefits against environmental sustainability'; and 'it's about minimising impacts and developing quality'.

Without a working definition of sustainability it is difficult to judge which TMP projects are most sustainable. For example, how sustainable is repairing and hardening the surface of a major footpath compared to installing environmental interpretation? Many of the initial projects were environmental improvements to infrastructure along roadsides, at viewing points, car parks and in village centres. Recently there has been a move towards arguably more environmental projects such as developing low-level walks, litter management and a peregrine watch scheme involving a video link from the birds' nest to the Forest Park Visitor Centre. Clearer thinking about sustainability at an earlier stage might have helped in prioritising projects and in assessing their value and impact.

The TMP has tried to encourage individual businesses to 'go green' themselves. A workshop was held and followed up by a series of 'energy audits' for businesses. However, apart from a few laudable exceptions most businesses do not perceive the benefits either to their overheads or for marketing purposes. This is obviously no different to other parts of the country, but it is discouraging if even businesses in such special environments and which have been given encouragement, do little to address their local and global environment.

Hughes (1996) draws attention to the marketing of sustainability as a commodified tourism product in Scotland. In the case of the Trossachs the TMP has involved relatively little marketing other than a new promotional leaflet for the Trail which does not particularly stress sustainability. It has, however, tried to encourage more environmentally friendly practices amongst partners and businesses and highlighted the developing market for environmental holidays. Most businesses are unlikely to respond until the market is more significant – a 'Catch 22' situation.

The scenery of the Trossachs is the major attraction for visitors. Most appreciate it in aesthetic or nostalgic terms (as described by Hughes) and/or as a setting for a particular activity such as walking or cycling. However, only a small (but perhaps increasing) minority consciously view their visit in environmental terms. Thus any attempts to promote green tourism directly to either businesses or visitors may be largely premature. This is why the public sector often avoids direct references to sustainable tourism and stresses other benefits such as cost savings for businesses or, for tourists, the attractions of wildlife or alternative forms of transport.

A fundamental problem for the TTTMP is that the whole initiative has been developed around a car trail. As one interviewee said 'a car based trail is not sustainable but that is the nature of tourism in the area – it's very difficult to totally change that'. This emphasis led to many of the early TMP projects being car-orientated such as signposting and car park improvements. Partners reflected on the seemingly impossible task of separating visitors from their cars. Eighty-seven per cent of visitors arrive by car and 80 per cent continue to use their cars to travel around the area. Few use the Trossachs Trundler bus service despite the significant efforts and funds invested by the partners. It has served more as a marketing icon than a serious contribution to sustainability. Interviewees felt more efforts should be made to encourage sustainable activities and alternative forms of transport such as cycling once people reached the area.

The area is well placed to target the growing wildlife tourism market. The Development Officer expressed a vision of recreating the Trossachs of Rob Roy's day – at least in environmental terms. He envisaged rejuvenated villages, healthy woodlands and an abundance of wildlife including currently rare species such as the capercaillie, pine marten and wildcat which visitors would be keen to observe. He acknowledged that there was still a long way to go. However, the landownership situation in the Trossachs provides more opportunity for co-ordinated management than in many other areas. Not only do Forest Enterprise and West of Scotland Water own vast areas but two other large estates are keen to work with the TMP. Thus the TMP may be able to co-ordinate joint working towards a sustainable environment.

This brings us back to monitoring and evaluation as it is impossible to judge progress towards sustainability without effective measures. The Task Force aims to develop sustainability indicators following on from the work on monitoring and evaluation mechanisms currently taking place amongst TMPs (e.g. St Andrews 1995). They aim to draw on the work of the Department of Environment and other agencies to develop sustainability indicators for TMPs which could be adapted to monitor each local situation (TETF 1996b: 10). This work is critically important and, if successful, could be used not only to monitor and evaluate sustainable tourism projects but also in their design and the prioritisation of their activities.

The future

The future for the TTTMP is not yet clear. A number of interviewees felt that the TMP was shifting towards less expensive 'software' or 'green' projects.

This was partly due to the evolving nature of the partnership and partly because funding was drying up. The final remaining major infrastructure project, the Trossachs Discovery Centre, is currently being built in Aberfoyle. Once it is complete it will form a focal point for the whole trail and Trossachs area. It is seen as the linchpin in the trail in terms of communication with the public, providing a central focus for interpretation and a key source of information about walks, activities and transport.

The TTTMP has been extended for two years beyond its original timescale but its future may be dependent upon the arrangements developed for the whole Loch Lomond and Trossachs area. Following the government instigated working party in 1992–3 a new management structure for the whole area has been discussed. Most recently the new UK government has announced that a Loch Lomond and Trossachs National Park is likely although this may not come into being for a number of years. What happens in the meantime is unclear.

This pattern of development of a TMP is by no means typical. Whilst all partnership initiatives develop and evolve over time (Waddock 1989) there is no fixed route to follow. The Task Force stresses this and outlines alternative exit strategies for TMPs (TETF 1997a: 10). TMPs are simply a mechanism. As one interviewee stated: 'the TMP is just an approach – it doesn't matter if the initials disappear – it's based on sustainability and partnership'. It is these principles of sustainability and partnership which must be adhered to and maintained into the future.

Conclusion

Partnerships are attracting increasing attention as they become ever more popular as mechanisms to galvanise, organise and implement tourism development. MacLellan (1996) made the valid point that too much attention is often paid to organisational structures involved in initiatives rather than to the outputs of the projects themselves. Whilst it is important to get the processes right and in particular to agree membership, a strategy and means of implementation in the early stages, it is more important to focus on the outputs and impacts of TMPs. TMPs and similar initiatives can have a significant impact in an area and speed progress towards sustainable tourism. They can smooth the implementation of complicated projects by tackling the 'choke points' which cause delays. TMPs focus energies on the improvements both large and small that are required. They add value and lead to better outcomes.

Hughes (1995) feels that public-sector agencies have taken a far too physical, scientific approach to the environment and sustainability. He feels that there is a need to incorporate the people element more. In the Trossachs the TMP partners had two groups of people in mind – the community and the visitors. Visitors were probably viewed as most important initially and the poor visitor experience the Trossachs offered was seen as central. The TMP has realised that it did not incorporate the second group – the community – initially and is currently trying to redress this. However, this will never be easy particularly when it is retrospective and when the area has a number of disparate communities. Incorporating the community and monitoring their attitudes should be a priority for partnerships from the start.

This chapter has also highlighted the problems of establishing effective monitoring mechanisms and how important these are for developing and managing sustainable tourism. The Trossachs demonstrates a co-ordinated approach using a broad range of measures. Whilst there are obvious weaknesses, it does demonstrate the range of techniques that can be used to monitor tourism in the area and the type of pragmatic approach that can be achieved involving several agencies all interested in different elements to assemble an overall programme.

However, there are limits to the amount that can be achieved by initiatives such as TMPs without adequate funding. The TTTMP has successfully implemented most projects for which agencies were able to identify funds. However, there are a number of key areas for which no funding could be identified such as the environmental monitoring. No additional funds from the Task Force were available even for demonstration projects. One is forced to conclude that if central government is serious about sustainability, additional funding should be identified to assist the process, as the tourism industry and tourists themselves are unlikely to convert *en masse* without encouragement.

The other fundamental problem which remains is that of sustainability itself. Partnerships may profess to be genuinely committed to the principle but if they have not agreed a working definition or a clear vision of how to achieve sustainability, progress towards sustainable development will be limited. Reliable and relevant sustainability indicators are urgently needed to monitor and evaluate mechanisms, outputs and impacts effectively. Important lessons have been learnt from the experience of initiatives such as the Scottish TMPs. The public sector may now be in a position to develop more sustainable tourism projects – better defined, more inclusive of communities, properly monitored and more likely to achieve something approaching sustainability.

Finally, it is important not only to incorporate the community into the process but also to identify what local communities see as sustainability. Whilst the opinions of individuals will vary enormously, it may be possible to work towards a collective, negotiated vision of how and to what extent tourism should be developed and managed in a local area. TMPs and similar initiatives may be best placed to incorporate community interests and act as facilitators to enable and empower local people to articulate their views.

Notes

1. The Task Force contains the following members: Scottish Tourist Board, Scottish Enterprise, Highlands and Islands Enterprise, Scottish Natural Heritage, Forest Authority, Scottish Sports Council, Historic Scotland, Scottish Office Industry Department, Convention Of Scottish Local Authorities, National Trust for Scotland, Scottish Landowners Federation, Scottish Wildlife and Countryside Link, and the Area Tourist Board network.
2. The Forum includes the TMP partners, community and business representatives, local councillors, plus the Scottish Landowners Federation, the Scottish Wildlife Trust, the National Farmers Union and the Royal Society for the Protection of Birds.

Acknowledgements

The author would like to thank all the interviewees and to express her appreciation of the assistance given by David Warnock, Trossachs Trail Development Officer and Duncan Bryden, Tourism and the Environment Manager during the research process.

References

Arnstein, S. R. (1971) A ladder of citizen participation in the USA. *Journal of the Town Planning Institute*, 57, 176–82.

Caffyn, A. (1998) Tourism, heritage and urban regeneration – community participation and power relationships in the Stirling Initiative. *The Environment Papers*, 1(3), 25–38.

Dargie, T., Aitken, R. and Tantram, D. (1994) *Trossachs Tourism Management Strategy – environmental monitoring report.*

Gold, J. R. and Gold, M. M. (1995) *Imagining Scotland – tradition, representation and promotion in Scottish Tourism since 1750.* Aldershot: Scolar Press

Hughes, G. (1995) The cultural construction of sustainable tourism. *Tourism Management* 16, 1: 49–59.

Hughes, G. (1996) Tourism and the environment; a sustainable partnership. *Scottish Geographical Magazine*, 112, 2: 107–13.

MacLellan, R. (1996) The effectiveness of sustainable tourism policies in Scotland. Conference proceedings 'Sustainable tourism: ethics, economics and the environment', Cumbria, April 1996.

Scott, P. (1993) *Trossachs Tourism Management Strategy.*

Scottish Tourism Co-ordinating Group (STCG) (1991) *Tourism and the Scottish Environment – a sustainable partnership.* Edinburgh: STB.

Scottish Tourist Board (STB) (1994) *Scottish Tourism Strategic Plan.* Edinburgh: STB.

St Andrews TMP (1995) Tourism Management Plan Monitoring report.

System 3 (1997) *Trossachs Trail Tourism Management Programme Economic Monitoring Survey.*

Tourism and the Environment Task Force (TETF) (1993a) *Going green – a practical guide for tourism businesses wishing to develop an environmentally friendly approach.*

Tourism and the Environment Task Force (TETF) (1993b) *Tourism and the Scottish Environment Guidelines for the development of tourism management programmes.*

Tourism and the Environment Task Force (TETF) (1996a) *Review and future directions.*

Tourism and the Environment Task Force (TETF) (1996b) *Action plan 1996–9.*

Tourism and the Environment Task Force (TETF) (1997a) *Guidelines for the development of Tourism Management Programmes.*

Tourism and the Environment Task Force (TETF) (1997b) *Positive Impact newsletter.*

Trossachs Trail Tourism Management Programme (TTTMP) (1995) *The first two years 1993–5.*

Waddock, S. A. (1989) Understanding social partnerships: an evolutionary model of partnership organisations. *Administration and Society*, 21, 1: 78–100.

7 Establishing the common ground

Tourism, ordinary places, grey-areas and environmental quality in Edinburgh, Scotland

Frank Howie

Introduction

Mainstream tourism development has been dominated by an assumption that the everyday or the commonplace is what the tourist wants to escape from. Plog (1972), Cohen (1974) and Smith (1977) each identified groups who sought the 'ordinary' in their travels – Plog's 'allocentric' who is happy to board with local residents; Cohen's 'experiential' traveller who looks for meaning in the lives of others; Smith's 'explorer' who desires to interact with his hosts and accepts local norms. In the 1970s these groups were marginal to the business of tourism. Today, guidebooks explain how to meet the 'real locals' (Rough Guides 1996) and companies targeting the independent traveller are amongst the fastest growing sectors of the industry.

Poon argues that best practice in tourism at any given time should be seen in the context of the prevailing paradigm, '. . . an "ideal" pattern or style of productive organisation or best technological "common sense" that prevails at the time' (Perez 1983 in Poon 1993: 73). Mass production – in tourism as in car manufacturing – was 'common sense' from the 1930s to the 1980s. Best practice may now lie in her context of 'New Tourism' where the motivations of previous minorities contribute strongly to a new mainstream.

Donegan (1997: 5) writes critically of what travel writing has become – 'pointlessly dangerous scrapes . . . of derring do adventurers' and '. . . haughty essays on the strange habits of the natives'. He asserts readers have grown tired of such self-indulgence noting the high popularity of Bill Bryson: 'I am just a tourist who writes books. I am one of us, a person who is slightly out of his element, lost and worried.' His product is 'the travel book which is extraordinary only in its ordinariness'. Norman Lewis (1997: 293), the veteran travel writer, expresses a similar sentiment: 'I am looking for the people who have always been there, and belong to the places where they live. The others I do not wish to see'.

A central argument of this chapter is that there is an evolving common ground between certain 'everyday' elements of the quality of life of host communities and the sought-after experiences of a growing number of 'new tourists'. This is not to preclude 'improvements' on the status quo – a significant outcome of the 'Earth Summit' (WCED 1987: 43) was the replacement of a former environmentalist objective of the 'steady state' with a commitment to growth within a framework of sustainability. Local communities and local authorities have a key role in implementation under 'Agenda 21'. The integration of the highly localised and the strategic approaches is essential: neither party should dominate since neither is infallible (Shoard 1996).

In support of this perspective a review is presented of contemporary developments in Edinburgh, now in a process of dramatic change – a historic-cultural-tourist city, but also one that is a living, working place. Three main themes are explored: 'grey-area' tourism; disenfranchised communities; environment and quality of life.

'Grey-area' tourism refers to contemporary, controversial topics, including 'moral panics', frequently highlighted by the media, such as sex tourism and begging. Issues raised include: moral acceptability; threats/challenges/shocks to the 'assumed' status quo; location of possible developments; local community, as opposed to strategic, perspectives.

Disenfranchised communities are communities excluded from the benefits resulting from tourism. Issues include: 'spreading the load' – reducing adverse effects of tourism by wider distribution; attitudes towards tourism development; nature of 'new tourism' resources, availability to the community, economic realities.

The third theme addressed is the convergence of attitudes held by local residents and tourists on certain key 'environmental' and 'quality of life' issues. Issues raised include: tourist expectations; host-community expectations; convergence or divergence of attitudes.

These themes are addressed within 'caselets' which together give an impression of the dynamics of the community–tourism relationship for the city as a whole.

Urban tourism and historic-cultural-tourist cities

Large cities are arguably the most important type of tourist destination, offering a diversity of resources (Law 1993: 1). Many a small town also tries 'to float its own little bit of bait upon the sea of tourist consumerism', though Boniface and Fowler (1993: 77) assert that 'Towns and cities are actually much more interesting than the tourist is often allowed to appreciate'. The 'bait' offered may have little chance of success and there are many examples of failure to regenerate a sound economy through the alleged catalysing effect of tourism after more than a decade of economic development initiatives. Go *et al.* (1992) suggested that small towns generally have insufficient 'pull' to attract tourists, but can serve as a medium to satisfy consumers' 'push' motives, such as exploration, evaluation of the self, and personal development. They suggest that destination attributes, resource endowment and potential impact are the basis on which communities should pursue specific tourist types – ethnic, cultural, historic, environmental and recreational, the first four being most controllable at this level. They identified the main barrier a community faces is 'the complacency of its leaders and residents with regard to what they have to offer tourists'.

It may be the less contrived local character itself, which is the attraction of an increasing number of contemporary tourists. Ashworth and Tunbridge (1990), specifically concerned with the tourist-historic city, cautions that it is value judgements that often determine whose choice of 'local' is favoured in development decisions. Hall and Jenkins (1995: 74), in a broader discussion of tourism development policy, describe tourism development in Monterey, California, where, 'History is "flattened" and conflicting histories are suppressed, thereby creating a simplified, generalised image which is consumed by the visitor'.

Special obligations often prevail in a historic-cultural-tourist city generating tension between locally perceived disbenefits of major events and other tourism-related developments and strategic gains to the city. There is a wider range of stakeholders. This is notably the case in Edinburgh, the focus of this chapter.

New contexts

Sustainable development

Evolving understanding of sustainability emphasises the inseparability of environmental, community and economic dimensions in planning and development. Sustainability as a framework for development might be visualised as a 'pediment' of sustainability supported by three 'columns' – the '3Es' (Howie 1996: 194). E1 represents ecology, the fundamental scientific principles underlying concepts of carrying capacity, rates of consumption of renewable resources, limits of acceptable development, etc. E2 represents economic considerations – the alleged 'economic realities' which determine decision-making in the 'real world'. E3 represents ethical (or equity) considerations.

Ethical considerations are the primary focus here. These include the increasingly acknowledged 'rights' of resident communities, e.g. rights of participation in decision-making, rights to a share of the benefits of development such as jobs and enhancement of quality-of-life as through environmental improvements. Tourists' rights are also acknowledged, and as well as the familiar trade aspects the experiential quality of tourism is increasingly important. Read's (1980: 193–202) acronym 'REAL' tourism referred to the motivations of a growing segment of the market, 'special interest tourists', though today's 'New Tourists' might reasonably expect travel experiences which are '**R**ewarding, **E**nriching, **A**dventuresome and **L**earning enhancing'. Krippendorf (1987) noted comparable expectations arising from the motivational change in the western tourist market towards an emphasis on the environmental and social context within which tourism occurs and the 'humanisation' of travel. On a more general level Hughes (1995) suggests, 'As consumers, individuals feel entitled to certain quality standards in the consumption of water, air and scenery'. Prentice *et al.* (1994) identified the formation of local contacts and friendships in a destination as an important factor in creating 'endearment' to a destination, suggesting that it may be the endearment to those friends rather than other destination attributes that encourages repeat visits.

The three 'columns' supporting the pediment of sustainable development rarely meet classical ideals. They are generally of unequal strength according to the circumstances and prevailing climate of opinion. Whether they should be equal in strength is debatable and lies at the heart of discussions of sustainable development. In certain circumstances scientific principles might be paramount, e.g. emphasising strict conservation management within a World Heritage Site. Elsewhere a community's rights to appropriate development of 'its' neighbourhood may be prioritised, perhaps conflicting with a strategic perspective on 'appropriate' tourism development.

The sustainable city

'It is not cities as such which are necessarily bad, then, but the ways in which they are built and used' (Haughton and Hunter 1994: 14). Cities are ecosystems in their own right – they require sustainable management, particularly given the global trend towards urbanisation. In addition to the natural, economic and socio-cultural dimensions, intangible quality of life elements are essential. The contemporary perspective also rejects the 'steady state economy' (Ecology Party 1976), where huge inequalities of wealth and opportunity exist, the only practical way to attain a more sustainable world is through sustainably managed growth. The focus on people – social justice – rather than the earlier, almost exclusive focus on the bio-physical environment, is characteristic. The spectrum of opinion in the sustainable development debate continues, however, to range from wholehearted belief in technology, to a 'deep green' philosophy demanding fundamental restructuring of human nature, centring on smallness of scale, co-operative rather than competitive practices and self-sufficiency (O'Riorden 1991).

In practice, initiatives lie between the extremes, responding to prevailing attitudes and opportunities. To merit the term sustainable city, policy-making, planning and development must be founded on quality of life issues, at scales ranging from the local neighbourhood to the 'city-region' of Patrick Geddes (Howie 1986: 15–17) – while recognising the finite resources of the Earth itself. Management of the 'spirit of place' is an under-developed skill. It addresses the factors that counter a sense of 'placelessness' – the casual eradication of distinctive places and the making of standardised landscape (Hinch 1996). Globalisation is a powerful force working against many aspects of sustainable development, tending to create homogenisation in environments and life-styles.

Cities are more resilient to cultural impacts of tourism as a consequence of their sheer size and also because of acceptance, or at least tolerance, of differing behaviour patterns of visitors, since comparable differences may exist within the cities' own resident populations. There are, however, limits to tolerance (Doxey 1975) as urban resources may be used differently by both residents and tourists and destinations generally rely heavily on 'friendly locals'.

The mass tourist has generally been held to blame for the adverse impacts of tourism. MacCannell (1976) however, referred to certain groups' desires to escape the 'fronts' of tourism and explore the 'backs'. Craik (1995) notes that cultural, eco-, rural, indigenous and adventure tourism are increasingly intrusive and dependent on the destination community. Hinch (1996) identifies local festivals as one of the most tangible ways in which the cultural dimensions of a city are expressed, but they are particularly vulnerable to increasing pressure to justify the public funding they may receive according to their success as tourist attractions. A dual function may be successful, but it may result in loss of a sense of local 'ownership'. What was a 'local festival' may become a contrived event for tourists with resultant loss of authenticity. It may then be staged elsewhere – perhaps in a more 'convenient' locality – resulting in a loss of cultural capital from the neighbourhood. 'Just looking' may appear a harmless aspect of tourism but can be oppressive. In certain communities it might even be hazardous to tourists. Formalised tourist trails as a response to tourist interest in 'ordinary' parts of a city may intensify reactions to the gaze.

Post-modernism: authenticity, the ordinary and the community

Cohen (1979) argued against Boorstin's (1964) view that all tourists were taken in by the artificialities created by the tourism industry; also against MacCannell's (1976) implication that all tourists might be compared to pilgrims seeking authenticity. Different tourist types are recognised today, but it is generally accepted that there is greater interest in authenticity in the tourist experience. A post-modern perspective asserts that authenticity is not found in reality but in interpreted representations of reality (Urry 1992). Raban (1974) in his novel, *Soft City*, asserts:

> For each citizen the city is a unique and private reality: and the novelists, planners or sociologists (whose aims have more in common than each is often willing to admit) finds himself dealing with an impossibly intricate tessellation of personal routes, spoors and histories within the labyrinth of the city. A good working definition of metropolitan life would centre on its intrinsic illegibility: most people are hidden most of the time, their appearances are brief and controlled, their movements secret, the outlines of their lives obscure.

Can there be a true community viewpoint in the 'soft city', rather than temporary agreements on certain elements of individual constructs of reality? Such a 'community consensus' may be valuable as evidence of support for a strategic, 'rational' perspective on the development needs of the city, but inevitably fails to represent all in 'the community'. Murphy (1985) in his seminal text, *Tourism: A Community Approach*, argued powerfully for the 'distinctiveness' that can arise from a community's heritage and culture, though Taylor (1995) suggests that to take this further is to risk taking a 'highly romanticised view of communal responsiveness'. Prentice (1993) raises similar doubts, noting that while a community might agree on general support for job creation through tourism, strong segmentation of views might arise over a more conspicuous expenditure on tourism development. Shoard (1996) suggested that the pendulum has swung too far and that 'community' has become a god word – parish pump politics should not automatically be given priority over other legitimate regional and national points of view. Simmons (1994: 106) notes, 'the public's knowledge of tourism appears, at best, to be barely adequate to instil confidence in the soundness of their contribution'.

There are numerous, often conflicting views within a given community on development proposals for 'its' locality, reflecting age-group, income bracket, incomers or long-term residents, business and residential 'community'. In microcosm this is the debate visible at national and regional levels concerning heritage. 'Periodic cultural misunderstandings and expressions of value differences [. . .] are cause for some pessimism over the acceptance of the concept of a plural heritage for a plural society' (Ashworth and Tunbridge 1990).

In the final paragraph of *Soft City*, Raban (1974: 250) explains his title:

> . . . we need to hold on tight to avoid going soft in a soft city . . . In the city one clings to nostalgic and unreal signs of community, taking forced refuge in codes, badges and coteries . . . We hedge ourselves in behind dreams and illusions, construct make-believe villages and make-believe families.

It might be concluded that while 'the tourist' has been subjected to close analysis, 'the community' has received less critical analysis in tourism debate. Murphy (1988) did much to correct this imbalance. Despite these reservations, public participation is rightly a component of statutory planning and tourism decision-making – it is the community that will live with the consequences of development.

The importance of the ordinary

Graburn (1983) argued that the essence of tourism was in regular breaks that reverse or 'invert' many of the norms of everyday life. This has been the logic behind much of tourism development to date – 'a set of contrived experiences grounded in "pseudo events" ' (Boorstin 1964; 1987). However, it could be argued that in a context of globalisation and increasing standardisation of products and urban environments, the developer should no longer overlook the fact that for a growing number of 'western' tourists the desired contrast between the tourist's 'centre' and 'the other' may lie in the everyday experience and environment of the local community in the destination area.

The spirit of place or genius loci has been described as what gives life to people and places and determines their character or essence (Norberg-Schulz 1980). Lynch (1980) asserted that it is this 'environmental image [. . .] that gives its possessor an important sense of emotional security'. The novelist, Laurence Durrell (1969) wrote: 'As you get to know Europe slowly, tasting the wines, cheeses and characters of the different countries, you begin to realise that the important determinant of any culture is after all the spirit of place.'

The celebration and enhancement of the sense of place is the focus of the work of the charitable trust, Common Ground (1993) which:

> . . . is working to encourage new ways of looking at the world to excite people into remembering the richness of the common place and the value of the everyday. [. . .] Places are not just physical surroundings, they are a web of rich understandings between people and nature, people and their history, people and their neighbours. [. . .] Little things (detail) and fragments of previous lives and landscapes (patina of time) may be the very things which breathe significance into the streets or fields.

Common Ground is concerned with the community and not the tourist – 'Try to define these things for others or at a grand scale and the point is lost' – yet these are the details sought by the new tourist.

'Community' is central to a contemporary perspective on tourism. The rights of local communities must be integral to considerations of sustainable development. Further, a strong sense of locality should be recognised as a valuable tourism resource. As a consequence, a closer interrelationship between 'host' and 'guest' is desirable. This, however, is not easily achieved in a highly emotive arena. A case study, comprising a number of 'caselets' in Edinburgh is examined to provide some guidance.

Edinburgh: the city, the tourism industry and the new context

The city and its tourism industry

Edinburgh's population is approximately 450,000. Tourism is a major industry, alongside financial services and information technology industries, supporting around 22,000 jobs. The city is the most visited in Britain, outside

London. In 1996 domestic visitors spent 3.5 million bednights in the city and overseas visitors 5.4 million bednights, reflected in spending by UK visitors of £165 million and by overseas visitors of £251 million, and the trend is upwards. The main reasons for visits are the historic town itself (mentioned by 42 per cent of British and 51 per cent of overseas visitors) and its status as the capital of Scotland (mentioned by 21 per cent of British and 41 per cent of overseas visitors). Specific events/purposes, including the festivals account for less than 10 per cent of visit motivations. Friendly/helpful people was a factor that impressed 23 per cent of British and 29 per cent of overseas visitors. UK visitors tend to be in the ABC1 classes. Edinburgh's Old Town and New Town was designated a World Heritage Site in 1996 (ELTB 1998).

There is a traditional view of the city as beautiful, but suffers from '. . . the stuffiness which is often regarded as Edinburgh's Achilles heel' (Rough Guides 1996: 48). A distinct change may be under way: 'Edinburgh is trendy, and the publicity from the book/movie *Trainspotting* has made the city even more trendy and popular' (*Dagbladet* 28 June 1996). The change is not universally approved of. The editor of *The Scotsman* newspaper commented, 'Trainspotting, a depressing film . . . how untypical it was of Scotland's great capital . . . Little did I realise that the Trainspotting culture had invaded the very heart of our capital' (Neil 1997).

Edinburgh is a generally affluent city, but the urban poverty typical of all major cities is present and 100,000 people have incomes at or below the minimum state benefit level (CEC 1996). The benefits tourism generates are not fairly shared – some communities complain of too much tourism (Old Town) while others, including the peripheral estates and older inner-city residential areas, are disenfranchised. A wider perspective on tourism and its potential benefits suggests there is common ground: 'environmental' objectives, nominally for the benefit of residents, are contributing to a 'green identity', arguably a positive and significant factor in destination choice. A European perspective reveals other historic–cultural cities in direct competition with Edinburgh – the unique selling proposition or, the distinctive 'spirit of place', takes on new importance, encouraging conservation of the natural, built and cultural environments.

A new context for development is evolving in the city. 'Our goal for 2010 is to be recognised as a city which has successfully balanced economic growth with quality of life and environment' (Begg 1997). *The Environmental Strategy* (CEC 1997) declares the vision for Edinburgh as a sustainable city, to be achieved through the process of Local Agenda 21 by working in consultation and partnership with all sectors of the community. The Lord Provost's Commission on Sustainability, a one-year project, established in 1997 and drawing on high-level expertise in the public, private and voluntary sectors, has a remit to gather evidence and undertake research to find a more sustainable way of life for the city and advise on key policy changes necessary to achieve it. These represent a move beyond the theory of sustainability in urban development, to its practice and implementation.

Tourism at neighbourhood level – an Edinburgh case study

Raban (1974) adopted looking-and-listening as an appropriate research method for his novel, *Soft City*. Bryson (1995: 244) comments on other well-established travel writers, Paul Theroux and Jan Morris, deriving much of their insights into a locality as it is experienced locally through conversation with its inhabitants. The approach was adopted for this work, while also drawing on personal familiarity and previous professional work in the locations discussed, plus the findings of public meetings and other local authority initiatives. People were engaged in conversation about their attitudes towards a number of contemporary tourism-related developments in Edinburgh, referred to here as 'grey-area' tourism and quality-of-life issues. The approach is ethnographical, relying on involvement in the community; also, informal and largely opportunistic. However, as Raban notes, '. . . there is no single point of view from which one can grasp the city as a whole. [. . .] For each citizen, the city is a unique and private reality'. The qualitative findings are a basis for subsequent, more detailed study.

Leith, the former Port of Edinburgh has experienced 'yuppification' over the last decade. Upmarket flats occupy former run-down housing and industrial areas and new, well-respected restaurants draw on the fishing and maritime associations and the claret trade with France of former times. The new Scottish Office is located there and there are ambitious plans for an 'ocean liner terminal', and international shopping and leisure destination, with the Royal Yacht Britannia as a principal attraction.

A century ago, Robert Louis Stevenson wrote about 'the underworld howffs of Leith' (Desebrock 1983). Irvine Welsh brought Leith to a wider audience with his novel, *Trainspotting* (1994) and subsequently proposed that Edinburgh, as an aspiring European capital city, adopt the Amsterdam approach to the sex industry: '. . . a good way to promote tourism would be to turn Leith into a social policy experimental area like Amsterdam, with drugs and sex liberalisation to the fore'.

The sex industry is an example of what is referred to here as 'grey area' tourism; there is no clear view on how it should be handled. A 'controlled but co-operative' policing approach is adopted. Contemporary examples include saunas where prostitution is known to be practised – in areas of Leith and other 'less-affluent' neighbourhoods in the city. 'Lap-dancing' – naked, erotic dancing, for individual customers – has recently become available in more up-market neighbourhoods, while nude dancing is on offer in several pubs near the heart of the Old Town. The existence of this contemporary 'other Edinburgh' is generally known though submerged beneath the dominant image of the city.

There are arguments for localisation of sex tourism in areas where the safety of sex workers and the welfare of local residents can be assured. There is no clear community view in the neighbourhoods where this example of 'grey-area' tourism might be located. The general pattern is one of initial hostility to proposals from local residents, often taken up by the press, then, if permission is granted, a 'settling-in' period after which the matter disappears from public debate if not view. Circumstantial evidence suggests that the more 'up-market' establishments are frequented by tourists as well as residents.

Ryan and Kinder (1996) in their study of Auckland, New Zealand, argue that sex tourism should not be perceived as 'a form of deviance', its general treatment in the tourism literature: 'tourism is part of the entertainment business; pornography is too'. They criticise the hypocrisy of 'official' responses to sex tourism and prostitution, a widespread urban phenomenon, noting that 30 per cent of the value of the sex industry comes from 'out-of-town clients and others who could be described as tourists'. They emphasise, however, that they are addressing the issue from a perspective of client-sex worker in a western society and that their conclusions do not apply to other societies where questions of economic dominance and power may be different.

As Edinburgh rises in status as a European capital city (CEC 1998), shaking off a former 'prim' image, it may be noted that Amsterdam is abandoning its former reputation of 'tolerance and liberalism, the red-light district, and the gay scene', the heterogeneous image giving way to a more polished image (Dahles 1998).

A 'Trainspotting Trail' has been proposed which would lead tourists through the areas of Leith and Edinburgh that feature in the book (Gray and Hannan 1997a, 1997b). Arguably this would reflect the success of 'literary trails' elsewhere in the UK, appealing to tourists who seek the counter culture, and bringing some local expenditure or employment. *The Scotsman* newspaper, in a perceptive editorial (17 June 1997), questioned the ethics of encouraging tourism in areas suffering from multiple deprivation, or of glamorising tragic lives. Disenfranchised communities may desire local investment and development, but might be unaware of the cultural impacts of apparently harmless tourism. 'Just looking' can be oppressive (Urry 1992). Arguably, in certain communities, it might even become hazardous to the tourists as local 'irritation' increased (Doxey 1975) .

These exemplify the prevailing attitude towards this end of the spectrum of 'grey area' tourism – the tourism industry avoids official statement, despite the economic and community significance of the ongoing development.

Edinburgh's Old Town is the gem in the crown of the city's tourism industry. In the 1960s an inner-city motorway scheme was proposed, necessitating relocation of a community from sub-standard housing in the St Mary's Street area – which would be demolished – to new flats in a peripheral estate. The process was begun, but after much controversy was abandoned by the 1980s in favour of a policy of 'conserving' the remaining residential community as well as the physical environment. In fact, the community structure has changed as young professionals have moved in, the rehabilitated flats being generally unsuitable for young families. This residential gentrification was not the direct result of tourism development or conservation, since there was positive discrimination towards retention of the original community and (limited) constraints on free-market profit-making. Different values placed by different social groups on conserved properties would appear to be the main reason for community change (Ashworth and Tunbridge 1990: 256). The Old Town has retained the fine examples of domestic, Scots Baronial architecture, undoubtedly contributing to the ambience prized by tourists, but the scheme (partially) failed in one of its objectives, of retaining the residential community. Indeed new residents are more vocal in their criticism of the down side of tourism such as competition for car-parking spaces, loss of local-needs shopping to tourist boutiques and restaurants and the growing sense of intrusion. The local business community is ambivalent towards further pedestrianisation, fearing (against evidence) a loss of trade (Lopez 1996). The 'rights' of tourists were – incidentally – respected. A less car-dominated transport policy and strengthened conservation ethos is evolving in the city, arguably permitting an enhanced tourist experience. Sections of the Royal Mile may be permanently pedestrianised while the conserved townscape is an important element of the city's visual quality and built heritage. Indeed a 1980s-built hotel adopted the architectural style, though purists have dismissed it as a pastiche.

Such adverse impacts at neighbourhood level deserve attention alongside the overall positive picture emerging of the area as a whole (Parlett *et al.* 1995). Sadly, certain equally important parts of the historic South Side of the city were swept away in the 1970s before more sensitive attitudes prevailed (Peacock 1974; 1976).

Completion of the new financial district, Edinburgh Exchange, west of the city centre, will affirm Edinburgh's standing as Britain's second biggest financial centre. It may precipitate a migration of offices and businesses from the New Town, leading to a reversion of the latter to its former residential character, where a strengthened community may have significant implications for tourism, transportation strategies and general city centre management policies.

There is, however, an established community in the immediate vicinity of this new financial 'quarter'. Conferences are worth £300 million to the Scottish economy (Munro 1992). The recently completed Edinburgh International Conference Centre (EICC) is central to an objective of placing the city firmly on the international conference circuit: 'Conference Square, between the EICC and the Sheraton Hotel will be the linchpin development due for completion for the new millennium and will create a central civic hub and bring life and vitality to the city centre' (Anderson 1997).

The October 1997 Commonwealth Heads of Government Meeting (CHOGM) illustrated the difficulties in integrating strategically important events into local neighbourhoods. The local residential and business community gave a largely negative response to this major event, alleging their needs were inadequately considered in the face of the major developments taking place in their neighbourhood, any debate being largely tokenism (Arnheim 1969). Businesses complained of major loss of trade due to road closures for security reasons, local residents complained of noise and disturbance round the clock. Since completion of the conference centre the character of the neighbourhood is changing. Pubs note changing clientele – some resisting this, others responding to a more affluent new customer. Proprietors of local shops and cafes, formerly supplying a local market realise the potential of increased property values and the new high-spending market.

As Edinburgh achieves new status as a capital city and as an internationally important conference destination and financial centre in addition to its established role as a historic-cultural-tourist city, there is the likelihood of further conflict between local aspirations and strategically important objectives. For residents of the city, there are increased job opportunities, an enhanced range of amenities and entertainment and potentially higher environmental quality; on the negative side, fear of displacement of local homes and businesses and for some a sense of loss or exclusion from their city.

Shopping is a significant component of the tourism product of Edinburgh and Princes Street is one of the foremost retailing streets in Europe. The street has been compared unfavourably to other major cities, despite the world-famous views to the castle and skyline of the Old Town, on account of the loss of architectural harmony through insensitive new frontages and inappropriate infilling, also litter and 'tackiness' of certain short-term leaseholdings. This 'Princes Street problem' is not new, being identified in a City Development report of 1943 (CRBE 1943). The threat from out-of-town shopping centres is a further factor. Few residents now live in the street but the possible re-emergence of the New Town as a residential area through the movement of offices to the new financial district will create invigorated local communities in this area. The 'grey area' of begging on the streets has polarised opinion, ranging from a demand for 'zero tolerance' to benevolent inaction and reference to 'the dead hand of municipal socialism' (Neil 1997).

Local festivals – Edinburgh's Hogmanay

There are nine major Edinburgh festivals involving direct expenditure of £44 million for Edinburgh and the Lothians out of a total of £72 million for Scotland (Scotinform 1991). Festivals are a major economic factor – their cultural impacts are more difficult to assess, though £9 million direct expenditure and 2,043 actual jobs in Edinburgh and Lothian is estimated. These are major events with a clear tourism role; the (city) community's benefits are jobs and income. Cultural disbenefits may be arising from the growing popularity of events developed out of existing community celebrations and a resultant loss of ownership and sense of exclusion. There may also be similar loss to the tourist who seeks out the 'real' spirit of place of a destination.

In Edinburgh various local festivals are or have the potential to become worthwhile components of the tourism product of the city, notably because of the known visitor profile of the city. An example is the young and less promoted 'Beltane' Festival, a quasi-pagan celebration of Spring. Other local community festivals have potential for development, though there is now awareness of the potential losses. Hogmanay has traditionally been celebrated with family and friends at home and by 'first-footing' or visiting neighbours after the stroke of midnight. In Edinburgh a traditional 'public' gathering place to hear the bells ring in the New Year has been at the Tron, off the Royal Mile. Popular opinion has it that 'Edinburgh's Hogmanay' was initially a move by the city authorities and other organisations to 'reward' the citizens for putting up with tourism the rest of the year. Certainly it was recognised as a local celebration, though one that could benefit from more activities – few commercial premises remained open for revellers after midnight. The third Hogmanay Festival of 1995–96 resulted in half a million people participating in city-centre events and concerts and £32 million to Scotland's economy. It also resulted in 350 injuries and acceptance that serious overcrowding took place. This represented a third more people than the previous year and a 40 per cent increase in injuries. The 1997–98 event saw continuation of the Princes Street parties and concerts, though the street and a wide surrounding area including the High Street were closed to all but ticket holders. A forecasted crowd of 300,000 to 400,000 was reduced to 180,000. While tickets were free, there was clear dissatisfaction by Edinburgh residents at a sense of being 'barred' from their own city centre.

This may be interpreted as a first acceptance in Edinburgh of the reality of limits to tourism development. Carrying capacity of the central area was here determined according to safety considerations. There was, however, informal evidence of deterioration in the participant experience and a growing sense of disenfranchisement from a local celebration. Van der Borg (1992) examined Venice, an extreme case due to its highly constrained nature, but which makes salient points. Based on earlier research (Van der Borg 1992), carrying capacity – as pertaining to maximum visitor numbers – was determined according to 'net availability', the capacity not used by residents and commuters of the most important facilities and infrastructure for tourist use, and found to be 25,000 visitors per day. This threshold contrasts with the ordnance on public security figure of 100,000 visitors per day used by the local police to take action to restrict access to the historic centre by closing the bridge from the mainland. Venice's tourism policy continues to follow demand aggravating the recognised urban crisis of the city centre.

An Edinburgh response to both safety and experiential quality thresholds might lie in zoning and decentralisation of events. Initially the 'official' event in Princes Street was for families and tourists – young people would continue to go to the informal, traditional – and rowdier – Tron gathering. Since rock concerts are now the major attraction at Princes Street there may be a case for appropriate support for local community gatherings at other points in the city, such as the traditional Tron. While some minor 'lawlessness' is perhaps a part of the celebrations, many locals say that this area had become too crowded and rowdy for families with children and older people.

Edinburgh villages

As in many cities, the growth of Edinburgh has seen the engulfing of former outlying neighbourhoods and villages and the development of peripheral estates. Historic, cultural and industrial heritage resources are present in many such neighbourhoods but it would be unrealistic to regard them individually as tourist destinations. The concept of a 'villages of the city' trail has some merit in this connection, while certain areas have individual potential for localised tourism.

Portobello lies to the east of the city, beyond Leith. From its Victorian heyday to the mid-1960s Portobello's long beach and its promenade was a popular day-trip destination for Edinburgh citizens and a centre for longer stays for tourists from elsewhere in the central belt of Scotland. In line with other cold-water resorts the town's tourism has suffered a long decline. A community 'visioning' project revealed an interest in reviving the area as Edinburgh's seaside. Residential conversions of former guest houses and a changed population structure as a commuter settlement precludes regaining former status, but this is not a community objective. Rather, low-intensity tourism is desired as a means of helping finance environmental improvements and conservation of certain local buildings.

Gorgie-Dalry is a neighbourhood just west of the city centre with an established tradition of community action, encouraged by active local authority support and a plethora of voluntary bodies. Tourism has never featured significantly in these processes and projects. The area is home to the city's only city farm, a resource with undetermined potential for tourism. It might be argued, however, that the focus should rather be on the more 'ordinary' aspects of the area as a traditional working-class, inner-city neighbourhood with its small shops, little trade premises and occasional, idiosyncratic small speciality businesses. Industrial heritage in the shape of breweries and a former rubber-works exist, though much of the built structure is gone. There is a scepticism but also an openness to the potential for local tourism. Social tourism has been noted – local people hosting 'matched' individuals and families from other European countries, in turn visiting their guests in their own countries. The need for an appropriate, preferably locally-based organisation to explore this further is recognised.

Craigmillar is an archetypal peripheral estate on the eastern edge of the city with characteristic problems of high unemployment, lack of local amenities, and high incidence of social problems. The long-established Craigmillar Festival Society has done much to foster arts, crafts, educational and environmental projects for the community. Tourism has not been considered until recently when the development potential of a historically significant local resource, Craigmillar Castle has been established. Long a ruin, the associations with Mary Queen of Scots are strong. Ironically – or fortuitously – this comes at a time when Linlithgow Palace, the birthplace of Mary, is playing down its association with her as 'old hat, over-romanticised and negative' (Jarman 1997). As part of Scotland's Millennium Forest project, a major planting scheme is also under way. A community 'visioning' has revealed interest in cycle and walk ways and a country park proposal. There is also considerable interest in the concurrent developments in the cultural and natural environments, led by schoolchildren enthused by local teachers. The strong 'community' focus has not yet been strongly linked with tourism potential.

Wester Hailes is a 1960s peripheral estate to the west of the city. A major tourism-related project now under development is likely to offer considerable potential for the area. When the estate was constructed in the 1960s, a section of the Union Canal which formed part of the water link across the whole country was culverted under the new estate as it was considered of no commercial value and a danger to children. In the 1990s the recreational/tourist potential became clear and in Wester Hailes this section of the 'Millennium Link' will be restored. The £32 million project has been supported by the Millennium Commission 'thanks to the support of communities along the length of the canal', an essential criterion for Commission recommendation: '[. . .] it will bring real economic, social and environmental regeneration to the canal corridor' (Millennium Commission 1997). Local pride in the award is evident, though jobs in construction and local recreational use is seen as the community gain. Local tourism opportunities – based on encouraging passing barges to stop in the area for servicing and berthing – is as yet unrecognised.

These 'disenfranchised' communities generally view tourism development as taking place elsewhere in the city. They are aware of the benefits the city receives from the industry – employment and improved amenities – but do not believe that their communities benefit directly. There is some feeling that money spent on city-centre tourism development is at the expense of their communities. In the peripheral estates where unemployment is high there is a fairly negative attitude towards jobs in the tourism industry and a general preference for 'real' jobs.

Conclusion

Local communities offer a potential complementary focus for tourism development, based on new tourist interest. There may also be a case for compensatory tourism development where tourism is unlikely to be a substantial local industry but can be a training ground for acquisition of work skills and encourages disenfranchised communities lacking in substantial tourist resources to regain pride in their neighbourhoods, aided by a realisation that certain tourists are interested in the very 'ordinariness' of this common ground, which at best reflects the diversity of the human experience. Who these tourists are and whether they can successfully fraternise with host communities outwith the established tourist areas are key questions for further research.

References

Anderson, D. (1997) in Burrel, I. 'Conference Square', 30 July 1997, Edinburgh: *Edinburgh Evening News*.

Arnheim, S. R. (1969) 'A Ladder of Citizen Participation', *American Institute of Planners* 35: 216–24.

Ashworth, G. J. and Tunbridge, J. E. (1990) *The Tourist-Historic City*, London: Belhaven Press.

Begg, D. (1997) *Moving Forward II*, October 1997, Edinburgh: City of Edinburgh Council.

Boniface, P. and Fowler, P. J. (1993) *Heritage and Tourism in the Global Village*, Routledge: London.

Boorstin, D. (1964; 1987) *The Image: A Guide to Pseudo Events in America*. 25th Anniversary Edition. New York: Athenium.

City of Edinburgh Council (1996) *Edinburgh's Economic Future*, Edinburgh: City of Edinburgh Council.

City of Edinburgh Council (1997) *Environmental Strategy*, Edinburgh: City of Edinburgh Council.

City of Edinburgh Council (1998) Edinburgh's Economic Future, http://www.ebs.hw.ac.uk/CEC/ECONREV/EconomicFuture/object6.html/

City and Royal Burgh of Edinburgh (1943) *Report of the Advisory Committee on City Development*, 27 October 1997.

Cohen, E. (1974) 'Who is a Tourist? A Conceptual Clarification', *Sociological Review* 22: 527–53.

Cohen, E. (1979) 'Rethinking the Sociology of Tourism', *Annals of Tourism Research*, 6: 18–35.

Common Ground (1993) *Local Distinctiveness*, London: Common Ground Publications.

Craik, J. (1995) 'Are there Cultural Limits to Tourism?', *Sustainable Tourism*, 3, 2: 87–98.

Dagbladet (1996) Edinburgh is Cool. *Dagbladet*, 28 June.

Dahles, H. (1998) 'Redefining Amsterdam as a Tourist Destination', *Annals of Tourism Research*, 25, 1: 55–69.

Desebrock, J. (ed.) (1983) *Picturesque Old Edinburgh by Robert Louis Stevenson*. Edinburgh: Albyn Press.

Donegan, L. (1997) 'Profile: Bill Bryson', *The Guardian*, 27 October 1997.

Doxey, G. V. (1975) 'A Causation Theory of Visitor-Resident Irritants: Methodology and Research Inferences'. San Diego, California: *Proceedings of the Travel Research Association 6th Annual Conference*, pp. 195–8.

Durrell, L. (1969) *Landscape and Character. Spirit of Place: Letters and Essays on Travel*, London: Faber and Faber Ltd.

Ecology Party (1976) *Manifesto for a Sustainable Society*, Sheffield, Ecology Party Third Annual Conference.

Edinburgh and Lothians Tourist Board (1998) *Tourism in Edinburgh and the Lothians*. Edinburgh: Edinburgh and Lothians Tourist Board.

Go, F. M., Milne, D. and Whittles, L. J. R. (1992) 'Communities as Destinations: A Marketing Taxonomy for the Effective Implementation of the Tourism Action Plan'. *Journal of Travel Research*, Spring 1992: 31–7.

Graburn, N. H. H. (1983) 'The Anthropology of Tourism', *Annals of Tourism Research*, 10, 1: 9–33.

Gray, A. and Hannan, M. (1997a) 'Trainspotting author floats plans for Leith sin city', *The Scotsman*, 15 February 1997.

Gray, A. and Hannan, M. (1997b) 'Take a ride on the wild side – the incredible Trainspotting experience', *The Scotsman*, 17 June 1997.

Hall, C. M. and Jenkins, J. M. (1995) *Tourism and Public Policy*, London: Routledge.

Haughton, G. and Hunter, C. (1994) *Sustainable Cities,* London: Jessica Kingsley Publishers.

Hinch, T. D. (1996) 'Urban Tourism: Perspectives on Sustainability', *Sustainable Tourism*, 4, 2: 95–110.

Howie, F. (1986) *The Patrick Geddes Heritage Trail, Old Town, Edinburgh. Environmental Interpretation*. Manchester: Centre for Environmental Interpretation.

Howie, F. (1996) 'Skills, Understanding and Knowledge for Sustainable Tourism' in Richards, G. (ed.) *Tourism in Central and Eastern Europe: Educating for Quality*, Tilburg: Tilburg University Press, pp. 183–206.

Hughes, G. (1995) 'The Cultural Construction of Sustainable Tourism', *Tourism Management*, 16, 1: 49–59.

Jarman, D. (1997) 'Mary Queen of Scots takes back seat', *Edinburgh Evening News*, 11 June 1997.

Krippendorf, K. (1987) *The Holiday Makers*, Oxford: Heinemann Professional Publishing.

Law, C. M. (1993) *Urban Tourism: Attracting Visitors to Large Cities*, London: Mansell.

Lewis, N. (1997) *The World, The World*. London: Picador.

Lopez, N. (1996) *Attitudes of Old Town Residents Towards Tourism*, unpublished dissertation. Edinburgh: Queen Margaret College.

Lynch, K. (1980) *Managing the Sense of a Region*, London: MIT Press.

MacCannell, D. (1976) *A New Theory of the Leisure Class,* London: Macmillan.

Millennium Commission (1997) '£32 million award for canal project', *Millennium Link Newssheet* No. 3 May 1997.

Munro, J. (1992) 'Mixing Business with Pleasure', *Edinburgh Economic and Employment Review*, 3, 3: 13–14.

Murphy, P. E. (1985) *Tourism: A Community Approach*, London: Methuen.

Murphy, P. E. (1988) 'Community Driven Tourism Planning', *Tourism Management*, 9, 2: 96–104.

Neil, A. (1997) 'The Other Edinburgh', *The Spectator*, 23 August 1997.

Norberg-Schultz, C. (1980) *Genius Loci: Towards a Phenomenology of Architecture*, London: Academy Editions.

O'Riorden, T. (1991) 'The New Environmentalism and Sustainable Development', *The Science of the Total Environment* 108: 5–15.

Parlett, G., Fletcher, J. and Cooper, C. (1995) 'The Impact of Tourism on the Old Town of Edinburgh'. *Tourism Management*, 16, 5: 355–60.

Peacock, H. (ed.) (1974) *Forgotten Southside*, Edinburgh: Edinburgh University Student Publication Board.

Peacock, H. (ed.) (1976) *The Unmaking of Edinburgh*, Edinburgh: Edinburgh University Student Publication Board.

Perez, S. (1983) 'The new tourism' in A. Poon *Tourism, Technology and Competitive Strategies*, Wallingford, Oxon: CAB International.

Plog, S. C. (1973) 'Why Destination Areas Rise and Fall in Popularity'. *Cornell HRA Quarterly*, November: 13–16.

Poon, A. (1993) *Tourism, Technology and Competitive Strategies*. Wallingford, Oxon: CAB International.

Prentice, R. (1993) 'Community-driven tourism planning and residents' preferences'. *Tourism Management*, 14: 218–27.

Prentice, R. C., Witt, S. F. and Wydenbach, E. G. (1994) 'The Endearment Behaviour of Tourists Through their Interaction with the Host Community', *Tourism Management* 15, 2: 117–25.

Raban, J. (1974) *Soft City*, London: Fontana.

Read, S. E. (1980) 'A prime focus in the expansion of tourism in the next decade: special interest travel' in B. Weiler and C. M. Hall (eds) *Special Interest Tourism*, London: Belhaven Press.

Ryan, C. and Kinder, R. (1996) 'Sex, Tourism and Sex Tourism: Fulfilling Similar Needs', *Tourism Management*, 17, 7: 507–18.

Rough Guides (1996) *Scotland: The Rough Guide*, (2nd edn), London: Rough Guides Ltd.

Scotinform (1991) *Visitor Survey: An Economic Impact Study of Nine Edinburgh Festivals*, Edinburgh: Scottish Tourist Board.

Shoard, M. (1996) 'When Home Rule's Wrong', *Countryside*, No. 78 March/April.

Simmons, D. G. (1994) 'Community Participation in Tourism Planning', *Tourism Management* 15, 2: 98–108.

Smith, V. L. (1977) *Hosts and Guests: The Anthropology of Tourism*, Blackwell: Oxford.

Taylor, G. (1995) 'The Community Approach: Does It Really Work?', *Tourism Management*, 16, 7: 487–9.

The Scotsman (1997) 'Keeking at the Schemies', 17 June 1997.

Urry, J. (1992) 'The Tourist Gaze and the Environment', *Theory, Culture and Society*, 9: 1–26.

Van der Borg, J. (1992) 'Tourism and Urban Development: The Case of Venice, Italy'. *Tourism Recreation Research*, 17, 2: 46–56.

Welsh, I. (1994) *Trainspotting*, London: Minerva.

World Commission on Environment and Development (The 'Brundtland Commission') (1987) *Our Common Future*, Oxford: Oxford University Press.

8 Local Agenda 21

Reclaiming community ownership in tourism or stalled process?

Guy Jackson and Nigel Morpeth

Introduction

Amongst the laudable aspirations of the Rio Earth Summit in 1992, the challenge of Local Agenda 21 (LA21) required local government globally to find ways of framing policy goals which would incorporate the central tenets of sustainable development and would also draw communities into a participative, collaborative policy making process. In assessing the extent to which we have realised these objectives, it is clear that tangible evidence is required that sustainable tourism as a policy goal can be translated into implementable policies, which will transform conceptually robust and well-rehearsed theory into action. More specific questions in the current context are whether LA21 is being adopted as such a policy mechanism, and to what extent it is impacting on tourism and generating community involvement.

This chapter explores the extent to which tourism policy within UK local authorities has been embraced within the Local Agenda 21 policy process and has resulted in participative interaction with communities. We find little compelling evidence to suggest that tourism, as a local authority service area, has been drawn within a Local Agenda 21 policy envelope in the UK, but also that the complex composition of heterogeneous communities and change in the interface between communities and local authorities creates difficulties for operationalising the type and level of community involvement envisaged in Agenda 21. In the face of slow local authority progress generally in the UK in realising the sustainable development goals of Agenda 21, particularly applied to tourism, we highlight evidence that shows that sustainable tourism development projects do have the characteristics for LA21 implementation. We advocate both wider recognition of the significance of LA21 for local authorities in the UK, and also the desirability of including tourism more substantially within its implementation.

What is Local Agenda 21?

The 1992 Earth Summit (the UN Conference on Environment & Development – UNCED) saw 179 countries endorse Agenda 21, a cross-national agreement on working towards 'sustainable development'. This is a term often taken for granted, but in this context, in the spirit of the Brundtland Report (emanating from the 1987 World Commission on Environment and Development), is taken to mean economic development which is sympathetic to safeguarding the economic possibilities of future generations by, for example, preventing or controlling the adverse impacts which often accompany economic development, such as irreversible degradation or over-exploitation of resources, be they natural, human, cultural, etc. Amongst other things, Agenda 21 argues that we will only achieve sustainable development through planned, democratic, co-operative means, including community involvement in decisions about the environment and development. As such, Agenda 21 has central relevance to the field of tourism and community development.

One of the key aims of UNCED in moving from ideas to policies for sustainable development, has been to integrate environmental and development concerns, and make them inseparable. But within this ambitious process of ensuring a long-term sustainable development future, it saw many of the problems and solutions to be at the local level. Thus, the role of local authorities and local action are focal. It is they who have the task of implementing the vision of integrated policy action regarding local environmental, developmental and social issues.

In addition to national sustainable development strategies, Agenda 21 calls on the local authorities of signatory nations to initiate local programmes and partnerships (Local Agenda 21 initiatives) that will bring about sustainable development at the local level. Thus, it is largely at the local level and through local government that the ambitions of the Rio Earth Summit were to be implemented.

What LA21 also does within this mandate is to require local government to make community involvement central in the implementation of strategic sustainable development initiatives and programmes. Chapter 28 of Agenda 21 requires every local authority to consult its citizens on local concerns, priorities and actions regarding the environment, development and other (e.g. social) issues, to encourage local consideration of global issues, and to encourage and foster community involvement. Local Agenda 21 statements reflecting this consultation were to be prepared by local authorities by 1996.

Is this relevant to tourism?

For tourism, the hope is clearly that Local Agenda 21 will help catalyse the implementation of the principles of sustainable development within tourism development planning and management. Despite the fact that Agenda 21 makes only scant reference to tourism *per se*, its significance *for* tourism and the tourism industry, and the significance *of* tourism to areas of overt concern (transport, industry, agriculture, etc.) are fairly evident. This has been reinforced, for example, by the EU in its *Green Paper on Tourism* (DGXXIII

1995) and in the 5th Environmental Action Programme: *Towards Sustainability* (1993–2000).

The concept of sustainable development is increasingly evident in tourism policy and practice at a variety of levels, from local to supra-national. For example, the European Commission *Green Paper on Tourism* argues that it is vital for the future of the European tourist industry for it to operate according to the broad principles of sustainable development, because of the significant potential environmental, economic, cultural, political, social and community costs and benefits accruing from tourism. Reflecting the Brundtland Report in sustainable development thinking, there is clearly a need to recognise the importance of ensuring that tourism does not threaten the natural and human resources available for both present and future generations.

From this more general concept of sustainable development, then, has emerged 'sustainable tourism'. Central to this is the notion that neither natural nor human resources will be allowed to be damaged significantly or irreversibly through tourism activity, and that they will be managed in ways considered to be sustainable (see e.g. Eber 1992; Hunter and Green 1995). It recognises that there is a need to maximise the benefits from tourism whilst minimising the costs, and to constrain tourism development within limits considered to be sustainable. The principles, issues, cases for and debates surrounding sustainable approaches to tourism development have been explored in detail variously and differently elsewhere, and have been summarised not least by a consortium of European Assocation for Tourism and Leisure Education (ATLAS) member universities, examining within the EU 'Action Plan to Assist Tourism', the principles of sustainable development and its application to tourism policy and management in different European contexts (Bramwell *et al.* 1998). These debates then are not re-examined here – except that we utilise the integrated and expanded concept of sustainable tourism development promoted in that work, to move beyond purely economic and environmental concerns, to encompass social, cultural and managerial issues of relevance to communities.

Local Agenda 21 and community

It appears increasingly evident that in attempts to implement the community involvement elements of Agenda 21, some efforts are floundering due to the imprecise and nebulous perceptions of community held by some local authority service providers. These represent some of the many pitfalls of oversimplifying the complex requirements of sustainable development agendas, and this becomes more evident if we recognise even some of the detailed work that exists on defining or clarifying the distinctions between and within communities.

Butcher *et al.* (1993) considered how the term community 'lacks definitional precision' and that community policy can best be viewed as 'a fragmentary process with a rag bag of programmes' (1993: 3). They emphasise how groups of people might share locality and common interests, but are not necessarily functioning 'communities'. There is good evidence from social policy research (see e.g. Butcher *et al.* 1993; Cooper and Hawtin 1997) that there is much more heterogeneity amongst people living in close geographical proximity, than appears to be recognised in the initial LA21 vision. These issues surrounding the complexity of 'community' raise concerns about the implementability of LA21 in its envisaged form. Too often the term 'community' has been used with 'loose currency' to describe a group of people who live in the same locality, with assumptions made about the homogeneity of community members and the consistency of their likely response to participative initiatives.

As Butcher *et al.* (1993) also identify, common agendas and other elements of 'social glue' are required for 'community' to exist, and to realise opportunities for action on these issues. The extent to which a community dynamic can be achieved through LA21, or any other vehicle, is further complicated by a recognisable transition taking place, certainly in the UK, from traditional communitarian values to what Butcher *et al.* term 'possessive individualism' (1993: 16). Cooper and Hawtin (1997), Miller and Ahmed (1997) and Stewart (1994), amongst others, have also commented on this type of process, highlighting the implications for community development of the rise of market-led individualism. The dynamic of communities, where it existed, has been changed in many developed nations by the wider political and economic environment, and particularly through the rise of the consumer society. Such issues and impacts are evidenced well in the UK, where community dynamics have been affected particularly by the significant economic restructuring and the market orientation of the previous Conservative government (Miller and Ahmed 1997). This has been further reinforced by changes to the operation of key functions within UK local authorities, reflecting the consumer-driven philosophies (these issues are revisited later). The corollary of this is that competing stakeholders are likely to relegate consensual and participative initiatives to a lower priority than competitive and sectional interests, and in the UK there is relatively little evidence to date of LA21-driven community participation initiatives being implemented robustly by local authorities in the face of such constraints.

Butcher *et al.* argue that it is more important than has often been recognised to employ methodological rigour in raising empirical questions about the nature of communities under investigation. This appears particularly the case for policy developments such as LA21. As they note, we should not make assumptions about people living 'cheek by jowl' in the same locality being necessarily part of a 'community' (1993: 15). Assumptions about community should be scrutinised by empirical questioning.

What is not open to question, however, is the significance of tourism to many communities, through economic benefit, visitor volume or other forms of intrusiveness. This is manifest, not only in terms of its impacts on local economies and environments, but on the people themselves. Whilst often necessary and beneficial, we know too well that the impacts can also be negative. The Federation of Nature and National Parks in Europe (FNNPE), for example, in their report *Loving Them To Death* noted how 'the demands of tourists can either fossilise or destroy traditional ways of life, potentially altering the socio-economic balance of rural communities' (1993: 3). Yet, despite our growing knowledge of social and cultural impacts, tourism's interactions with, and impacts on, the physical environment have so often been the primary focus of work on sustainable tourism, at the expense of a full recognition of tourism's interactions and impacts on communities (Craik 1995; Eber 1992; Lovel and Feuerstein 1992).

Despite the tendency to equate sustainability almost exclusively with environmental issues, sustainability must encompass *inter alia* the maintenance of different ways of life, the solidity of local social structures, and the viability of local communities (see e.g. Craik 1991; Bramwell *et al.* 1998). Tourism planners and managers must recognise (or be influenced to recognise) the contribution of people, lifestyles and local social structures to their primary product. What then does Local Agenda 21 offer us in terms of guidelines or hope that future development planning, and particularly tourism development, will go forward sustainably, rather than with the destructive swath that has come to characterise it in many locations?

LA21's significance to tourism: the UK experience

Whilst there is increased awareness of the need for environmental and other resource conservation within tourism and other forms of economic development, radical solutions and large-scale transformations in the values of industry and communities are not practical in most contexts. Arguably this is one reason why LA21 initiatives offer hope for the implementation of sustainable practice, with their concern to create local strategies for sustainable development, involving community self-determination. Theoretically, LA21 can and should be being utilised within economic development policies and planning for tourism and other economic vehicles, to maintain development that has economic, community and other social benefits, but which is increasingly planned and managed in such a way, or using initiatives, that ensure sustainability into the future. Concerns include ensuring that resources are not over-exploited or depleted irretrievably, that positive economic and social benefits are equitably shared, and that stakeholders are consulted, even empowered, in decision-making.

What then is the reality of the situation at present? Is LA21 acting as a mechanism for progress towards sustainable tourism development, community empowerment, etc.? The immediate answer is that it is probably too early to say – although there are some interesting trends and divergent views on its success to date and its potential, and overall progress appears slower than was envisaged at Rio, particularly in the UK.

Some attempts have been made to implement LA21 initiatives in tourism planning, often (as is desirable) within wider economic development planning. However, the extent to which local authorities across Europe have embraced LA21 clearly varies widely – often reflecting the extent of central government involvement in environmental concerns, and the significance of green politics in each country. The International Council for Local Environmental Initiatives (1994, quoted in Whittaker 1995), for example, suggests that across the globe hundreds of local authorities have made 'strong commitments' to the LA21 process.

In the UK, for example, a nation-wide study of local authorities for the Local Government Management Board (Tuxworth and Carpenter 1995) suggested that 41 per cent of local authorities are committed to LA21 initiatives. However, other studies (e.g. Pattie and Hall 1995; Leslie and Hughes 1997) have found less take-up and as yet relatively little awareness of the significance of LA21 to policy development and practice, particularly related to tourism, and relatively little action taken. Where action towards the implementation of sustainable tourism development and practice has taken place, it has often been implemented outside the auspices of LA21.

The Local Government Management Board (LGMB)[1] has recently produced a key document which is a review of the first five years of Local Agenda 21 in the UK (LGMB 1997). Charged with monitoring the progress on LA21 implementation, the LGMB reveals generally slow progress in local authorities drawing up action plans for community involvement in the LA21 process, despite the spirit of LA21 to rekindle local democracy and partnership. From the document it is evident that tourism as a key area of economic activity, social significance and an environmental presence, has generally been ignored as an area for LA21 implementation. The community participation element also appears problematic in the UK, both for local authorities who lack the resources and the will to implement consultation effectively, and for the local communities themselves, who lack confidence in the process and in their ability to influence decisions.

There is no doubt that in the UK, as well as some of the broader changes taking place within communities, and the wider socio-economic environment, there are also a number of other significant factors which are reducing the speed, scope and effectiveness of LA21 initiatives. Amongst the more significant are lack of enthusiasm from central government for the process (despite the rhetoric) and, related to this, the lack of resourcing for the process and its non-mandatory nature at a time of high pressure on public-sector resources.

Related to points made earlier about the economic restructuring and market orientation of the UK's previous government, and its impact on the ability of communities to respond to participative initiatives, the same period of transition has seen a privatisation of local government service areas (Agyemen and Evans 1994; Patterson and Theobald 1996). The interface between local government and communities has changed, with a consequent change from a concept of 'community' to a focus on local 'consumers'. The privatisation of local government services, allowing market forces ascendancy in service delivery, e.g. through Compulsory Competitive Tendering (CCT), has arguably resulted in a more centralised form of local authority administration and decision-making which mitigates against the type of participatory practice advocated for LA21 initiatives.

Additionally, and linked to this process, a number of analysts (see e.g. Hill 1994) have drawn attention to the increasingly complicated interface between citizens and government at the local level. Citizens increasingly have to interact with a complex framework of public, semi-public and private organisations, rather than their elected government and its officers. Recent years have also seen the increasing influence of central government in agenda setting for local authorities in the UK, with the 1980s and 1990s seeing systematic reductions in the powers and functions of local authorities (Stewart 1994). Again, this arguably makes the LA21 process of enhancing local democracy more difficult and mitigates against the widening of participatory practice. Leslie and Hughes (1997) also point to the reorganisation of local government in the 1990s, increasing the size of local authority districts in many cases, increasing the problems of direct contact and participatory involvement with communities.

Other constraining factors within the implementing authorities include uncertainty about actions to be taken, difficulty in establishing consultation structures and procedures, and scepticism about the value of the process. Recurring problems in public consultation which have long been evident, and remain, are those of communities' lack of enthusiasm and low expectation of the process, how to provide up-to-date information to the public on which informed decisions can be made, and ensuring the involvement of underprivileged groups (see *inter alia* Almond and Verba 1977).

The existence or development of other 'sustainable' initiatives, such as environmental management programmes etc., outside LA21, can also be seen to be limiting the number of specific LA21 initiatives. Herein lies a more positive view of what is happening in many local authorities towards the implementation of sustainable practice in areas of economic development, including leisure and tourism. There is evidence that in the UK, local authorities are responding, albeit at greatly differing rates, to other national and supra-national calls to implement 'sustainable' practice in economic development processes, outside LA21. Many, for example, are responding to the EU 5th Action Programme, and to other national sustainable development initiatives, promoted in the UK, for example by the Countryside Commission, national tourist organisations, Rural Development Commission, etc. Whilst tourism-specific schemes may be rare, some local authorities have adopted broad scale approaches, such as implementing environmental management systems and monitoring programmes to enhance sustainable practice across a range of development areas. LA21 initiatives have been incorporated within a number of broad scale initiatives; for example the 'Environment Cities' programme (e.g. Leeds), environmental forum partnership schemes (e.g. Sheffield), eco-management and auditing (Bristol). Some, though still relatively few, authorities have established LA21 officer posts.

Outside the UK it is easier to find evidence of sustainable development initiatives (integrating development and environmental issues) implemented both within and outside LA21. Some of the most significant community consultation exercises in LA21 schemes are found in Northern Europe. Whittaker (1995), for example, reviews the participatory planning schemes of The Hague and Gothenburg, which involve continuing rounds of information campaigns, exhibitions, seminars, discussion groups, questionnaires, etc., making local communities central in the development of blueprints for the future of these cities.

However, even observing more widely, in general it is difficult to find many examples where effective community consultation has formed a key part of the implementation process of LA21 or other development initiatives. In fact, one of the major stumbling blocks in the implementation of LA21, particularly in the UK, appears to be the community participation element that is central to it. So far, examples of significant community participation, empowerment, etc., in major tourism development initiatives appear particularly rare.

Most of the progress in terms of the implementation of sustainable practice in tourism in the UK has been achieved as a result of the *Maintaining the Balance* report of the government's 'Tourism and the Environment Task Force' from the early 1990s. This was initiated originally to examine the nature and scale of problems caused by visitor numbers, other tourism-related environmental issues, and to identify the benefits of tourism to, e.g., rural areas and historic towns. One of the key recommendations was that tourist boards, in conjunction with other agencies and local authorities, should set up 'pilot projects' to test the solutions to local problems of developing and operating tourism sustainably. Several were established and include examples of the involvement of local communities in decision making, community participation in local action, and other examples of sustainable practice (see *Sustainable Rural Tourism: Opportunities for Local Action*, Countryside Commission 1995).

However, despite the fact that, as with LA21, attempts to foster community initiatives or at least maximise the benefits to local communities are at the heart of some of these projects, and within many of those that have subsequently emerged, the great majority of these sustainable tourism initiatives in the UK have been implemented without reference to the LA21 process. In fact, in terms of assessing the overall impact of LA21 on tourism within the UK to date, there appears relatively little evidence of the implementation of LA21 initiatives generally, and still less involving tourism, despite its increasing recognition as a significant economic agent and of its potentially damaging nature on sensitive environments.

In view of the fact that LA21 implementation plans were, on the Rio timescales, to be complete by 1996, we should by now clearly have witnessed more significant implementation. These observations from our involvement in a range of policy initiatives in this field, with several types of local authority, are supported by some available quantitative assessments. Leslie and Hughes' (1997) work shows, in contrast to the more optimistic interpretation by the LGMB (1997), that in many UK local authorities awareness of LA21 and its significance is limited. By 1995 only 35 per cent of local authorities had established or were developing programmes or strategies for the implementation of LA21, and some of these were clearly only extending existing environmental or greening strategies for their council operations.

Leslie and Hughes' work is also useful in supporting the impression that even fewer authorities have included tourism within their LA21 strategies. In fact, only 5 per cent of authorities appear to have LA21 tourism initiatives and Leslie and Hughes conclude that there is, as yet, relatively little awareness of the significance of LA21 to policy development and practice in the UK, particularly related to tourism, and relatively little action yet taken. In hindsight, and from further analysis, it is evident that this circumstance is at least partly the result of, or is reinforced by, the analytically blunt review documents published by the quasi-governmental organisations charged with, or at least interested in, catalysing the implementation of LA21. Despite an economic and environmental significance which is major in some areas, tourism has either been under-emphasised or ignored in the guidance literature on LA21 (see e.g., LGMB 1997; Countryside Commission, English Heritage, English Nature 1996). For their part, the organisations involved in extending the adoption of sustainable practice in tourism settings (e.g., ETB, Countryside Commission, Rural Development Commission, etc.) have been slow to integrate this potentially significant initiative of LA21 within their work with local authorities on sustainable tourism developments and practice.

In the UK then, the situation appears to be one where, as yet, little progress has been made by local authorities in implementing Local Agenda 21, despite the passing of the date by which implementation plans were to have been complete. There remains much ignorance of the initiative's background and significance. More specifically, there appears to be very little evidence that tourism, as a key area of economic activity, social significance and an environmental presence, has been recognised as a potential area for LA21 implementation. What is also disappointing is that the global initiative of Local Agenda 21, to realise and implement the principles of sustainable development at the local level, which can clearly incorporate tourism initiatives and give greater impetus to the process of implementing sustainable practice in tourism, has been largely ignored by the key agencies with responsibilities for, or interests in, tourism in the UK.

Despite this, and while clearly there are problems in translating the laudable aims of Brundtland and Rio into action at the local level, realising in particular the goals of Agenda 21's Chapter 28 for community involvement, we can find evidence of progress in isolated schemes that give hope for the longer term realisation of the aims of sustainable development practice in tourism and for the viability of community involvement in local development.

Case study: Local Agenda 21 and the creation of a national cycle network in the UK

We attempted to find case study evidence that there are sustainable options for tourism development that could be widely implemented with both community participation and involving the LA21 process. Thus, if conceptually robust sustainable tourism initiatives and appropriate policy mechanisms both exist, we can highlight even more the inertia which appears to have materialised in implementing this LA21 vision in the UK, certainly relating to tourism.

'Cycle tourism' appears to represent a potential exemplar of sustainable tourism which is ripe for implementation through LA21, and the present research was prompted by evidence that cycle tourism is becoming an increasingly significant element within rural sustainable development projects within the UK. The 1995 *Sustainable Rural Tourism* report (*op. cit.*) highlights 21 projects which exemplify different aspects of sustainable tourism practice, and within this a pattern emerged in that eight of the projects positioned cycling as a tourism form which is being developed in a variety of rural locations to encourage a different type of focus to tourism development.

Additionally, the civil engineering and sustainable transport charity 'Sustrans', which has been the catalyst for the creation of a national cycle network in the UK, has made specific reference to the growing significance of both utilitarian and leisure cycling to LA21 initiatives (see e.g., 'LA21 and the National Cycle Network', Sustrans 1996a). Also relevant here is the potential for greater community involvement in local development highlighted by Sustrans, and this has recently gained greater attention in the UK through the elevation of 'Sustrans' to a national profile, due largely to their award of £42.5 million from the Millennium Commission in 1995 for their work in the establishment of 2,500 miles of the 6,500 mile National Cycle Network. Sustrans consider that this network can make a key contribution to the process of sustainable development and that the LA21 process can play a pivotal role in realising the goals of wider access to a more diverse and sustainable transport network.

The basis of the National Cycle Network is a linked series of traffic-free paths and traffic-calmed roads through the main urban centres and linking to all parts of the UK. They estimate that this network has the potential to carry over 100 million journeys each year (by bicycle, on foot or using wheel chairs) with 60 per cent of journeys for utilitarian purposes and 40 per cent for leisure. The route would be a 10-minute cycle ride from 20 million households (Sustrans 1995; 1996a).

Sustrans' work provides one of the best examples from the UK of community involvement in development. The Sustrans approach has included not only interface with local authorities and quasi-governmental agencies, but has also involved a significant community dimension. Examples include attempts to attract community participation and ownership of the projects, and a tradition of attracting volunteers from local communities with local knowledge of routes, to help in both planning and construction. There is also now a considerable body of subscribers to Sustrans, as well as volunteer assistance. This community involvement represents a clear platform of community-centred initiatives on which local authorities could build. The LA21 sustainable development ethos is also reflected in Sustrans' work to help communities derive the environmental benefits of reduced traffic, not only related to promoting tourism without increasing car-borne visitors, but also in working with local authorities to produce sustainable transport policies locally.

Within this process, Sustrans has mobilised a diverse range of co-operating local authority departments and helped combine their sectional interests in, for example, planning, transport, housing, economic development, education, leisure and recreation, tourism, etc., through the medium of the National Cycle Network. For the UK, this provides a significant example of how sustainable development issues and their planning can be established in a co-operative and integrated fashion, with participatory elements, fuelled under the influence of Local Agenda 21 thinking.

The UK 'Trans-Pennine Route' from Hull to Liverpool is currently being developed as part of a partnership arrangement between local authorities. Its charitable status has enabled Sustrans to play a catalytic role in bringing together funding partners to commit money to financing elements of this and other routes. Sustrans' vision goes beyond a strategy for the UK, with a proposed 'Atlantis Route' from the Isle of Skye in Scotland to Cadiz in southern Spain. This route would, for example, mobilise the resources of local authorities within the 'Atlantic Arc' region. Sustrans has also lent its support to the development of the 'Nord Zee Route' from northern France to the Netherlands (Sustrans 1996b).

In the UK, the mobilisation of a diverse range of policy communities has been over-laid by a national response to cycle initiatives by central government through the 1996 'National Cycle Strategy' (DoT 1996). Central government has also stipulated that policy planning guidelines, particularly within the sphere of transport policy, should acknowledge the requirement of equitable planning proposals to incorporate cycle initiatives.

In their 1996 document 'Local Agenda 21 and the National Cycle Network', Sustrans acknowledge the work and support of the Local Government Management Board. Sustrans have sought to combine a range of 'Sustainable Themes' into the local authority and community domain in creating the national network. These were positioned as complementary elements of the benefits that are likely to accrue to communities from the creation of a National Cycle Network. 'Sustainable Theme 12', for example, identifies that opportunities for culture, leisure and recreation are readily available to all, and in 'Theme 8' that access to facilities, services, goods and other people, is not achieved at the expense of the environment or limited to those with cars.

Through joint working between local authorities and communities, despoiled land has been turned into amenity space. Additionally, Sustrans consider that Local Agenda 21 processes, in tandem with the creation of the National Cycle Network, have the capacity to assist in the rebuilding of communities, in the sense that they consider the *overall* transport needs of an area, enabling greater expression of choice, and more specifically can overcome transport poverty, providing people without access to motor cars an opportunity to travel.

The basis of the National Cycle Network, including the role of local authorities and their citizens, was essentially to promote utilitarian access. However, in terms of tourism, Sustrans has also promoted the Network as an opportunity to reduce the costs and negative impacts of car-borne tourism, and the development of a tourism form that has the potential to generate £150 million in tourism receipts annually in the UK and create an additional 3,700 jobs (Sustrans 1995).

Sustrans' objective of increasing tourism without increasing car-borne visits is currently achieving fruition through the 'C2C Route', a 170-mile coast to coast cycle route across the north of England. Sustrans estimates that this attracted 15,000 largely tourist cyclists in its first year (1995). This sustainable tourism initiative is seen by many as a good exemplar of how tourism to under-utilised areas can be achieved, whilst minimising negative impacts and maximising benefits. The route was awarded the British Airways 'Tourism for Tomorrow Award' in 1995 and the economic benefits to communities in the area, whilst unquantified, have been significant. Claims have been made for the route as a contributor in maintaining marginal rural communities in west Cumbria and the north Pennines.

It could be argued that Sustrans, in highlighting the significance of cycling and cycle tourism to the policy processes of Local Agenda 21, has provided local authority departments, particularly those involved in leisure and tourism, with a tailor-made opportunity for sustainable development activity and opportunities for community involvement in an emerging and growth area of tourism demand.

However, in our experience, even where LA21 policy has been developed, progress towards integrating sustainable tourism and wider sustainable development initiatives, even in this fast developing and apparently ideal example area of cycling and cycle tourism, has generally been slow. As noted earlier, LA21 schemes have tended to be dominated by small scale initiatives for 'greening' local authority operations, despite the fact that the vision of Rio went much further. Even where cycle tourism initiatives are being implemented, seemingly ideally suited to inclusion within LA21 programmes, this integration has not occurred, providing more evidence to support Leslie and Hughes' findings of inertia, in terms of a recognition of the significance of LA21 in UK local authorities, and especially in its application to tourism. Our enquiries with a number of local authorities and with Sustrans themselves show that while Sustrans have highlighted the relevance of cycle tourism initiatives to LA21, in most cases cycle tourism schemes, even those which include community involvement, are being implemented outwith the LA21 process.

Conclusion

So what are the results of the proposed implementation of the LA21 policy mechanism by the mid- to late 1990s for the UK? Has community ownership and involvement in tourism planning been reclaimed, or has the LA21 process stalled?

It is clear that the take-up and implementation of LA21 has been slow in the UK. Certainly, much of the evidence at present tends to indicate that the Local Agenda 21 mechanism is failing to have as significant an impact as had been envisaged. It is equally clear, by looking even at one case study type, that the LA21 policy process is a potential mechanism for sustainable tourism policies and implementation, and further that these can include a meaningful interface with community self-determination, especially where the complexities and the heterogeneity of communities are researched and recognised.

It is acknowledged that, not only are LA21 initiatives at an early stage of development within the UK for the reasons described, but that tourism did not have a particularly high profile within LA21 in the initial Rio vision. However, like others, we conceive that tourism is sufficiently significant to be relevant, if not important, in some areas to the implementation of Local Agenda 21, which itself has significant potential as a policy vehicle towards achieving sustainable development, particularly through the public-sector arena.

The current realities for local authorities in the UK are that despite a requirement to consider new approaches to policy implementation (e.g. acknowledging LA21), the local government environment has become sufficiently centralised, and deregulated with the privatisation of key policy areas, that the long-termism associated with sustainable development is commonly inconsistent with the short-term realities of vehicles such as Compulsory Competitive Tendering. As a result, the vision of local authorities and community involvement as central in the implementation of sustainable development initiatives, is presently not being realised.

This review has evidenced some of the other problems with local authority implementation of LA21 in the UK, which include lack of resources, the non-mandatory nature of these programmes, lack of impetus from central government, the increasing difficulties of local authorities interfacing meaningfully with communities as a result of, for example, the increasing size of some districts, the increasingly complicated interface between local authorities and their communities, as well as the requirement to treat residents as consumers, not communities. All these suggest that there are likely to be problems in the UK (although not there exclusively) in delivering the vision of Local Agenda 21. Questions arise as to whether LA21 as a policy mechanism has enough support from those charged with its implementation to stimulate this type of co-operative action between modern local authority service providers and local communities.

In the UK, there also appear additional problems for tourism. Although there is evidence of sustainable practice emerging and this will undoubtedly spread, Leslie and Hughes (1997) have confirmed with quantitative evidence that tourism is rarely viewed as a legitimate area for LA21 programmes, even where these have been implemented. In the UK, LA21 implementation, such that it is, has been largely subsumed within a narrowly defined environmental and operational greening focus. Leslie and Hughes highlight what has also been our experience, which is that in most local authorities, the implementation of LA21 and tourism policy development is operationally separate, with responsibility located in separate departments and rarely attracts cross-departmental collaboration. The result to date, despite the relevance we have highlighted, has been an almost comprehensive separation of LA21 and tourism policy.

In highlighting cycle tourism initiatives as a policy area exemplifying the logic of convergence across this organisational and policy divide, it is also argued that community benefits can be accrued through a strategic response by local authorities to combining initiatives on cycling and tourism, through the policy envelope of Local Agenda 21. Equally, we recognise the evidence that suggests that policy initiatives for cycle tourism, as with other forms of sustainable development, are emerging independent of LA21 as pragmatic responses to available resources. This, in fact, gives further support to the observations and concerns of Leslie and Hughes (1997) regarding the centrality of LA21 in local authority operations in the UK, and particularly within tourism planning and development.

However, sustainable tourism developments do offer genuine hope for marginalised communities to play a meaningful role in delivering tourism initiatives appropriate to their specific localities and regions, as has been evidenced through a number of the schemes reported in the 'Sustainable Rural Tourism' report and through the work of Sustrans. There is evidence that policy for implementing sustainable development can, as envisaged, be realised through collaboration between local authorities and communities. We have found that several cycle tourism initiatives now being implemented in the UK also fit this role, which was identified within the original LA21 vision. In reality, however, most of these initiatives are being delivered without detailed reference to, or inclusion within, LA21 programmes.

The policy challenge ahead is for local authorities to respond to the LA21 vision, within the constraints under which they operate, and in the face of what remains inconsistent awareness of the significance of LA21 and the need for sustainable development. For LA21 to be implemented more comprehensively, local authorities (certainly in the UK) need to adopt more integrated approaches which transcend traditional departmental divides, and which include elements such as tourism, transport policy, etc. This must replace the notion, prevalent in UK local authorities which have LA21 schemes, that they are merely an extension or alternative packaging for existing 'greening' initiatives for local government operations.

Note

1. The Local Government Management Board provides services and support to local authorities in England and Wales. Governed by a Board elected from the representative Local Government Association and funded from grants, subscriptions and revenue earning services, the Board's work includes contracted activities and support for the range of local government organisations. Examples of work include national pay negotiations, developing vocational qualifications, developing good practice guides, and monitoring initiatives. It was the LGMB that was invested with the task of promoting LA21 implementation among UK local authorities, producing guidance, monitoring progress, publicising the experiences of implementing authorities and promoting good practice. A number of LA21 guidance publications and monitoring reports provide examples (see LGMB 1993; 1994; 1997).

References

Agyemen, J. and Evans, B. (1994) 'The new environmental agenda' in J. Agyemen and B. Evans (eds) *Environmental Policies and Strategies*, Harlow: Longman.

Almond, A. and Verber, B. (1977) 'Theoretical perspectives on planning participation' in A. Thornley, *Progress in Planning*, 7 (1), London: Pergamon.

Bramwell, B., Henry, I., Jackson, G. and van der Straaten, J. (1998) 'A framework for understanding sustainable tourism management' in B. Bramwell, I. Henry, G. Jackson, A. Prat, G. Richards and J. van der Straaten (eds) *Sustainable Tourism Management: Principles and Practice*, Tilburg: Tilburg University Press, 2nd edn, pp. 23–71.

Butcher, H., Glen, A., Henderson, P. and Smith, J. (eds) (1993) *Community and Public Policy*, London: Pluto Press.

Cooper, C. and Hawtin, M. (eds) (1997) *Housing, Community and Conflict: Understanding Resident 'Involvement'*, Aldershot: Arena.

Countryside Commission, Department of National Heritage, Rural Development Commission, English Tourist Board (1995) *Sustainable Rural Tourism: Opportunities for Local Action*, Cheltenham: Countryside Commission.

Countryside Commission, English Heritage, English Nature (1996) *Ideas into Action for Local Agenda 21*, Peterborough: English Nature.

Craik, J. (1991) *Resorting to Tourism: Cultural Policies for Tourism Development in*

Australia, Sydney: Allen & Unwin.

Craik, J. (1995) Are there cultural limits to tourism? *Journal of Sustainable Tourism*, 3, (2), 87–98.

Department of Transport (1996) *The National Cycle Strategy*, London: UK Department of Transport,.

DGXXIII (1995) *Green Paper on Tourism*, European Commission, DGXXIII, Brussels.

Eber, S. (1992) (ed.) *Beyond the Green Horizon: Principles for Sustainable Tourism*, Godalming: World Wide Fund for Nature.

Federation of Nature and National Parks of Europe (1993) *Loving Them To Death? The Need for Sustainable Tourism in Europe's Nature and National Parks*, Grafenau: FNNPE.

Hill, D. (1994) *Citizens and Cities: Urban Policy in the 1990s*, Hemel Hempstead: Harvester Wheatsheaf.

Hunter, C. and Green, H. (1995) *Tourism and the Environment: A Sustainable Relationship?*, London: Routledge.

ICLEI (1994) *Strategic Services Planning: Sustainable Development – A Local Perspective*, ICLEI, Toronto.

Leslie, D. and Hughes, G. (1997) Agenda 21, local authorities and tourism in the UK, *Managing Leisure*, 2, 143–54.

Local Government Management Board (1997) *Local Agenda 21: The First 5 Years – Review*, London: LGMB.

Lovel, A. and Feuerstein, B. (1992) After the carnival – Tourism and community development, *Community Development Journal: An International Forum*, 27 (4), 353–60.

Miller, C. and Ahmed, Y. (1997) Community development at the crossroads: a way forward, *Policy and Politics*, 25 (3).

Patterson, A. and Theobald, K. S. (1996) Local Agenda 21, Compulsory competitive tendering and local environmental practices, *Local Environment*, 1 (1), 7–19.

Pattie, K. and Hall, G. (1995) The greening of local government: A survey, *Local Government Studies*, 20 (3), 458–85.

Stewart, M. (1994) 'Between Whitehall and Town Hall: The realignment of urban regeneration policy in England, *Policy and Politics*, 22 (2), 133–45.

Sustrans (1995) *The National Cycle Network: The Bidding Document to the Millennium Commission*, Bristol: Sustrans.

Sustrans (1996a) *Local Agenda 21 and the National Cycle Network: Routes to Local Sustainability*, Bristol: Sustrans.

Sustrans (1996b) *LFI Noordzeeroute*, Amersfoort: Sustrans.

Tuxworth, B. and Carpenter, C. (1995) *Local Agenda 21 Survey 1994/5, Environment Resource and Information Centre*, University of Westminster, London.

United Nations Conference on Environment and Development (1992) *Agenda 21; Chapter 28*, UNCED, Geneva.

Whittaker, S. (1995) *Local Agenda 21 and Local Authorities: European Experience*, Proceedings of the 1995 European Environment Conference, University of Nottingham (UK), 11–12 September 1995.

World Commission on Environment and Development (1987) *Our Common Future*, (The Brundtland Report), Oxford: Oxford University Press.

Part 3

Developing community enterprise

9 Fair trade in tourism – community development or marketing tool?

Graeme Evans and Robert Cleverdon

Introduction

The concept and practice of 'fair trade' has been developed mainly in relation to the production of foodstuffs and handicrafts and is often linked to aid programmes in developing countries. Examples include 'CaféDirect' coffee, organic chocolate and other foodstuffs and crafts goods produced by rural communities in southern countries, in collaboration with western aid agencies and charities (e.g. Oxfam, Traidcraft). The fair trade movement which is now established in Europe and North America, seeks to improve the working conditions, production and marketing of goods and services in these communities, through premium pricing, training and investment with the goal of minimising economic leakage, widening the distribution of economic benefits and guaranteeing price stability and more sustained income (Barratt Brown 1993). This is a response to the otherwise fickle price variations and trade, and poor working conditions and commercial exploitation, which can lead to commodification, mass production and in the case of tourism, over-development, mass tourism and an unsustainable destination 'life cycle'.

This chapter discusses the notion of fair trade and its possible application in the case of tourism services based on an action research project which looks to the principles of 'fair trade' and tourism branding in the north and its operation in southern communities. This project is a joint venture between the Centre for Leisure and Tourism Studies (CELTS) at the University of North London, the Voluntary Services Overseas (VSO) and Tourism Concern – representing academic/research, overseas aid and tourism campaigning and advocacy perspectives. The experience and mechanisms developed for branded 'fair trade' products are assessed and the position of tourism in local economic and community development is analysed. This includes the development of fair trade policy guidelines and benchmark measurements of what constitutes a fairly traded tourism product, and which may be adopted and promoted by tour operators, travel agents, tourist boards and destinations themselves – with parallels to fair trade goods and to 'green' or 'environmentally friendly' products and 'ethical investments'.

Fair trade or free trade?

In the shadow of the North American Free Trade Agreement (NAFTA) and the Uruguay round of the General Agreement on Tariffs and Trade (GATT 1993), control and regulation in developing countries of investment, ownership, and intervention to protect indigenous industry (especially rural-based) and culture, seems likely to be sacrificed in the name of free trade. Under GATT principles a foreign-owned company operating in a World Trade Organisation (WTO) member state has to be treated in the same way as a domestic company. The fear is that protectionism by US and European powers and transnationals will ensure that financial and other key services remain out of reach of developing countries, whilst technology, brands and intellectual property rights will be enforced in these same developing countries. The developing world is therefore likely to lose out as a result of full free trade (in the context of welfare economics a 'Pareto loss' through the imposition of comparative advantage), despite the fatal attraction of post-Fordist industrial development exploiting cheap labour and infrastructure costs (Goldin *et al.* 1993; Page and Davenport 1994). This is nowhere more apparent than in the border industrial areas of Mexico, or the all-inclusive resorts of the Caribbean and Pacific.

The inclusion of tourism and services within GATT (namely the General Agreement on Trade in Services – GATS) has for the first time raised the issue of tourism exchange at international trade level, rather than the traditional bilateral relationships (e.g. airline routes, tour operations, visas). The European Union (EU) has for the first time shown interest in this aspect of free trade, commissioning studies into its likely interpretation and impact on tourism activity, having brought the operation of tour operators in the EU under tighter regulatory control through the 'Travel and Tourism Directive' in 1992. Services now account for over 60 per cent of world production and 20 per cent of international trade – cross-border trade in services, including tourism, totals over $900 billion annually. In the implementation of the 1993 GATT round, developing countries have been reluctant to give up 'protection' of key industry sectors, however national governments have become increasingly aware of the advantages from low labour costs and more modern structures in transport and communications and policies have shifted away from import protection. This is particularly so in the case of tourism services (Table 9.1) where a remarkable 100 per cent of developing country WTO members offered tourism to the free trade regime allowing freedom of foreign investment, ownership control and acceptance of expatriate (i.e. western) professional/managerial staff (e.g. in hotel management).

The tension between 'free trade' as embodied in supranational and free market movements, implemented through the World Trade Organisation (WTO) and its members, and the notion of 'fair trade' is further confused by the use of the term 'fair trade' as an extension of consumer protection

Table 9.1 Developing countries' services by sector (percentage of countries offering 'free trade')

Region	*Tourism*	*Transport*	*Communi-cations*	*Construction*	*Distribution*	*Finance*	*Health*	*Recreation*	*Business*
Africa	100	47	53	33	13	53	7	7	60
Latin America	100	50	50	41	32	73	32	41	77
Asia	100	50	61	56	11	50	33	22	94
Total LDCs	100	49	55	40	20	60	25	25	78
EU, US, Japan	100	100	100	100	100	100	100	100	100

Source: *World Trade Reform* – Page and Davenport, Overseas Development Institute (1994)

(not producer protection). This is epitomised in the Office of Fair Trading (OFT) in the UK which mirrors the free trade objectives through preventing anti-competitive behaviour and ensuring consumer rights and protection. Fair trade 'deals' are now familiar in the brochures of major tour operators. This is interpreted in terms of the consumer and the contractual relationship and quality assurance of the travel product being purchased and it says nothing about the fair trade relationship with host destinations or the objectives of fair trade promoted here. In fact the adoption of consumer 'fair deals' by operators may run counter to a fairly traded approach.

Fair trading

Fair trading is therefore a complex and problematic issue. However, companies which attempt to trade fairly do have a number of similar features. All are committed to the support of their producers as well as the promotion of their products. For instance, the European Fair Trade Association's (EFTA) mission is as follows:

- to support efforts of partners in the south who by means of co-operation, production and trade strive for a better standard of living and fairness in the distribution of income and influence
- to take initiatives and participate in activities aimed at establishing fair production and trade structures in the south and on the global market.

Fair trade organisations also seek to promote development towards self-reliance and empowerment by establishing fair trade relations, i.e., a trade relationship that is more equal and open than, for example, that between the normal western tour operator and host destination. They also endeavour to include consumers in this process by providing them with information about the origin and culture of the goods and services provided so that they may make informed choices (similar to environmentally friendly, 'eco' or 'green' product information, now familiar in supermarkets and in ethical investment portfolios). In the case of Oxfam, the major charity in the UK dedicated to poverty alleviation (with a turnover of £100 million, 3,000 staff and 30,000 volunteers), their fair trade aim is:

> to work developmentally with small scale producers in the South, especially women, supporting their efforts to earn a sustainable livelihood by providing: access to appropriate marketing outlets (including Oxfam shops and direct mail); and services that increase producer's business skills, trading capacity and quality of life. (Oxfam 1997: 1)

A recent collaborative project with development aid charities in the UK (Oxfam, Traidraft, Twin Trading, Equal Exchange) involves the import of coffee under the 'CaféDirect' brand, from co-operaive farmers in Mexico, Peru and Guatemala, which guarantees price stability. In Mexico the 'CESCAFE' co-operative represents 34 collective growers, all small family farms, in the Chiapas region. A premium of 25p is paid by consumers of this product, which is the first to be sold in major supermarkets, and with greater local processing and reduced distribution costs, growers receive more than four times the normal income from other brands (Jones 1993). This brand has already captured 3 per cent of the retail coffee market, an experience repeated in the Netherlands, Germany, Belgium and Switzerland. Links with independent tour operators increasingly involves collaboration with guide training, eco-tourism and craft and food product promotion, such as the Belize Eco-Tourism project which works with the Toledo Soil Association in Mayan village-based tourism, and with organic chocolate producer, Blacks, who have also achieved supermarket status for their product. These regions have also been the subject of encroaching tourism and transport development as well as growing crafts trade with fair traders and through local intermediaries, with tourists.

Fair traders are also questioning and reassessing ways of production and trade in services at every level. This involves exploring mutually supportive trading relationships and encouraging environmentally and economically sustainable production and services (Jones 1993; Traidcraft 1993). It can also involve empowering the workforce so that the trade relationship is not an exploitative one, but is of mutual benefit. Apart from the financial and practical aspects of trading, there are cultural issues of fairness which require consideration. For instance, the problem of turning individual artistry into mass production, so that what was once a pleasurable combination of work and craft, merely becomes 'cash-crop' employment, or the deleterious effects of servile employment in the hospitality industry drawing key workers from other work and responsibilities (van den Berghe 1992). Examples can be seen where entire village communities lose a generation of men to hotel and catering work in tourist resorts and where incumbent communities are displaced by tourism and related development, such as in Mexico (Maya), Kenya (Masai) and Egypt (Bedouin).

Local[1] and indigenous communities (Goodland 1982) can develop a degree of self-sufficency through job creation and land ownership, the development of production and producer services and fair trade. When combined with the experience of local economic development in response to globalisation, particularly in urban areas, fair trading relationships between northern consumers and intermediaries, and between tourist destinations in the south, present particular advantages over the free-market system prevalent in international tour operations, where 'market failure' through the unsustainable life-cycle, short-term contract and trading horizons and small profit margins is perpetuated (Josephides 1993).

Fair trade in tourism services

As already stated, the fair trade movement and development of minimum standards, certification (e.g. social audits, performance indicators) and more transparent trade relations has to date largely been concentrated in the area of commodities, such as foodstuffs and handicrafts. The following factors have emerged in these cases, which will need to be considered in terms of their applicability and transferability to fair trade in tourism:

- Most of the Northern fair trade organisations are non-profit oriented (either charities/aid organisations, community businesses and co-ops).
- Their image is one of social/moral/ethical responsibility.
- Their products are easily definable as trade commodities (e.g. coffee, crafts).
- Collective community trading structures (e.g. farmers, crafts villages) already existed for organisations to start trading with minimum start-up costs.
- The emphasis is on fair trading as a *development tool* rather than commercial activity or as a campaign tool (an exception in the UK is the BodyShop company creating a niche market for fair traded cosmetics).
- Fair price, advance payments, premium and long-term relationships are essential minimum requirements for fair trade. This provides security for producers and creates an atmosphere of trust with banks who are more inclined to give credit at affordable interest rates.
- Quality of the product is critical – it must compare favourably with other competitors in order to sell and be accepted by consumers/distributors.

Developing fair trade in tourism presents several difficulties, not limited to the concentration and domination of the mass market by a small number of operators and transport carriers in the north. Larger hotel chains are also evident in the Caribbean, Mexico, Pacific Asia and Southern Europe, leaving little scope for fair trade in the mass-tourism market. Tourism is, of course, a perishable commodity which is also increasingly substitutable as destinations become more homogenised. Prices are also driven down by factors outside the influence of local tourism businesses, such as relative exchange rates, international competition and unfair trade (Josephides 1993). Where generator countries dominate inbound tourism, as is the case in Malta and Cyprus, a single tour operator can demand exchange rate subsidies from the host government, further increasing economic leakage and restricting more sustainable and quality tourism development – a vicious circle.

Some conditions are, however, present within the tourism industry which may be used as a starting point for developing a fair trade in tourism market. To some extent existing alternative and independent tour operators already practice many of the fair trade guidelines and aims. Certain types of more sustainable and eco-tourism claim to respond to the problems of economic leakage, environmental damage and local working conditions, and may not be commercial operations but more special ventures by individuals pursuing their interest in a particular country or destination region. By definition these are likely to be small-scale until larger operators move in, as has been the case for safari tours, outdoor pursuits and cultural trails/heritage tours. The growing independent tour movement is also organised in the UK through the Association of Independent Tour Operators (AITO) and other groups, and the interest of established aid and development agencies who are already active in fair trade. This offers an opportunity for the fair trade approach to tourism to be tested and developed through a receptive independent tour operator group who may be willing to extend their business into a more full fairly traded one.

Such approaches are therefore likely to require collaboration and joint ventures between tour operators, aid and development organisations, and developing relationships with host organisations, businesses and communities. The latter raises a number of problematic issues around 'who' are the hosts? Intervention in economic development is complex and fraught – risking claims of favouritism, uncompetitiveness and special interests, particularly where the political economy is moving towards a less interventionist position at national government level. A key balance to strike is between the entrepreneurial approach and activity of small businesses and the need to develop small and medium-sized enterprise (SME) networks and co-operation, including public–private partnerships in establishing local fair trade associations (Buhalis and Cooper 1998). The role of intermediaries in tourism and north-south trade generally is also fundamental – they act as buffer, barrier and middleman between local economies/producers and end-consumers. This role can be benign, beneficial or exploitative (Evans 1994a). In tourism the consumer travels to the place of production, but nevertheless in southern communities a range of intermediaries are involved in tourism production, administration and finance. Often these key intermediary functions are run by a business élite, or dominant ethnic and social groups, and failure to engage local economies with national policies and political and trade systems will therefore render any fair trade measures largely ineffectual.

Local tourist-oriented businesses may not be represented or have influence over most of these power groups. Representation and networking of small enterprises is therefore key to gaining greater power and becoming noticed (Evans 1996). The experience of SME networking in urban economies (Curran and Blackburn 1994; Blackburn *et al.* 1990) and of collective and credit union initiatives in developing countries may be of interest here. It will therefore be important for fair trade initiatives in the north to identify the 'tourism production chain' in destination areas. Identifying and supporting key SME leaders will also be important both in championing the fair trade 'cause' and local trade network and in developing communications and exchange between fair trade operators and the host destination over time. The thorny issue of pricing requires from the outset a more transparent analysis and exchange of cost information, as well as testing the 'willingness-to-pay' and price sensitivity of consumers to a fair traded tourism product.

Given the complexities of adapting fair trade to the tourism situation, two approaches are put forward here (Figure 9.1), which although not exclusive, are ideally both required to be in place if the full benefits of fair trade are to be achieved. On the one hand, establishing consumer awareness of fair traded tourism may help existing alternative or more sustainable tourism activity increase its trade and possibly income through premium pricing. Unless appropriate local economic development and trading relationships are established, however, this is unlikely to generate fair trade benefits at a local level. On the other hand, to some extent the support of local economic development through networks, self-help and co-operation may produce some improvement to small providers (SMEs) and suppliers, but without public-sector support the situation will not move beyond this level or achieve distributive social or economic objectives.

An integrated and two-way tourism and local economic policy will therefore be required in order to protect and develop small enterprises through access to marketing budgets and promotional campaigns, tourist information at generator tourist offices/boards, transport gateways and at the destinations themselves. Combining an integrated tourism/local economic development policy and planning regime with a fair trade niche market in generator countries will ensure the optimum benefits of full free trade through the production chain.

The most difficult aspects of achieving significant local benefits and reducing leakages rests with the control of accommodation and transport contracts by multinational companies and their access to higher-volume

1. NORTH (consumers and operators)

Consumer product – Targeted initially at the more aware, 'alternative' tourist; niche market per fairly traded coffee, crafts etc. Requires fair trade 'branding' – through certification/guarantees/standards/guidance and monitoring, and adoption by associations (e.g. tour – AITO, retail – ABTA) as well as through the media – print, guides and broadcast (e.g. campaigns, awareness, market/consumer research).

⇕

2. SOUTH (host-producers and local operators)

Local economic development – Fair Trade relationship with tour operators and other fair trade organisations (e.g. crafts, cultural, aid); preferential relationship with regional/national host government and institutions (financing, credit, planning, employment, training, promotion). Requires use of small firm (SME) networks; joint promotion and marketing, training, technology sharing and transfer (e.g. IT, marketing).

Figure 9.1 Models and approaches to a fair trade relationship.

markets through advertising and distribution relationships. It should be recognised, therefore, that the economic benefits accruing through a fair trade tourism product will still be marginal in terms of the total price of an international tour package. Approximately 10–15 per cent of the retail price comprises retail/commission fees, insurance and generator services and profits and 30–35 per cent goes to international air transport providers. Less than half of the retail price is therefore available at the destination, of which accommodation may account for half of this sum. Where hotels and resorts are foreign owned or managed and no local purchasing policy or diversification exists (e.g. in many small island states), the actual spending available for the local economy may be less than 20 per cent of the total tour price. The extent of national and local involvement in transport and accommodation is therefore key to minimising economic leakage, increasing economic diversity and maximising opportunities for fair trade in the local economy. This requires developing the soft infrastructure of skills – training, technical and professional services (computing, communications, marketing) – as well as priority use of local versus imported supplies. In St Lucia, for instance, where 'all-inclusives' and cruise-ship stop-overs dominate international tour operations from the US and UK, legislation now requires hoteliers to provide access for local traders in their shops and crafts outlets, and there is an initiative to encourage the purchase of locally grown foodstuffs and beverages.

Consumer demand for fair trade

A goal, but one with uncertain likelihood of short-term success, will be the adoption of 'fair trade' as defined by major tour operators, akin to environmentally aware and 'ethical' mainstream products, such as non-animal tested cosmetics and investments. This point will only be reached if and when commercial operators perceive and respond to consumer demand for fair trade as a clear market requirement, i.e., tourists will prefer fairly traded tourism products to unfair ones. Transparency in pricing is the most problematic obstacle here since commercial operators will plead commercial confidentiality, backed up by 'fair trade' [*sic*] competition policy and regulations. Premium pricing could still operate even in the case of major tour operators, however, without publicly opening up cost structures, for example by advertising a percentage or cash premium sum to local tourism businesses and other fair trade benefits to host communities. This might be akin to corporate social responsibility reporting. How far premium pricing can actually develop outside those small, niche markets which are already prepared to pay a higher price, remains to be seen (see Forsyth 1995). This will of course be dependent upon other demand and price factors, notably relative exchange rates, the availability of substitutes, airline pricing, generator economies and the perceived value and quality of the fair trade product.

The extent of consumer demand for commodity-based fair trade products was tested in research carried out in 1993 by the NOP market research company for Christian Aid (NOP 1993). The research clearly indicated that trading fairly with a Third World country is a better way to help that country develop than giving aid. Of the 1000 survey respondents 86 per cent agreed with a statement on 'trade not aid', whilst 85 per cent agreed that workers in these countries are exploited and do not get enough for their produce. The same proportion wanted to see fairly traded products available in their supermarkets and 68 per cent agreed that they would pay more for such items if they could be sure that farmers and workers received a fair return (agreement amongst women was strongest at 73 per cent – women are still the prime purchaser of holidays for what the marketing experts label the 'decision-making unit'). Premium prices for coffee ranged from 25p on a £1 purchase and 32p on a £2 one, again women were more likely to pay more than men (29p and 36p respectively). A further NOP survey for the SuperMarketing organisation in 1994 indicated that 10 per cent of the sample (1,004) had bought a fair trade branded product within the last month, with highest awareness in the south of England and in Scotland and amongst the highest socio-economic groups and people aged 55–64 years. Over half of the sample expect fair trade products to be of the same quality as other products, and when given the offer of either giving to charity or buying fair trade product, the majority of those in the highest socio-economic groups and 25–35 year olds chose trade.

The impact of fair trade branded goods does offer some insights into consumer attitudes and behaviour that might be transferable to fairly traded tourism. The link with aid/charity is not one that has been developed or exploited within tourism (with a very few exceptions), although initiatives by development aid organisations, notably VSO, Friends of the Earth and the World Wildlife Fund (WWF), in adding tourism development to their assistance schemes and campaigns, suggests that there is potential to do so. The growth of global trade exhibitions which bring together alternative tour operators, fair trade and overseas development organisations, offer opportunities for both joint promotion and collaboration, as do mail-order activities and initial Internet marketing ventures. The opportunity presented by the World Wide Web to remoter and developing tourism destinations is only starting to be recognised (Buhalis *et al.* 1998). Websites linking small operators and destinations to consumers and retail travel agents offer access to marketing and distribution systems controlled by NTOs, tourist boards and importantly can mean that local organisations can present their own image and information, not the imposed images chosen for tour brochures and national tourism marketing (Evans 1994a). The control over the marketing and images presented of destinations and host communities is another aspect which therefore needs to be considered as part of a fair trade tourism strategy.

Monitoring and benchmarking fair trade activity

Fair trade branding involves both an ethical stand and a measurable economic position. In the latter case, reduction of economic leakage, enhanced local employment, continuity of trade (including controlling the effects of seasonality) and environmental improvements are obvious indicators. The 'wealth indicator' at a local level might also include inward investment, amenities, social housing, as well as more quantifiable quality of life indicators such as health, family/community retention and economic diversity. The qualitative measurements might include conditions of employment (e.g. age minima, equal opportunities, pension, holidays, health and safety, training); the preservation of cultural and environmental assets and practices (e.g. revived crafts, festivals, observances). The ethical aspects of fair trade are reflected partly in the goal of fair trade companies and intermediaries (including host governments) and the transmission of these to consumers and ultimately to operators and governments. The Fair Trade Foundation (1995), for example, has agreed a set of minimum standards of employment for producers, as follows (and see Daly and Cobb 1989). As well as the rights of workers, conditions for smallholders/SMEs include:

- a genuinely democratic producer group
- a fair price
- credit terms
- a long-term trading commitment.

The monitoring and verification required both to establish a fair trade 'mark' and maintain this through regular reviews, requires an evaluation system that meets the expectations both of consumers and travel suppliers in the north and is acceptable, feasible and cost effective in the south. In other spheres the application of social and environmental auditing and more sustainable 'wealth' indicators developed by the World Bank and the New Economics Foundation (NEF, and see Lingayah *et al.* 1997) provide some models and mechanisms which may be used in validating fair trade production. The extent of management information, legislation (e.g. health and safety, environmental health/control), accounting (e.g. payroll, taxation, insurance) and reporting that a tourist destination and local economy can offer, will determine the degree of external verification required. The more that internal systems, audits and standards already operate, the greater the degree to which self-evaluation, internal audit and certification can take place, and be accepted in the north. The Catholic Institute for International Relations (CIIR) in collaboration with the NEF present these options diagrammatically in Figure 9.2.

These options accept that a phased approach to benchmarking will be taken in most situations, which will also vary according to the scale and area a particular activity or production type covers. The size of this task should

Scenario 1 'Company-based monitoring and external verification' Tour companies set up own codes of conduct which they apply to contractors/local suppliers. Companies engage an external verification team.	*'Optional aspects'* • graduated scale to allow companies and contractors to improve over time • local NGOs as verifiers • verifiers accredited by an independent body in which all stakeholders are represented • use of hallmarks, labels • standardisation of codes.	*Scenario 2* 'Company-based monitoring and institutional certification through external verification' Companies within a sector (e.g. hotels) sign up to a common code of conduct. Tour companies which sign up are members of a foundation. They carry out their own monitoring which is verified by verifiers engaged by the foundation.
	Scenario 3 'Supplier certification' Tourism suppliers are certified against an agreed scale of labour conditions by a national or regional council which includes representatives of all stakeholder groups. Tour companies meet their own codes of conduct by promoting certified tourism service suppliers (e.g. hotels, guides, transporters, caterers).	

Source: CIIR (1997)

Figure 9.2 Open trading: options for effective monitoring of corporate codes of conduct.

not, however, be underestimated. The involvement of consumers themselves in verification has generally been under-developed, risking institutionalisation and over-professionalisation of the evaluation process. This is a dangerous path if customer credibility and positive behaviour is to be maintained – they are the essential stakeholders. Tourism, unlike the fair trade in goods, involves tourists as both consumers and participants and their experience will be invaluable in contributing to the fair trade approval system. For example, the largest German tour operator, TUI, uses its customers to a greater extent than UK operators. Through client questionnaire and feedback and constant monitoring, they evaluate environmental conditions in all their destinations and if standards slip, pull out of resorts with unacceptable conditions. In the UK, the US-influenced litigious trend is evident, fuelled by the consumer fair deal and EU protection legislation. This creates both an oppositional environment between consumers, travel agents and tour operators and between operators and host suppliers.

Fair trade in crafts and cultural tourism

As well as coffee and other crop production linked to eco-tours, another area where fair trade relationships are well established is in crafts production and distribution and tourism has naturally evolved as a mechanism through which consumers and travellers exchange pre- and post-tour information and purchases. This has extended to fair trade crafts exhibitions and trade fairs in the UK, Germany and Sweden, to cities and hubs in destination countries (trade shops/centres and markets), as well as via specialist retail travel and tour operators in generator countries. Detailed accounts of these are provided in Evans (1994b and 1996). Cultural tourism involving cultural trails and itineraries, live performance/festivals and traditional heritage sites and venues, also provide a link between tours that are led and controlled by hosts (e.g. as guides, museum curators, traders, artists and producers) and fair trade tour operators. In this growing specialist market, premium pricing, reduced leakages and greater involvement of host intermediaries are more feasible, given the niche market concerned and value added presented by such experiences. As ethnic groups rediscover the value of identity and cultural practices and their appeal and added value to tourists and fair trade organisations, perhaps they can avoid the more derogatory associations with tourist 'souvenir art' (Cohen 1988; 1993), with tourists being hassled in markets, beaches and heritage sites, and with what MacCannell (1984) refers to as 'staged authenticity'.

Agenda 21 and fair trade in tourism

The environmental impact of production is also an important aspect of fair trading. An obvious facet of this is in craft production with the use of recycled raw materials (for example glass, aluminium, natural dyes in Mexico and natural tools and fuel in Lombok, Indonesia – Evans 1996). Perhaps less well acknowledged are the ways in which Third World trading can affect the actual structure of a community. Small businesses and workshops run by families, such as in Lombok, provide work for members of isolated villages. Their ability to maintain a sustainable trade may often continue the life of these communities, arresting the drift of people towards the cities and resorts, and help keep vital craft skills and cultural expression alive, for the benefit of those people, and as an externality, for the public good. The success of crafts and cultural tourism aligned with tour operators who are close to a fair trade approach and relationship, offers one starting point for developing a full fair trade and tourism model as outlined above. Potential destinations which have been identified include the Gambia and southern India, where putative advocacy organisations and tourism operators are already established, although lacking in influence with government or in international markets.

Agenda 21 is primarily directed at governments and educators, although key sections address business and trade unions (Leslie and Muir 1997). Tourism activities that involve intensive use and impact on land-use, the built environment and traffic pollution, and noise, will need to meet such environmental sustainability measurements, including the estimation of acceptable carrying capacity at tourist sites, including festivals, events and 'honeypot' attractions at peak times. Design and energy considerations will also arise in tourism developments, particularly since many facilities are often poorly designed, both functionally and aesthetically, and are energy hungry and inefficient (particularly water, heating, lighting and ventilation in hotels, airports and facilities such as golf courses). As well as the environmental and physical impacts raised by a sustainable development approach, a section of Agenda 21 relating to social and economic dimensions focuses on strengthening local economies, changing consumption patterns and also on strengthening the role of local communities in their environment and provision of amenities (Quarrie 1992). Fair trade in tourism has particular significance for achieving Agenda 21 and other sustainable policy objectives, and environmental and amenity indicators are an obvious element of the validation process for certifying fair trade tourism services.

Conclusion – a future for fair trade in tourism

Experience to date suggests that extending the fair trade branding to tourism operation and development in the south should involve both niche marketing through receptive alternative and independent tour operators and agencies in the north, and local economic development and assistance initiatives in the south, supported by targeted expertise, investment and promotion. Tourism is not a commodity good and actions taken by fair traders will need to take into account the fragile relationships and problems associated with the international tourism system, and of the move towards 'free trade' and the overriding economic imperatives of host governments themselves.

Specific goals within the fair trade and sustainable tourism rationale and ethic are as follows:

- in collaboration with southern organisations, to demonstrate to consumers and industry in the north what fair trade in tourism might look like;
- through this objective, to start to develop markets in the north for fair trade in tourism operations.

Two areas have therefore been identified, which require further investigation in order to develop fair trade in tourism in practice:

> *Transparency*
> This will require analysis of the tour operator system and related contractual and market relationships, including pricing, integration, transfer pricing, costing, cash flow and payment terms. It will extend to standards, international conventions and employment conditions in destinations.
>
> *Stakeholders*
> This will require analysis, delineation, definition and mapping of the interest groups that will benefit from fair trade. This includes the role of intermediaries, networks and relationships with NTOs, host governments, NGOs and operators in the north. A specific assessment will be made of the representation, image and advertising of destinations in the north and by NTOs, and research into consumer demand for fairly traded travel and tourism.

This analysis and the models presented will need to be tested in case studies based on selected destinations and tour operators, and if successful, may lead to the development of markets in the north for fair trade in tourism products and operations with interested tour operators and agencies. The longer-term aim of moving mainstream tour operators towards a fair trade approach will look towards models of good practice and consumer research and awareness programmes; however, once viable fair trade tourism is established in the industry through specialist and niche companies, the interest of commercial operators is likely to rise.

Note

1. From initial discussions (April 1997, London) with representatives from southern communities, the delineation of 'local' in terms of host, benefits and impacts was not accepted – 'interest groups' – those stakeholders affected and/or involved was preferred. The notion of 'host community' was felt to be a northern construct. This is problematic – 'local' in terms of geographic, community (e.g. village), ethnic and other groupings are definable entities and economic impact measures tend to use such definable areas (e.g. input–output, multiplier). Tourism, however, cuts across administrative and other physical boundaries and the links between interest groups are therefore important, rather than location (which might encompass imported labour, expatriates and exclude displaced or peripheral workers and communities).

References

Barratt Brown, M. (1993) *Fair Trade: Reform and Realities in the International Trading System*, London: Zed Books.

Blackburn, R., Curran, J. and Jarvis, R. (1990) 'Small Firms and Local Networks: Some Theoretical and Conceptual Explorations', 13th National Small Firms Policy and Research Conference, Harrogate, November.

Buhalis, D., Min Tjoa, A. and Jafari, J. (1998) *Information and Communication Technologies in Tourism 1998*, New York: Springer.

Buhalis, D. and Cooper, C. (1998) 'Competition or Co-operation? Small and medium-sized enterprises at the destination' in Laws, E. *et al.* (eds) *Embracing and Managing Change in Tourism*, London: Routledge: 324–46.

CIIR (1997) *Open Trading: Options for Effective Monitoring of Corporate Codes of Conduct*, Lonson: New Economics Foundation/Catholic Institute for International Relations.

Cohen, E. (1988) 'Authenticity and Commoditization in Tourism', *Annals of Tourism Research*, 15: 371–86.

Cohen, E. (1993) 'Introduction: Investigating Tourist Arts', *Annals of Tourism Research*, 20(1): 1–8.

Curran, J. and Blackburn, R. (1994) *Small Firms and Local Economic Networks, The Death of the Local Economy*, London: Paul Chapman.

Daly, H. E. and Cobb, J. B. (1989) *For the Common Good: Redirecting the Economy towards Community, the Environment and a Sustainable Future*, London: Merlin Press.

Evans, G. L. (1994a) 'Whose Culture is it Anyway? Tourism in Greater Mexico and the Indigena', in Seaton, A. (ed.) *Tourism: State of the Art*, Chichester: Wiley, 836–47.

Evans, G. L. (1994b) 'Fair Trade: Crafts Production and Cultural Tourism in the Third World', in Seaton, A. (ed.) *Tourism: State of the Art*, Chichester: Wiley, 783–91.

Evans, G. L. (1996) 'Small Crafts Producers and Cultural Tourism in Developing Countries', Proceedings of *Globalisation and the SMEs*, 23rd International Small Business Congress, Athens.

Fair Trade Foundation (1995) *The Fairtrade Mark*, London.

Forsyth, T. (1995) 'Business Attitudes to Sustainable Tourism: Responsibility and Self Regulation in the UK Outgoing Tourism Industry', *Sustainable Tourism World Conference*, Lanzarote.

Goldin, I., Knudsen, O. and van den Mensbrugghe, D. (1993) *Trade Liberalisation: global economic implications*, OECD/World Bank, London: HMSO.

Goodland, R. (1982) *Tribal Peoples and Economic Development – Human Ecological Considerations*, Washington: World Bank.

Jones, J. (1993) 'Fair Trade', *Green Magazine*, December, West Sussex: 12–15.

Josephides, N. (1993) *Proceedings of Sustainable Tourism Conference*, London: English Tourist Board.

Leslie, D. and Muir, F. (1997) *Local Agenda 21, Local Authorities and Tourism: A United Kingdom Perspective*, A Report Prepared for Tourism Concern, Glasgow Caledonian University.

Lingayah, S., MacGillivray, A. and Raynard, P. (1997) *The Social Impact of Arts Programmes – Creative Accounting: Beyond the Bottom Line*, Working Paper 2, Stroud: New Economics Foundation and Comedia.

MacCannell, D. (1984) 'Reconstructed Ethnicity: Tourism and Cultural Identity in Third World Communities', *Annals of Tourism Research*, 11(1): 375–91.

NOP (1993) *Attitudes to Fair Trading in Consumer Goods*, London: National Opinion Polls.

Oxfam (1997) *Oxfam Context, Oxfam Trading*, Oxford: Oxfam.

Page, S. and Davenport, M. (1994) *World Trade Reform*, London: Overseas Development Institute.

Quarrie, J. (1992) *Earth Summit*, London: Regency Press.

Traidcraft (1993) *Social Audit*, Gateshead: Traidcraft.

van den Berghe, P. L. (1992) 'Tourism and the Ethnic Division of Labour', *Annals of Tourism Research* 19(1): 234–49.

10 Tourism, small enterprises and community development

Heidi Dahles

Introduction

A major development in the Indonesian economy since the 1980s has been the expansion of non-oil sectors and especially the tourist industry. Tourism has benefited from the government policy of deregulation which is intended to facilitate private-sector activities, particularly in the export sector (Booth 1990). Examples of deregulatory measures are tax incentives for big companies, cutting tariffs, simplifying export procedures, eliminating permits, and introducing tax holidays for newly established companies. Measures designed specifically to benefit the foreign tourist sector have included the partial abolition of visa requirements, the granting of additional landing rights to foreign airlines in the major ports of entry, the establishment of more international airports, the reduction in the number of licences required to build new hotels, and the definition of new tourist destination areas outside the islands of Java and Bali. The high priority given to tourism in national development policy generated a rapid growth in tourist arrivals and in earnings from tourism which has become a major source of foreign exchange.

The lower echelons of the Indonesian economy, however, have not benefited from these measures. The government, in fact, counters the deregulatory measures at the top with more regulation and control below. Although Indonesia has one of the largest informal sectors in the world (Evers and Mehmet 1994), the Indonesian government does not regard micro entrepreneurs as a force in economic development, in general, nor in tourism development in particular. Instead they are seen as an obstacle (Clapham 1985). It seems that tolerance towards small-scale economic activities is declining. Particularly, the 'informal' sector is currently undergoing a period of formalisation. The ease of entry into this sector has been reduced due to a multiplicity of permits and licensing required by various levels of government authorities. Evers and Mehmet (1994: 3) indicate that the labour-absorption capacities of the informal sector in Indonesia have declined during the 1980s and will decline further, owing to the sixth national Five-Year Development Plan that aims to increase the productivity and income levels of informal-sector participants. As a consequence, the position of small-scale enterprises in Indonesia is changing. They have to operate under harsher conditions and conform to increasing government regulations, or face destruction. The Indonesian government has attempted to shift petty traders into multi-storey buildings and formalise trading arrangements (Guinness 1994), to demolish the squatter housing in Jakarta (Jellinek 1991), and to banish pedicabs in many large cities in Java (Van Genugten and Van Gemert 1996; Kartodirdjo 1981) and *bemo* (motorised pedicabs) from Jakarta (Jellinek 1991). Food sellers are losing their businesses because of the authorities' reluctance to issue licences, and lottery sellers and street guides have to hide because of occasional police raids (Dahles 1996, 1998b). The recent blows that tourism in Bali has suffered have been blamed on the informal sector, leading to harsher government measures against hawkers, food vendors and street guides (Bras and Dahles 1998).

Outside the formal markets, traditional marketing arrangements still flourish and as soon as the police turn their backs, hawkers, lottery sellers, and street guides carry on with their business activities as usual. The small-scale enterprises are a vigorous and visible element in the tourist sector, employing a large proportion of the labour force. Manifold opportunities for informal employment are emerging within the industry and should be considered when examining the economic impact of tourism. Much of the employment generated by tourism is in the form of self-employed, small-scale entrepreneurs. The employment effects of this small-scale sector are often excluded in the assessment of tourism employment because available employment data are not accurate (Cukier 1996). As research has shown, 'traditional' modes of production do not disappear because of modernisation, but exist alongside the capitalist economy and often even grow and become more important in the process (Tokman 1978; Evers 1981; Drakakis-Smith 1987; Hart 1993; Verschoor 1992). Empirical evidence has shown that capitalist development does not absorb the traditional economy but that both simply exist side-by-side in a dual system (Gilbert and Gugler 1992; Drakakis-Smith 1987). Instead of the informal sector being a hindrance to economic expansion, some governments in developing countries have started to regard this sector as very productive, offering employment opportunities to the poor and less educated with little state intervention. The same thing applies to tourism: in developing countries, where the presence of the informal sector is well established, it continues to increase and diversify as the tourism industry develops (Cukier 1996: 55).

Beyond general discussions of the impact of transnational organisations, the tourism literature is remarkably uninformative on the interrelationship between tourism development and small businesses. However, more recently, the tourism literature shows an emerging interest in small entrepreneurs in tourism in developing countries (Samy 1975; Cohen 1982; Britton and Clark 1987; Smith and Eadington 1992; Wilkinson and Pratiwi 1995; Dahles 1996, 1997, 1998a, 1998b; Bras and Dahles 1998). This new interest is partly related to investigations into the potential of tourism for sustainable developments. New forms of tourism are required that consist of smaller-scale, dispersed and low-density tourism developments located in and organised by communities where it is hoped they will foster more meaningful interaction between tourists and local residents (Brohman 1996). These forms of tourism depend on ownership patterns which are in favour of local, often family-owned, relatively small-scale businesses rather than foreign-owned transnationals and other outside capital. By stressing smaller-scale, local ownership, it is anticipated that tourism will increase multiplier and spread effects within the host-community and avoid problems of excessive foreign exchange leakages.

It may seem obvious that where tourism is thriving, it absorbs many people who would otherwise be unemployed. However, these emerging employment opportunities do not necessarily contribute to community development. As dependency theorists would argue, employment in tourism is not apt to contribute to community development, it rather increases dependency on foreign markets (Harrison 1992). Small entrepreneurs are not supposed to be appropriate forces in community development either, as they are regarded as exponents of capitalist thinking (Schuurman 1993). According to the classical economic definition, however, entrepreneurs are instruments for transforming and improving the economy and society, as they are regarded as innovators and decision-makers pursuing progressive change. Following this definition, entrepreneurs can thrive only under minimal state intervention and in a free market economy (Clapham 1985). Liberal market theorists believe that prosperity is the outcome of successful individual entrepreneurship. Local ownership implies that economic success for the entrepreneur results in benefits to the local economy. Tourist developments based on local entrepreneurship are much more likely to rely on local sources of supplies and labour and are much less likely to produce negative socio-cultural effects associated with foreign ownership. Local tolerance to tourism activities is significantly enhanced if opportunities exist for active resident involvement in the ownership and operation of facilities. Small-scale operations can also respond more effectively to changes in the marketplace and fill gaps overlooked by larger, more bureaucratic organisations (Echtner 1995). In Indonesia, as in many other developing countries, the informal sector consists of self-employed individuals and small family enterprises operating under strong government restrictions. The issue that has to be raised is to what extent the small-scale tourism enterprises actually constitute a vital, innovative force in the Indonesian tourism industry and whether they contribute to community development at the same time.

To understand the conditions under which this sector operates and how it copes with diminishing government tolerance, we need to understand the social meanings embodied in the production, exchange and consumption of the tourism product. As the activities of small-scale tourism enterprises do not unfold in a vacuum, we need to analyse their definition of the situation, the arenas of their actions, their routines of social interaction, social networks and the kind of resources they control. To analyse the processes by which small entrepreneurs manage their everyday social worlds, this chapter deals with three interrelated questions:

1. What are the characteristics of Indonesian government policy towards tourism in general and small-scale tourism developments in particular?
2. What characterises small-scale entrepreneurs in the Indonesian tourism sector, how do they operate and interact with other entrepreneurs in this sector in their local community and beyond?
3. What significance do small businesses in the Indonesian tourism industry have for community development?

Tourism, government policy and the dual economy in Indonesia

In many developing countries, including Indonesia, the state has discovered tourism as a source of modernisation. Tourism is supposed to contribute significantly to economic development in terms of measurable growth. International and domestic tourism has grown considerably since the 1980s. The number of foreign visitors increased by more than 200 per cent between 1988 and 1995, and the income from foreign tourism more than doubled between 1990 and 1994 (Table 10.1). The government estimates that in the year 2000 about 6.5 million foreign tourists will visit Indonesia, yielding $9 billion of foreign exchange earnings. Growth scenarios for the next ten years anticipate visitor arrivals to double and income from foreign tourism to triple (Parapak 1995).

The Indonesian government has put much effort into large-scale tourism development. Indonesia's strategic national and regional plans emphasise the construction of integrated resorts as a strategy for tourism development. The basic model for provincial plans has been the Bali Tourism Provincial Master Plan from the 1970s that designates limited zones for tourism development and is controlled through the state-owned Bali Tourist Development

Table 10.1 Number of foreign visitors and revenues

Year	*Number of visitors*	*Revenues (in US$)*
1988	1,254,000	1.0 billion
1989	1,569,000	1.3 billion
1990	2,734,000	2.1 billion
1991	2,964,000	2.5 billion
1992	3,205,000	3.3 billion
1993	3,455,000	4.0 billion
1994	3,731,000	4.8 billion
1995	4,030,000	

Source: Statistik Kunjugan Tamu Asing, Biro Pusat Statistik, Jakarta (1996).

Corporation. The most celebrated result of these government efforts to boost international tourism is Nusa Dua, an isolated area in south Bali boasting the highest concentration of carefully monitored five-star luxury resorts in Indonesia. The hawkers, food and souvenir stalls, masseuses, informal guides, and beggars that are so prominent in other Balinese tourist places, are absent from Nusa Dua as, owing to its isolated location, these groups are effectively kept out (Cukier 1996). Although the Nusa Dua concept of controlled development has been deficient in many ways (Wall 1996), it is being used by the government as a model for 16 Tourist Development Corporations in other provinces (Sammeng 1995). In doing so, it is hoped that 'quality' tourism will be encouraged, which is a euphemism for attracting rich tourists.

Large-scale tourism development accomplished with large-scale and partly foreign investments suited the Indonesian state under the Suharto regime which acted on behalf of the ruling élite. To protect the élite's interests and facilitate their access to land and other resources, the government's role in community development was characterised by an overtly anti-participatory mode. As Midgley (1986) points out, the anti-participatory attitude is found in states where the government sees community involvement as a threat rather than an opportunity. This anti-participatory attitude is reflected in tourism policy. The responsibility for managing tourism is distributed over two main levels of government, the national and the provincial/local levels. The national government is not willing to transfer authority for programmes that produce large financial benefits to provincial and local levels. This applies to the accommodation sector and associated travel agencies rated in the four- and five-star category, generating the government-targeted 'quality tourism'. Government agencies and offices have been established in the provinces to service the tourism industry and increase the number of tourist arrivals, particularly in the 'quality' category. Airlines and travel agents, hotels and tour operators based in Jakarta or overseas, control much of the tourist sector (Guinness 1994). As a consequence, a large percentage of the profits from tourist traffic do not go to the local people. Procedures relating to land acquisition, tenure and utilisation, and the management of star-rated hotels and resorts do not fall under the authority of the regional government. It may be clear that tourism policy in Indonesia does not include any perspective on community participation in tourism developments.

Critics point to large resort developments, particularly resort enclaves, as being major contributors to the negative impacts of tourism (Harrison 1992; Wood 1993). Not only are they often out of scale with the indigenous landscape and ways of life, it is argued, but they consume large quantities of capital which could be more usefully applied in other ways. The involvement of outside investors is necessary because of the large capital requirements which inhibit the participation of local people. The result is that profits are taken out of the community. Management positions often go to outsiders, as few local people have the appropriate skills. Local residents are often denied access to resources, such as beaches, which they previously used, and reap few benefits from the developments. In contrast, it is suggested that smaller developments are less disruptive, have more modest capital requirements which permit local participation, are associated with higher multipliers and smaller leakages, leave control in local hands, are more likely to fit in with indigenous activities and land uses, and generate greater local benefits (Wall and Long 1996).

The Indonesian government seems to realise that the preferences and tastes of foreign tourists are changing. The market only partly corresponds with the objectives of the state-controlled tourism industry. The needs and tastes of Western visitors are different from the Asian growth markets (Japan and the Newly Industrialising Countries), not to mention the domestic markets that emerge in many developing countries with growing prosperity and political stability. Moreover, there are indications that the Western market is becoming fragmented because of changing consumer tastes for the highly specialised and individualised programme and accommodation that resort tourism cannot supply (Urry 1990, 1995). While a couple of years ago the government was reluctant to facilitate low-cost tourism since that type of visitor is associated with drug-using hippies allegedly having a rather detrimental effect on local communities (McTaggert 1977), at present they have a different attitude. In a meeting of international tourism experts in Yogyakarta in 1995, the former Director General of Tourism, Mappi Sammeng, recognised that tourism trends in the 1990s indicate that many travellers are seeking alternatives to large-scale, beach-oriented resort development. The government agreed to encourage small-scale projects, especially in the outer islands, to support tourism developments in still underdeveloped destination areas. Small entrepreneurs are supposed to play a key role in enhancing local participation in tourism and, hence, contribute to sustainable development. However, given the priority that the Indonesian government directs towards resort development, one may doubt whether these statements made by the Director General in a meeting of tourism experts define a new direction in national tourism policy or simply an adjustment to an international audience.

Small entrepreneurs

An entrepreneur is regarded as a person who builds and manages an enterprise for the pursuit of profit in the course of which he or she innovates and takes risks, as the outcome of an innovation is usually not certain. Boissevain (1974) distinguishes between two distinct types of resources that are used strategically by entrepreneurs. First-order resources include resources such as land, equipment, jobs, funds and specialised knowledge which the entrepreneur controls directly. Second-order resources are strategic contacts with other people who control such resources directly or who have access to people who do. Entrepreneurs who primarily control first-order resources are called patrons; those who control predominantly second-order resources are known as brokers. While patrons manipulate the private ownership of means of production for economic profit, brokers act as intermediaries, they put people in touch with each other directly or indirectly for profit, they bridge gaps in communication between people. Entrepreneurs can become brokers if they occupy a central position which offers them a strategic advantage in information management.

If we categorise small-scale tourism businesses in Indonesia according to Boissevain's distinction between patrons and brokers, we find that most of the local owners of small businesses act as patrons, while many categories of people that mediate between them and their clients usually combine patronage with brokerage, i.e., exploiting a combination of first- and second-order resources. Other people operate as brokers, i.e., they make a living exclusively by manipulating second-order resources. To understand the way in which small entrepreneurs in the Indonesian tourism industry operate, we have to distinguish two categories: independent patrons and network specialists.

Independent patrons

Local owners of small-scale tourism businesses act as patrons since they control first-order resources. They own land, real estate, money, equipment, and other means of production. We find them in various tourism-related sectors and branches:

- the accommodation sector: the owners of homestays, *losmen*, and guesthouses;
- the tour and travel branch: the owners of small travel agencies that arrange excursions in the vicinity of their home base;
- the transport sector: the owners of pedicabs, horsecarts, and minibuses who rent their vehicles to drivers who in turn transport tourists;
- the restaurant sector: the owners of foodstalls, street cafes and small restaurants offering local and 'tourist food';
- the souvenir business: the owners of small shops, stalls and workplaces producing and selling souvenirs, woodwork, leatherwork, batik paintings, clothes, cheap jewellery, sunglasses, toys, and other gadgets tourists buy;
- the beauty service sector: predominantly self-employed women operating as specialists in massage, manicure, hairstyling, and make-up;
- the entertainment business: non-professional groups and individuals giving performances in traditional dance, theatre, shadow play, and gamelan music, and in popular genres of music, like the innumerable reggae bands playing in the streets and cafes at night.

In addition to these enterprises, there are large numbers of tourist-related businesses operating in the semi-legal or illegal sphere. These enterprises are diffcult to investigate and, due to lack of data, have to be left out of this analysis, i.e., escort services, brothels catering particularly to tourists, gambling and drug dealing. The latter has been associated with bars and discotheques. The small enterprises are established without large capital investments, sometimes in collaboration with foreigners, often on a provisional basis, operated with no or only a few employees, preferably family members, and not always in possession of appropriate licences and government permits. However, these enterprises are easily combined with several other economic activities providing the owner with considerable flexibility to move in and out of the sector.

One of the many small-scale entrepreneurs in tourism-related businesses are the owners of homestays or *losmen*. These are family-run accommodations in which local people take visitors into their homes in much the same way as bed-and-breakfast accommodations (Van der Giessen and Van Loo 1996; Wall and Long 1996). Homestays are most prominent in Bali, where locals started to take in foreign backpackers as early as the 1960s (Mabbett 1987). These are low-budget establishments, usually owned by local families. They consist of less than ten rooms, offer basic facilities, and are run by family members and perhaps a couple of non-family employees. Homestay owners are able to meet the relatively low standards sought by budget tourists. They require a relatively low initial capital outlay and this type of business is potentially accessible to any family with a spare room or the space to build one. Not all the enterprising families are Balinese; people come from all over Indonesia, especially from Java, to benefit from the Balinese success in the tourism market. There are also foreigners in business, many of whom came to Bali as tourists, became friends with local people, and stayed, some of them playing a tense game with immigration officers and hoping that their Balinese partners and frontmen will not decide to do without them (Mabbett 1987, 1989). However, returns to owners and investors are unclear. Operators generally lack formal management and pricing skills, and actual prices paid vary since bargaining is a part of most transactions. Most likely, homestay operation is a supplementary income for many operators. The conversion of traditional Balinese homes to tourist uses has been an important factor in enabling residents of Bali to become involved in tourism as entrepreneurs and not merely as employees (Wall and Long 1996). Successful involvement is a result of the willingness of local entrepreneurs to take risks and strategically use first-order resources such as the home, land, agrarian products, and family networks in order to establish businesses to meet a growing demand from foreign tourists. Although small-scale enterprises can succeed economically with relatively small numbers of tourists because of low overheads and limited leakages, the small-scale, low-occupancy rates and low prices restrict the economic benefits. As the management of a homestay does not require any specific knowledge and training, it is not a potential career for the young. As Wall and Long (1996), Van der Giessen and Van Loo (1996), and Long and Kindon (1997) observe for Bali, and Smith (1994) and Peeters and Urru (1996) observe for other destinations in developing countries, homestay operations may have implications for family life, particularly as most require considerable family input in their operation and management. In addition to homestay tasks, owners often take on other forms of employment and also maintain their social and religious responsibilities. Homestay operations increase household workloads, particularly for women and children who do the cooking, cleaning, laundry and shopping.

Small establishments find it hard to attract individual tourists and, because of the fragmentation of the industry, they are in a weak bargaining position *vis-à-vis* the large tour operators. This has led to suggestions that small enterprises in the hotel industry should get together and market their products jointly (DeKadt 1995). By establishing co-operative networks and banding together, small firms could afford the diagnostic and consultancy services beyond their reach as individuals; they could also market their products jointly, blending sectoral co-operation with enterprise competition. Networks and systems have become the fundamental survival route of small tourism firms because of the benefits they offer in terms of cost advantages, marketing, information access, and, ultimately, flexibility. Informal self-help organisations among small owners of accommodations, restaurants, and travel agencies do emerge, but do not yet represent a vigorous force. Independent patrons seem to fear the loss of their autonomy if they get involved in any organisation, including private local initiatives. However, there is informal co-operation among these entrepreneurs in order to enhance their profit. There are occasional examples of small enterprises co-operating with multinational firms, for example, local food producers selling their product directly to big hotels (Telfer and Wall 1996). Although economic considerations are the basis of co-operative efforts, personal networks, family obligations and friendship are necessary for mutual support. Homestay owners help other non-contesting tourism businesses in a particular area to market their product.They spread flyers containing information about excursions and trips organised by local travel agencies or the menus of local restaurants. Sometimes they offer brochures of other homestays or losmen in what will most probably be the next destination of their guests. There are boards with the business cards of local masseurs, beauty salons, transport services and souvenir shops. Many accommodations are visited daily by the owner's family or friends from a compatible business to check on new guests and offer their products. The most important force in the marketing of the local tourist product is the category of small entrepreneurs known as brokers. There are manifold relations of interdependence between local enterpreneurs and mobile operators of local networks.

Network specialists

Individuals occupying a central position in the tourism business can become brokers provided they possess considerable amounts of second-order resources, i.e., flexibility in terms of time, space and social contacts. Brokers are in the middle of the action. They have free access to tourists and to the businesses in the local tourism industry. They are informed about where and when tourists arrive, where and how long they stay, where they will go next, their activity patterns, their expectations and needs, and their spending power. Moreover, they are familiar with the local market and the opportunities to match demand with supply in a way that enables them to make a profit. There are several categories of entrepreneurs that fit this profile. Taxi drivers, pedicabmen, lottery sellers, street vendors and informal guides are the most flexible and mobile people in a tourist area. They usually have access to a large network and up-to-date information. Others are more limited in their freedom because of permanent or part-time jobs, but nevertheless have free access to tourists and connections in the local tourism industry that enable them to operate as brokers. Brokers are frequently employees of small accommodations and travel agencies, waiters and bartenders, formal guides, shop assistants and security men. Many brokers combine a number of 'informal' jobs to expand their network. To illustrate how the network specialists operate, I will briefly discuss the strategies of informal guides.

Indonesian beach resorts and other popular tourist destinations are swarming with young men offering their services to passing tourists (Mabbett 1987; Wolf 1993; Dahles 1996, 1998b). Often street guides operate in small, loosely structured groups of friends, sharing and controlling a hangout, but many operate on their own. The best hangouts – bus stations, post offices, restaurants, shops, discotheques and other tourist spots – are controlled by young men who occupy a relatively powerful position in the world of street guides. In some places, power might come from being a member of a local family (Dahles 1998b), in other places from being of Javanese origin (Cukier 1996). Within groups there is a loosely structured division of tasks. Success in the 'tourist hunt' is largely dependent on communicative abilities, outward appearance and mastery of foreign languages. Group members scoring high on these criteria usually take the initiative of approaching tourists. If they are successful, they receive the biggest share of the profit. If they fail, other group members try to take over. While the tourist experiences a series of approaches by different young men, these men often belong to the same group. If one of the group 'has a bite' the others follow him and his guest at a distance, observing which restaurants or shops are visited, what souvenirs are bought, and how much money is spent. After the tourists and their 'guide' have left, group members enter the shop, guesthouse or restaurant to collect the commission, which will be divided amongst them.

The street guides present themselves as 'friends', as someone who wants to help. Whatever the tourist desires, from an inexpensive place to eat and drink to drugs or a prostitute, guides know the right place and the right person. To cater to tourists' needs, they apply knowledge of a pragmatic kind: where to shop for souvenirs, where to stay or eat, how much to pay; and they possess an understanding of human nature in their ability to read a social situation and turn it to their advantage. They have general conceptions about tourist motivation, national stereotypes and tourist types (Dahles 1998b; Crick 1992). To understand the tourists' needs, the street guides have to study their background. Tourists find themselves being interviewed by these guides. This interviewing renders information which enables the guide to classify the tourist: Is he rich? Will he spend a lot of money on souvenirs? What kind of souvenirs? Is she travelling alone? Is she available as a sexual partner?

These young men usually combine work as a street guide with different kinds of economic activities. They frequently work as touts for several shops, restaurants, guesthouses and bars on a commission basis. Some sell toys, ice cream or cold drinks in the street; others work incidentally as barkeepers, waiters, security guards, bellboys in guesthouses, taxi drivers and even pedicab men. Sometimes they invest money in the bulk buying of goods to sell at a profit. Most of the time, they do all these things. They spread themselves thinly over a wide range of deals rather than plunge deeply into any one activity. Since tourism is a very precarious industry, small-scale peddling and guiding are flexible jobs which fit in with busy high-season and calm off-season trade. The most marked aspect of the street guides' work, i.e., accompanying tourists, is only a strategy to earn money. The guides' income consists mainly of the commissions they receive for taking customers to the small hotels, souvenir shops, bars and restaurants. The commission is a percentage of the selling price of the products and services purchased by the tourists. Guiding, as such, does not provide a substantial income. They have to be satisfied with tips from tourists, a meal, a drink or cigarettes. If they are lucky, they receive gifts of some value: western consumer goods like wrist watches, walkmans, radios, leather jackets, an invitation to accompany the tourist to the next destination, or even a ticket to the tourists' home country. If a tourist is reluctant to buy souvenirs and does not tip, the guide has no other choice than to walk away, since he has no right to charge a fee. Windfalls are made, but they may go for days without business. Street guides have high hopes and expectations of the tourists, and try to keep the relationship going even after the tourist has left for the next destination or returned home. They like to boast with notebooks full of addresses of their foreign tourist friends, business cards and letters from abroad.

Whereas the 'independent patrons' are closer to the formal end of the continuum of employment in the tourism industry since their enterprises require government licences and permits, brokers are closer to the informal end. Street guides do not require a licence as long as they do not pretend to be proper tourist guides and stay away from the formally guided tourist attractions where they would pose a threat to the professional guides. They therefore prefer to introduce themselves as 'friends' or 'students' who want to improve their language proficiency. In many tourist places, there is a booming trade in fake student ID and guide licences. Nevertheless, guides go into hiding when the news of an upcoming police raid is passed by word of mouth. However, these raids that are incidentally organised by the local police force together with the provincial department of Justice are not effective in controlling unlicensed guiding. The guides are long gone by the time the police arrive. Similar raids are held among pedicab men, street vendors, taxi drivers and stall operators. If they fail to produce a valid ID and licence, they are fined and have to stop doing business immediately. Goods and pedicabs are confiscated and stall holders have to pack up and move on. Souvenir sellers and masseuses are chased away by security guards and police (Bras and Dahles 1998). While network specialists have to deal with police raids that hinder their mobility and free access to tourists and social contacts, patrons have to cope with increasing regulations regarding the payment of taxes, registration and licences. There is a 10 per cent government tax added to all transactions in the restaurant and accommodation sector, and owners have to prove with a certificate that they have paid the tax. Local governments send officials to inspect the accommodations and restaurants and register the number of rooms or tables and the type of facilities. The idea is to establish a classification system (Van der Giessen and Van Loo 1996). However, as local people point out, the registration is meant to check whether the owners are evading taxes. As a consequence, many accommodation owners are vague about the number of rooms and the quality of the facilities, and prefer not to be listed in government-issued brochures on local 'places to eat' or 'places to stay'. Officials are bribed to turn a blind eye to an extension to a losmen, and only some of the tourists staying in the accommodation are officially registered. The government's measures do not seem to discourage people from establishing a business. Although we lack official figures, it seems that a growing number of Indonesians are attracted to tourist areas to benefit from tourism developments. Tourism is a catalyst of population movements. Young men in particular are attracted by the Western consumption patterns and lifestyles in tourist areas. Tourists enact their dreams of western consumerism and hold the promise of a better life. While poverty and the lack of economic opportunity are reasons to leave one's community, the promise of quick money and a better future pull people to tourist areas.

Conclusion

In the introduction to this chapter the issue was raised whether small-scale tourism enterprises have the potential to constitute a vital force in community development. It may be clear that small-scale businesses are both a reservoir of hidden unemployment as well as overt innovative and enterprising forces that are integrated in the local economy through extended networks. Small entrepreneurs are operating neither in the margin of the tourism industry nor the centralised state bureaucracy. They form an integral part of the industry and are becoming more and more important to the Indonesian government in its search for new forms of tourism to develop a product that is competitive in the global market. Small-scale entrepreneurs are neither representatives of a traditional, informal economy, nor do they fit in definitions of the completely modern, formal, capitalist sector. They participate in both economies. Depending on the kind of resources, some (i.e., the patrons who rely predominantly on private property) operate more in the formal sector, others (i.e., the brokers who rely predominantly on personal networks) participate more in the informal sector. The relations between the formal and informal spheres are characterised by mutual dependence. Patrons depend on brokers for the advertising and marketing of their accommodation, restaurant or shop. Brokers, in turn, depend on patrons for their commission and access to tourists. Patronage and brokerage actually constitute a safety belt that allows small entrepreneurs to operate in a rather flexible manner. For these reasons small entrepreneurs seem to be more successful and small enterprises to be more sustainable than the large-scale resorts that have recently experienced severe setbacks due to a collapse in major tourist markets.

Small entrepreneurs, patrons as well as brokers, depend heavily on networks based on personal friendships and family relations. The most effective forms of co-operation are those that meet the need for income security, insurance and protection, marketing, advertising and information. However, this does not mean that small entrepreneurs react passively to changing markets and to attempts of the bureaucracy to regulate economic life from above. They are enterprising, inventive, innovative and creative in the exploitation of new niches in the market as well as in the law. Small entrepreneurs are extremely flexible in using changing consumer preferences and government regulations to their advantage. Pedicab men benefit from the tourists' need for local transport and company, homestay owners from the budget travellers' need for an inexpensive place to sleep and/or to experience 'authentic' local life in a *kampung* (neighbourhood), masseuses from western beauty standards, and street guides from the disorientation of the unorganised tourists.

The hustle and bustle of small entrepreneurs which is dominating street life in Indonesia's tourist areas and big cities is not necessarily an indicator of a vibrant economy. In established tourist areas the competition among small entrepreneurs is tough, causing irritation among tourists and leaving the competitors with less and less income. In emerging tourist areas the business activities of local people are hampered or even destroyed by more experienced migrants from developed tourist areas. Tourism development cannot rely exclusively on free-market principles, but has to be supported and to a certain extent controlled by the state. If provisions are not made to increase local economic participation, this greatly increases the likelihood of the domination of Indonesia's tourism sector by transnational capital from the metropolitan core. Instead of focusing on regulatory measures, the Indonesian government needs to facilitate tourism development by making available public goods necessary for small entrepreneurs to generate a tourism product and to set out rules and legal measures without stifling entrepreneurial initiatives. In Midgley's (1986) terms, then, the Indonesian government has to adopt a participatory approach to support local tourism developments.

References

Boissevain, J. (1974) *Friends of Friends. Networks, Manipulators and Coalitions*, Oxford: Basil Blackwell.

Booth, A. (1990) 'The tourism boom in Indonesia', *Bulletin of Indonesian Economic Studies* 26, 3: 45–73.

Bras, K. and H. Dahles (1998) 'Women entrepreneurs and beach tourism in Sanur, Bali. Gender, employment opportunities and government policy', *Pacific Tourism Review*, 1, 3: 243–56.

Britton, C. and W. Clark (eds) (1987) *Ambiguous Alternatives: Tourism and Small Developing Communities*, Suva, Fiji: University of South Pacific.

Brohman, J. (1996) 'New directions in tourism for Third World development', *Annals of Tourism Research*, 23, 1: 48–70.

Clapham, R. (In co-operation with R. Strunk, H. G.H. Schaldach, G. Clapham) (1985) *Small and Medium Entrepreneurs in Southeast Asia*. Research Notes and Discussion Paper no. 49, Singapore: Institute of Southeast Asian Studies.

Cohen, E. (1982) 'Marginal paradises: bungalow tourism on the islands of Southern Thailand', *Annals of Tourism Research*, 9, 2: 189–205.

Crick, M. (1992) 'Life in the informal sector: street guides in Kandy, Sri Lanka' in D. Harrison (ed.) *Tourism and the Less Developed Countries*, London: Belhaven Press.

Cukier, J. (1996) 'Tourism employment in Bali: trends and implications' in R. Butler and T. Hinch (eds) *Tourism and Indigenous Peoples*, London: International Thomson Press.

Dahles, H. (1996) 'Hello Mister! De rol van informele gidsen in Yogyakarta, Indonesië', *Derde Wereld*, 15, 1: 34–48.

Dahles, H. (1997) 'Tourism, petty entrepreneurs and sustainable tourism' in H. Dahles (ed.) *Tourism, Small Entrepreneurs, and Sustainable Tourism. Cases from Developing Countries*, Tilburg: ATLAS.

Dahles, H. (1998a) 'Tourism, government policy, and petty entrepreneurs in Indonesia', *South East Asia Research*, 6, 1: 73–98.

Dahles, H. (1998b) 'Of Birds and Fish. Streetguides, tourists and sexual encounters in Yogyakarta, Indonesia' in M. Oppermann (ed.) *Sex Tourism and Prostitution. Aspects of Leisure, Recreation, and Work*, New York: Cognizant Communication Corporation.

DeKadt, E. (1995) 'Tourism policy management after structural adjustment', Plenary V: International Tourism, Development, and Policy-Making. International Conference on Cultural Tourism, Indonesian-Swiss Forum on Culture and International Tourism, Yogyakarta.

Drakakis-Smith, D. (1987) *The Third World City*, Routledge Introductions to Development, London and New York: Routledge.

Echtner, Ch. (1995) 'Entrepreneurial training in developing countries', *Annals of Tourism Research*, 22, 1: 119–34.

Evers, H.-D. (1981) 'The contribution of urban subsistence production to incomes in Jakarta', *Bulletin of Indonesian Economic Studies*, 17, 2: 89–96.

Evers, H.-D. and O. Mehmet (1994) 'The management of risk: informal trade in Indonesia', *World Development*, 22, 1: 1–9.

Gilbert, A. and J. Gugler (1992) *Cities, Poverty and Development. Urbanization in the Third World*, Oxford: Oxford University Press.

Guinness, P. (1994) 'Local society and culture' in H. Hill (ed.) *Indonesia's New Order.*

The Dynamics of Socio-economic Transformation, London: Allen and Unwin.

Harrison, D. (1992) 'International tourism and the less developed countries: the background' in D. Harrison (ed.) *Tourism and the Less Developed Countries*, London: Belhaven Press.

Hart, K. (1993) 'Markt en staat na de Koude Oorlog – De informele economie opnieuw beschouwd', *Derde Wereld*, 12, 2: 87–103.

Jellinek, L. (1991) *The Wheel of Fortune. The history of a poor community in Jakarta*, Sydney: Allen and Unwin.

Kartodirdjo, S. (1981) *The pedicab in Yogyakarta; Surabaya*, Yogyakarta: Gadjah Mada University Press.

Long, V. H. and S. L. Kindon (1997) 'Gender and tourism development in Balinese villages' in Th. M. Sinclair (ed.) *Gender, Work and Tourism*, London/New York: Routledge.

Mabbett, H. (1987) *In Praise of Kuta. From slave port to fishing village to the most popular resort in Bali*, Wellington: January Books.

Mabbett, H. (1989) *The Balinese. All about the most famous island in the world* (2nd edn), Wellington: January Books.

McTaggert, D. W. (1977) 'Aspects of the tourist industry in Indonesia', *The Indonesian Quarterly*, 4: 62–74.

Midgley, J. (with A. Hall, M. Hardiman and D. Narine) (1986) *Community Participation, Social Development and the State*. London/New York: Methuen.

Parapak, J. (1995) *The curricula in tourism education and training. The case study of Indonesia*, Department of Tourism, Post and Telecommunication, Republic of Indonesia. Education and Training for Industry Growth Conference, Jakarta.

Peeters, S. and J. Urru (1996) 'Homestays: Een glimp van het echte Javaanse leven? Een onderzoek naar de budget-accommodatie sector in Yogyakarta, Indonesië' unpublished MA Thesis, Department of Leisure Studies, Tilburg: Tilburg University.

Sammeng, A. M. (1995) 'Tourism as a development strategy', Plenary Address. Plenary V: International Tourism, Development and Policy-Making. The 1995 Indonesian-Swiss Forum on Culture and International Tourism, Universitas Gadjah Mada, Yogyakarta, Indonesia.

Samy, J. (1975) 'Crumbs from the table? The workers' share in tourism' in S. Tupouniua *et al. The Pacific Way*, Suva, Fiji: South Pacific Social Sciences Association.

Schuurman, F. (1993) 'Introduction: Development Theory in the 1990s' in F. Schuurman (ed.) *Beyond the Impasse. New directions in development theory*. London/New Jersey: Zed Books.

Smith, V. L. (1994) 'Privatization in the Third World: small-scale tourism enterprises' in W. F. Theobald (ed.) *Global Tourism: The next decade*, Oxford: Butterworth-Heinemann.

Smith, V. L. and V. Eadington (eds) (1992) *Tourism Alternatives: Potentials and problems in the development of tourism*, Philadelphia: University of Pennsylvania Press.

Statistik Kunjugan Tamu Asing (1996), Jakarta: Biro Pusat Statistik.

Telfer, D. and G. Wall (1996) 'Linkages between tourism and food production: an Indonesian example', *Annals of Tourism Research*, 23, 3: 635–53.

Tokman, V. (1978) 'An exploration into the nature of informal sector relationships', *World Development*, 6: 1065–75.

Urry, J. (1990) *The Tourist Gaze. Leisure and Travel in Contemporary Societies*, London: Sage.

Urry, J. (1995) 'Tourism, Travel and the Modern Subject' in J. Urry *Consuming Places*, London/New York: Routledge.

Van der Giessen, E. and M.-Ch. Van Loo (1996) 'Bali, a "paradise" with two faces. A study of low-budget accommodations in Kuta and Ubud on the island of Bali in Indonesia', unpublished MA Thesis, Department of Leisure Studies, Tilburg: Tilburg University.

Van Genugten, E. and H. Van Gemert (1996) 'Tukang Becak. A Study of becak drivers who operate in the tourist sector of Yogyakarta, Indonesia', unpublished MA Thesis, Department of Leisure Studies, Tilburg: Tilburg University.

Verschoor, G. (1992) 'Identity, networks and space: new dimensions in the study of small-scale enterprise commoditization' in N. Long (ed.) *Battlefields of Knowledge: the interlocking of theory and practice in social research and development*, London: Routledge.

Wall, G. (1996) 'Perspectives on tourism in selected Balinese villages', *Annals of Tourism Research*, 23, 1: 123–37.

Wall, G. and V. Long (1996) 'Balinese Homestays: an indigenous response to tourism opportunities' in R. Butler and T. Hinch (eds) *Tourism and Indigenous Peoples*, London: International Thomson Business Press.

Wilkinson, P. F. and W. Pratiwi (1995) 'Gender and tourism in an Indonesian village', *Annals of Tourism Research*, 22, 2: 283–99.

Wolf, Y. (1993) 'The World of the Kuta Cowboy. A growing subculture of sex, drugs and alcohol is evident among male youth in the tourist areas of Bali and Lombok as they seek an alternative to poverty', *Inside Indonesia*, 15–17 June.

Wood, R. E. (1993) 'Tourism, culture and the sociology of development' in M. Hitchcock, V. T. King and M. J. G. Parnwell (eds) *Tourism in South-East Asia*, London and New York: Routledge.

11 Gili Trawangan – from desert island to 'marginal' paradise

Local participation, small-scale entrepreneurs and outside investors in an Indonesian tourist destination

Theo Kamsma and Karin Bras

Introduction

> The day the people called 'a cloudy Tuesday' (18-4-1995) was the moment on which the cleaning team formed by the provincial government started their action. Houses, tourist accommodation, restaurants, souvenir shops were demolished in a short time. [. . .] The cleaning began at eight o'clock and four hours later 25 houses were broken down [. . .] The result of this action was not only felt by the local people themselves, but also by the tourists who were having a holiday on the paradise island. Tourists usually wake up at nine o'clock, but this morning they were woken up earlier because their bungalows were target of the demolishing action. [. . .] As a result of this many of them shortened their holiday, which normally lasts a week.
>
> (*Bali Post*, 20 April 1995).

After 'cloudy Tuesday', the provincial government proclaimed a four-day visit prohibition in order to keep journalists, activists and tourists away from Gili Trawangan. In spite of this prohibition, the Gili Trawangan case has had national as well as international press coverage (Kamsma 1996; *Bali Post* 1995; Suara Nusa 1995; *Inside Indonesia* 1993; Breda 1997). The problems concerning this small island located in the province of Nusa Tenggara Barat in Indonesia, are characteristic of the Indonesian land policy in areas that are regarded as promising tourist destinations. Small-scale local initiatives often lose ground in favour of a government-controlled larger-scale development. In the tourists' search for new idyllic places, as the Gili Trawangan case will show, locals are often pioneers in meeting new demand for tourist facilities. They start homestays, food stalls and transportation, or hire out snorkel gear, motorcycles or mountain bikes. The initial investments are small, but gradually the local entrepreneurs expand their businesses, as at several locations on Lombok, where the success of small-scale entrepreneurs did not remain unnoticed. In the last ten years land ownership in present and future tourist areas has felt the growing pressure of commercial interests of a new élite. Wealthy enterprises and private entrepreneurs from the Nusa Tenggara Barat region but also from Bali and Java, have bought large parcels of land. Apart from land speculation it is their goal to develop large-scale tourist resorts, like the planned Putri Nyale resort in the south of Lombok, where an area of 1,250 ha has to be transformed into a sun-sea-sand destination that will include a marina, golf courses, dozens of hotels and other attractions. Or like Senggigi in the north-west, the first area being developed for tourism, that already had a history of land speculation and disputes. Tourism development is of course not the only cause of land speculation, but as the Gili Trawangan case will show, tourism development on Lombok is strongly characterised by large disputes about land and the right to build tourism facilities. There is a strong local involvement in this potential growth sector. The introduction of tourism led to economic diversification. In the past the people of Trawangan made their living through agriculture, fishing and herding small livestock. Nowadays you can find shopkeepers, *losmen* (homestays) and restaurant owners and personnel, souvenir sellers, boatmen, massage ladies, diving instructors and local tourist guides on Trawangan. But interest is also emerging from external entrepreneurs who are willing to invest in large-scale resorts. It remains to be seen whether these new initiatives will not be at the cost of the locals of Trawangan.

In this chapter we want to discuss how, through a government orientation towards so-called 'quality' tourism, which leads to drastic regulations and the involvement of outside investors, local participation in tourism is put under pressure. The concept of local participation is an important issue in the debate on tourism development in Third World countries and is usually discussed in relation to the growth of mass tourism and its negative impacts on local communities. In Third World countries the growth of tourism is not often accompanied by the creation of local linkages to spread the benefits of growth in social, sectoral and regional terms (Brohman 1996; Murphy 1985, 1994; Simmons 1994; Harrison 1992; Shaw and Williams 1994). Brohman gives an outline of the shortcomings that are commonly associated with the Third World tourism industry. He states that normally the three most lucrative elements of Third World tourism – marketing and the procurement of customers, international transportation and food and lodging – are dominated by vertically integrated global networks. The technical, economic and commercial characteristics of mass tourism sectors tend to favour the development of large-scale, integrated, multinational enterprises (Brohman 1996: 54). Foreign capital profits from local natural resources, but because of the high rate of foreign ownership the profits made in tourism are not locally distributed. The relatively isolated position of tourism in the local economy reduces linkages with other economic sectors. The consequences are low multiplier and minor spread effects outside the tourism enclaves.

These discussions have led to a general conclusion that tourism should be seen as a local resource and that the desires of local residents should be the principal tourism planning criterion. Essentially, a tourist destination has to be regarded as a community and the local residents as the nucleus of the tourism product. Local support is indispensable for developing a sustainable tourism product (Brohman 1996: 60). The introduction of the concept of local participation as 'the direct involvement of ordinary people in local affairs' (Schaardenburgh 1995: 10–11) is therefore a viable one. As locals are influenced by tourism developments themselves, they have to participate in plan making, decision-making and implementation in order to control changes that affect their lives (ibid. 10–11). This means empowering people to mobilise their own capacities, be social actors rather than passive subjects, manage resources, make decisions, and control the activities that affect their lives. Local tolerance to tourist activities is significantly enhanced if opportunities exist for locals to be involved in tourism as entrepreneurs through involvement in the ownership and operation of facilities, and not merely as employees in the hotel or restaurant sector.

The result of discussions on local participation in tourism was a boom in community development strategies and projects (see for example Saglio 1979; de Kadt 1979; Smith and Eadington 1992; Schlechten 1988; Bras 1991, 1994) and the introduction of terms like responsible, sustainable, grass roots and rurally integrated tourism. No attention has been paid in these discussions to existing, locally based small-scale private enterprises (Smith 1994; Shaw and Williams 1994). Private initiatives in the accommodation sector or adjoining sectors like the souvenir industry are evident wherever tourists emerge. The literature, however, is uninformative about the contribution of small-scale private enterprises to local communities. How do these small-scale enterprises operate and why are they left out of the discussions on local participation?

Furthermore, notions of small-scale entrepreneurship and local participation have to be described against the background of the regional and national tourism policy in Indonesia. The national government emphasises the importance of tourism within regional development and has defined new target areas in its *Repelita VI* (sixth Five-Year Development Plan 1993–1998). The eastern islands, still regarded as poor and backward, constitute one of the areas that get more and more attention in tourism promotion campaigns directed by the government. An analysis of the tourism development on Lombok, and more specifically on Gili Trawangan, will introduce the actors who play a role in tourism development and will provide an insight into the different interests that are at stake. One group of actors are the locals of Trawangan who, by operating small-scale private enterprises, try to improve their living conditions and raise their standard of living. The growing competition and the up-grading activities put the existing tourism entrepreneurs under pressure. Will the locals of Trawangan still be able to meet the required quality standards a few years from now, or will they have to make room for star-rated hotels and a golf course?

Gili Trawangan

Together with Gili Air and Gili Meno, Gili Trawangan is one of the popularly called Gili's, three tropical 'bounty' islands on the north-west coast of Lombok. Until 1976 Gili Trawangan was uninhabited. The island, covered with dense mangrove woods, was considered to be impenetrable. In 1976 the former provincial governor of NTB, Wasita Kusuma, gave his four children the right of customary use (*HGU – Hak Guna Usaha*) of a part of the island to start a coconut plantation. Other companies like Pt. Generasi Jaya and Pt. Rinta also obtained these rights. Officially, the land remained state property. The Sasak, the local inhabitants of Lombok, were recruited to work on these plantations. Due to the difficult circumstances, the lack of infrastructure and diseases, the harvests failed. The companies withdrew from the island but some workers stayed. They obtained one or two hectares of land and built themselves a livelihood with fishing, agriculture and cattle breeding. Nowadays about 150 families or 350 people live on the island, a number which doubles in the high tourist season in June, July and August. In 1995 there were 31 low-budget hotels with an average of six rooms each (Dinas Pariwisata Prop. DATI 1 NTB 1995).

The story goes that in 1981 the first tourist, a German, spent a night on the island (Mucipto 1994). He wrote about his experiences on the island to the publishers of a German travel guide. Usually travel guide publishers (i.e. Lonely Planet) rely heavily on contributions from travellers. The author of the travel guide used the information and mentioned the island as a new unspoilt paradise. Since then thousands of tourists have visited the island. As at first there were no accommodations or restaurants on the island, the tourists were invited by the villagers to stay at their homes, full board and lodging being offered for one price. Later on three or four households built bungalows together. Their relatives were employed in the accommodation, other islanders supplied food, beverages and transportation to the island. Gradually the standard of living improved as more people came to stay on the island and the locals became more and more absorbed into the tourist industry.

All the accommodation, bars and restaurants are grouped along the east coast beach, close to where the boats from the mainland arrive. The local residences were also scattered along the east coast beach. Some locals have their own dwelling, others occupy a bungalow or live in an extension built next to the restaurant. A sandy path circles the island and divides the beach from the bungalows. The restaurants are often located on the beach, a strip of about 25 metres at its widest. Bamboo is generally used as building material for the dwellings, but many entrepreneurs have recently upgraded their bungalows and restaurants by using bricks for construction. The compact layout gives the area a friendly village atmosphere.

The accommodation and restaurants on the island are no longer strictly run by people from the island or the nearby mainland. Through the years people from other parts of Lombok, but also from other islands and even from other countries, came to Trawangan to work or to operate a business. The majority, however, still originate from the island or the nearby mainland, like Bangsal and Pemenang or the city Mataram. In spite of this influx of 'outsiders' tourism enterprises are identical in form, meaning small-scale, locally oriented and built with relatively small investments. Foreigners are able to exploit a business only in co-operation with an Indonesian partner. In practice these foreigners show up only once or twice a year during their holiday. They consider their enterprise as a nice holiday address rather than an important source of income.

Trouble in paradise

In 1991 a group of entrepreneurs from the west of Lombok – the Basri Group – demanded that the Trawangan people who occupy the eastern coastal strip, the touristically most lucrative part because of the coral reef in front of the beach, should clear the area. The business-group had bought up the rights to use the land (HGU) from Pt. Generasi Jaya, which had held these rights since 1976 to start a coconut plantation. But the Basri Group was not the only one that claimed these rights. Another group of businessmen associated with Ponco Sutowo (Jakarta Hilton) and Nellie Adam Malik (widow of a former vice-president) are negotiating with two companies which have had government permission since 1986 to develop 200 of Gili Trawangan's 380 hectares for tourism (McCarthy 1994). A third party is the Bupati from West Lombok. He claims that the former governor of NTB, Wasita Kusuma, had no right to grant the HGUs on Gili Trawangan to his children. These rights should have been given at *kabupaten* (district) level. The children of the former governor Wasita Kusuma are the fourth party. For their part, they of course claim that although they left the island years ago, their HGUs are still legally valid. The government takes the position that the land is state property and that no claims are legally valid other than those granted by the government. As soon as it became clear that the development of tourism on Gili Trawangan is profitable, the interested parties staked their claims. The price of the land in tourism areas – approximately $5000 per acre in 1996 – is still rising. An investment in tourism promises to be lucrative and selling land-ownership or user rights in these areas is highly profitable. As a consequence the different parties were and still are entangled in different lawsuits. Parties that have not been mentioned yet in these lawsuit entanglements are the local people of Gili Trawangan. They also sought legal aid to protect their interests. The locals consider the claims made by the Basri Group as legally invalid. Before selling the land-user rights Pt. Generasi Jaya, from which the Basri Group bought the HGUs, let the land lie fallow for more than two years. In that period the people of Trawangan started to use the land for their own benefit. They claim that they are now legally entitled to use the land. In addition they consider the plans of the Basri Group to build a tourist resort inappropriate, since the land-user rights were granted to exploit a coconut plantation. On the basis of these two issues the Trawangan people started a lawsuit in 1991, which they lost. An appeal against this decision in 1994 was also lost. Paradoxically, the misuse of the land has served as a pretext for the provincial government to justify the 'slash and burn' actions by the police and army (Mucipto 1994). After the verdict in 1991 the local police went to the island to confiscate locally owned bungalows. The locals ignored them, however, and continued to operate their bungalows. The police returned in 1992 to demolish the lodgings. At the same time local activists visited the Minister of the Interior, Rudini, to halt the demolition. The minister sent a radiogram to cancel the action. Nonetheless, in September 1993 all of the bungalows were demolished by force (Mucipto 1994). As an alternative the provincial government issued a spatial planning concept for Gili Trawangan that allows locals to own and operate lodgings of not more than five rooms, on one hectare of land, in a designated area of the island which is the least desirable as it is not near the coral reef. The regional government (*Pemerintah Daerah* (Pemda)) and the regional tourist department (*Dinas Pariwisata Daearah* (Diparda)) claim that local businessmen do not meet spatial management requirements and cannot provide adequate services and accommodations (Mucipto 1994). The locals for their part claim that if the provincial government treated the local entrepreneurs in the same way as they treat the outsiders, by helping them to get the necessary business permits and access to bank credits, there would be no problem meeting the requirements of the provincial government. Some of the islanders accepted the deal with the provincial government and relocated their business to the designated area of the island. Others, partly because they did not possess the necessary documents to claim a new spot, turned their backs on the island embitteredly. A small group persists in its struggle. Even after the last 'cleaning action' of the army they rebuilt their businesses on the old spot, albeit on a smaller scale.

Present-day Trawangan

During a visit to the island in the summer of 1996 it appeared that the spatial planning concept drawn up by the provincial government had been carried out. Surrounding the area on the east coastal strip, where the new resort is planned, is a barbed-wire fence. According to the spatial planning concept it is also forbidden nowadays to build any type of construction directly on a beach front. In 1994 the businesses were still built on both sides of the main road. On the sea side there were mainly little shops, travel agencies and other small businesses. On the land side were the hotels and restaurants. In 1996 all the constructions on the sea side had vanished and flower boxes had been installed instead. On the land side there had been a regrouping of the accommodation. Every entrepreneur who accepted the deal with the provincial government was offered a small piece of land on which he could build a five-room *losmen* or homestay.

In the early years the bungalows on the island, and also on Gili Air and Gili Meno, were rather standardised. They were all simple wooden constructions. Prices were also pretty much the same, approximately $2–4 a night. As there were no independent restaurants operating on the island, three meals were usually included. This has now changed because there is a more diverse supply of accommodation and restaurants. Bricks are used more and more for construction, and nowadays it is even possible to get a fan in your room. To keep the customers in the restaurants and bars there are video shows and parties are held at a different hotel every night.

Lombok, a new tourist destination

Although the province Nusa Tenggara Barat (Lombok and Sumbawa) hosts no more than 3 per cent of the international visitors to Indonesia, Lombok in particular is becoming an increasingly popular destination. For a long time Lombok was not noticed as a tourist destination of any importance, mostly because of the popularity of its neighbour Bali. In his book *The Island of Bali* Covarrubias described Bali as an island that has the lush and splendid greenery of Asia, while Lombok is arid and thorny like Australia (Cederroth 1981: 26). For many visitors Bali was the only island worth visiting. Travel guides almost immediately state that tourists who visit Lombok should not expect the cultural refinement of Java or the lively dance and music of Bali. However, with the growing demand for new destinations the interest for Lombok began to develop, and nowadays Lombok is described as a new paradise. Anticipating the popularity of Bali the regional tourism department on Lombok used to promote the island through the slogan: 'You can see Bali on Lombok, but you can't see Lombok on Bali'. This slogan referring to the Balinese who live on Lombok has been dropped, but Lombok still seems to have difficulties in becoming more than a nearby replacement for its famous neighbour Bali. In promotional material the island is still linked with Bali. A hotel in one of the tourism areas, for instance, uses a banner with the following text to promote its happy hour: 'Lombok as Bali. In Sixties watching the Eastern Bali with cold draught beer Bintang. Buy one and get two.' And the Garuda in-flight magazine started an article about Lombok with a comparison: 'Lombok? Sure, I've heard of it. The Bali of 20 years ago, isn't it? Without all that commercialisation. Sounds great!' This is a popular view of Lombok. Images of empty white-sand beaches without jet skis, traditional villages without tourist buses and simple bungalow-style guest houses without pizza bars. This image is, however, no longer correct. The number of foreign as well as domestic tourists has increased rapidly in the last few years, as has the number of hotel rooms, as is shown in Table 11.1.

In spite of the relatively rapid tourism development Lombok is still predominantly an island of farmers. Fifty per cent of the Gross Regional Domestic Product (GRDP) comes from agriculture (Lübben 1995: 56). Although there has been an average annual rise of GRDP of 8.54 per cent,

Table 11.1 Number of visitors and hotel rooms in Nusa Tenggara Barat 1988–95

Year	*Foreign tourists*	*Domestic tourists*	*Star-rated hotel rooms*	*Non-star-rated hotel rooms*
1988	44,846	55,475	340	382
1989	56,148	67,146	340	1,037
1990	107,210	76,817	386	1,216
1991	117,988	99,011	549	2,167
1992	129,997	102,040	819	2,278
1993	140,630	106,907	859	2,463
1994	157,801	120,279	1,025	2,682
1995	167,267	140,940	1,183	2,949

Source: Dinas Pariwisata DATI 1 Propinsi Nusa Tenggara Barat (1995).

Nusa Tenggara Barat still is the province with the lowest income (Dinas Pariwisata Prop. DATI 1 NTB 1995). The development of larger industries on Lombok is negligible, but the growth of home industry, on the other hand, is considerable. Especially in the handicraft home industry there is a growth in outlets within the country, but also internationally (Lübben 1995: 55–6). The growing importance of tourism and handicrafts within regional development is noticeable. The contribution of the industrial sector to GRDP is only 3 per cent and trade and tourism contributes 16 per cent. Jobs in the hotel and restaurant sector, in transport and in the travel business are regarded as a solution to the high level of unemployment, estimated at 35 per cent.

Almost all accommodation and most of the tourism activities are located in three main areas, namely Senggigi, a beach resort on the west coast; the Gili's, three small islands on the north-west coast and Kuta, a small beach resort in the south. Outside these areas tourists are still rare enough to attract a great deal of attention. The tourism master plan is characterised primarily by large-scale projects launched by Javanese or foreign entrepreneurs like the Putri Nyale resort which is planned in the south of Lombok. In the UNDP (United Nations Development Programme) which studied Lombok's tourism potential in 1987 (WTO), the three beaches in South Lombok (Putri Nyale, Seger and Aan) were designated as an area suitable for the development of an integrated tourism resort. Altogether 1,031 luxury rooms and 953 middle-class rooms were planned. The project area is about 6 km long, from Kuta village in the west to the end of Aan Beach in the east (WTO 1987: 51). The Lombok Tourism Development Corporation (LTDC) has bought this 1,250 ha area and is going to turn it into a sun-sea-sand destination that will include a marina, golf courses, dozens of hotels and other attractions. The LTDC is a joint venture of the private organisation PT Rajawali Wira Bhakti Utama, which holds 65 per cent, and *Pemda*, the regional government of Nusa Tenggara Barat, with a 35 per cent holding. The prediction is that LTDC will create approximately 15,000 jobs during a 10–20 year period. In addition to the 15,000 jobs directly involved in tourism, an estimated 6,000 to 10,000 additional jobs will be generated by new activities such as administration, health, education and handicrafts (McCarthy 1994: 60). It remains to be seen whether the local population will profit from this new resort. McCarthy states that luxury hotels require sophisticated staff and that even in Bali the majority of hotel workers are not local (1994: 80–1). Lübben (1995) concludes for Kuta that even in the present situation the locals have a very limited share in the tourism development. Kuta has at the moment mainly *melati* hotels, bungalow-style accommodation and small restaurants. The initiatives come from people from Praya, Mataram or Bali and seldom from people from the Kuta area (1995: 95–6). Therefore, it remains to be seen whether the local people are prepared for tourism that is dominated by the international tourist industry. The planning of the Putri Nyale resort has already had an influence in this area. Through the years big parcels of land have been compulsorily purchased by the LTDC and the local owners received only low compensation, often not enough to buy a new plot of land in another area. LTDC bought the land in this area from the local farmers at a price of $70 per acre. The same land is now sold to Indonesian and foreign investors for $3,500 to $5,000 per acre, almost fifty times the amount of money the locals received (Kamsma 1996). The latest developments are land evictions, in order to build the newly planned international airport to make the Putri Nyale resort more accessible in the near future. The airport is planned to come into operation in the year 2000.

Tourism policy in Indonesia

The fifth Five-Year National Development Plan (FYDP) (1988–92) anticipated that tourism would be one of the driving forces for regional growth. The early 1990s were years of extremely rapid growth. International arrivals increased from 1.1 million in 1987 to over 4 million visitors in 1994 (see Table 10.1). The position of Indonesia as a tourist destination rose from the 11th most important in Asia in 1986 to 6th in 1994. The target during the sixth FYDP (1994/95–1998/99) is to increase the number of foreign tourists by an average of 12.9 per cent per year. In the final year of the sixth FYDP it is expected that 6.5 million foreign tourists will visit Indonesia. This will generate foreign exchange earnings of approximately US$ 8.9 billion. At the moment tourism is the third most important non-oil based source of foreign income after the timber and textile industries. The prediction is that within the next decade tourism will overtake oil as the principal source of income (Sammeng 1995: 3–4). Since 1992 a group of Indonesian researchers has made an inventory of the touristic potential of each of the 27 Indonesian provinces. Their goal is to formulate a balanced long-term tourism development programme. In their research they use the resort life cycle model designed by Butler (1980). Each province is assigned to a particular stage in the life cycle. Bali and Jakarta are considered to be in the rejuvenation phase. These two destinations together account for 50 per cent of the total number of international visitors to the provinces of Indonesia. A relatively great effort will therefore be required to consolidate their position. As the national government has defined new target areas in the *Replica VI* [1993–98], Bali especially is losing its privileged position in national tourism policy which the island held for 25 years. Government-initiated campaigns like 'Beyond Bali' deliberately promote new destination areas in Indonesia (Aziё 1995). 'There is more to Indonesia than Bali' and 'Bali and Beyond' have become powerful slogans to attract international tourists. The growth in tourism, as far as the Indonesian government is concerned, has to be generated by new destinations like Manado in Northern Sulawesi (diving and snorkelling), Sumatra's Niass islands (surfing), Kalimantan (jungle-trekking) and Lombok (beaches).

Quality tourism and small-scale accommodation

Although the national government encourages large-scale and small-scale tourism projects, depending on the specific character of a destination area (Sammeng 1995), in practice priority is given to the development of areas like Nusa Dua and Sanur in Bali which are designated as luxury beach resorts with mainly star-rated hotels. This focus on resorts, with an obvious preference for the 'quality' tourist, makes it difficult for the locals to participate in tourism, other than being employees in the hotels or restaurants. There are, however, numerous examples of how locals gradually got involved in tourism at times when tourism was not yet a major asset and a destination did not yet occupy a place on the 'tourist map' (Mabbeth 1987; Hussey 1989; Cohen 1982, 1996). They occupy themselves with the recruitment for organised tours, they offer transport facilities or, most noticeably, they exploit restaurants and lodgings. Not much research has been done on the extent to which the operation of small-scale private enterprises is beneficial to local communities. Some examples that highlight their importance come from the accommodation sector on Bali. Indonesia has a wide range of small-scale accommodations managed by local people, such as *losmen*, homestays, *wisma*, *penginapan* and guesthouses. Homestays are most prominent in Bali where locals started to take in foreign backpackers as early as the 1960s (Mabbeth 1987). Research done in Kuta, Bali in the 1980s (Hussey 1989; Wall and Long 1996) showed that local entrepreneurs were able to meet the relatively low standards sought by budget tourists because of the availability of beaches, the proximity of the airport and, maybe most important, the absence of competition from professional developers in this area. Before 1970 there were two hotels in Kuta, by 1975 the area contained more than 100 locally owned accommodations and 27 restaurants (Hussey 1989). The activities of the homestay-owners are entrepreneurial in the sense that their successful involvement resulted from a willingness to take risks, that they were able to recognise changes in tourist demands and that there were possibilities to manipulate traditional resources such as residence, land, agrarian products and family social networks. Even in situations in which resources appeared limited, available resources were used effectively by local people if they saw an opportunity to engage in entrepreneurial activities (Wall and Long 1996). In the beginning the demand for rooms was high. Within a year and a half and with an occupancy rate of only 40 per cent local entrepreneurs were able to make their investments profitable (Hussey 1989). Nowadays homestays in this area can still survive with only small numbers of customers, but the benefits are generally also very limited. In most cases, however, operating a homestay is not the only source of income. It is regarded as a possibility to earn additional income.

The future scenario for Gili Trawangan

On Gili Trawangan the local population responded to tourism by offering accommodation and additional services. They slowly developed the infrastructure and the facilities on the island to the point where almost everyone is involved in the local tourist economy. In the absence of a concentrated effort by the government or foreign funds to 'implant' tourism in the area, the tourist business developed spontaneously and on a small scale through local initiative. The tourists who visit Gili Trawangan are mainly young travellers who are satisfied with a nice beach, good company and modest accommodation. They do not demand extensive services, which made it possible for the locals successfully to take advantage of tourist needs and participate. Gili Trawangan, together with the other two Gilis, is now the best visited site on Lombok.

The success of Gili Trawangan attracted outside investors, who put considerable effort into penetrating the area and dislocating the locals. At the same time the Gilis became incorporated in provincial development plans, which do not provide for support to small-scale entrepreneurs. The focus of the provincial government is on the development of 'quality' tourism. Therefore, priority is given to building resorts with star-rated hotels, swimming pools and golf courses. This emphasis on 'quality' tourism will lead to rising land prices and to landowner disputes, as is the case on Trawangan. Efforts to upgrade the area, rising costs and more regulations will make it very difficult for the small-scale entrepreneurs to continue their businesses as they used to. Their low-budget accommodation and other modest initiatives will no longer meet the required standard of services and accommodation asked for in the resort type of development.

National plans state that priority has to be given to entrepreneurship in tourism through training programmes, marketing and management, so that small business people can compete in an international environment (Sammeng 1995). But these plans seem to exist only on paper. The Trawangan case shows that small-scale entrepreneurs are marginalised rather than supported through government measures. The riposte of the local entrepreneurs is that if they are given the same support as foreign investors by means of credit facilities, education and training, they also would be able to meet the required international standards.

After the regrouping of the accommodation setting some of the owners occupy the least attractive sites on the island and are allowed to exploit only five bungalows. In view of the future upgrading plans it remains to be seen whether these locals will still be able to make a living from accommodating tourists. Another question is whether their target group, the low-budget tourists, will still find their way to Gili Trawangan when the accommodation and the other facilities and services are upgraded and become more expensive. This is what already happened in Senggigi. Because of its luxurious image Senggigi is nowadays infamous among the young, low-budget tourists. The low-budget accommodation has been replaced by the star-rated hotels, and new initiatives to exploit modest guesthouses or losmen are impossible because of the exorbitant land prices. Where in Lombok will the low-budget tourists go when the Gilis are no longer affordable? Which new paradise will be created?

Conclusion

What has been neglected in the discussion about local participation are the existing locally based, small-scale private enterprises in tourism (Smith 1994; Shaw and Williams 1994; Wall 1995). Current research either focuses on the economic, environmental, social and cultural effects of large-scale resorts and the linkages these resorts have with the local community in the area of labour and agriculture (Telfer and Wall 1996), or research is centred on community-based or community development projects which emerged as a reaction to the development of mass tourism. Surprisingly, small-scale tourism enterprises, mostly in the form of family-oriented businesses, hardly get any attention although locally owned enterprises are generally smaller in scale and offer greater direct economic payback and control. These enterprises are more likely to rely on local supplies and labour, and local ownership implies that economic success for the entrepreneur results in benefits to the local community. While it is easy to sympathise with arguments used to promote small-scale developments, it is difficult to find examples in the literature of successful, small-scale indigenous tourism developments (Wall and Long 1996) and their contribution to local communities.

Why are these small-scale initiatives neglected? One reason could be the still persistent, romanticised, western idea that a local community in a developing country is a unity that has to be saved from larger-scale disruptive forces and that local private entrepreneurship undermines this idea of homogeneity. Opportunities for local entrepreneurship are not necessarily equally accessible. In order to succeed, apart from the availability of resources like capital, land and labour, an entrepreneur has to have the ability to recognise niches in the tourism market, to innovate and to take risks. Furthermore, the income does not necessarily find its way back to the local community. The successful entrepreneurship of those who had capital or offered good employment, can lead to a widening gap between them and less fortunate community members (Smith 1994: 166). An important additional question is: how local is local? What is the precise identity of the 'locals' who supposedly benefit from small-scale tourism developments? New opportunities for employment such as those offered in the tourist industry often attract migrants who have different characteristics from those of the local population. Stress is placed on the community, and problems of rising land prices, social organisation and cultural values can arise (Cohen 1982, 1996; Wall 1995; Smith 1994; Schaardenburgh 1995).

It is inevitable that opportunities for local entrepreneurship are not equally accessible. Every destination is essentially a community (quoted in Brohman 1996: 60), but what is often overlooked in community development programmes is that every community is also essentially heterogeneous and dynamic because there is a constant coming and going of residents. It is an illusion to think that every member of a community should be and can be involved in tourism development equally. As often stated before, (Brohman 1996; Smith and Eadington 1992; Vickers 1989) tourism creates winners as well as losers among local residents.

Without centrally planned tourism development projects the locals of Trawangan were very capable of launching a local tourism industry. Although not every inhabitant is equally involved in tourism and although not all of them benefit in the same way, it is evident that the standard of living has improved considerably over the years. The profits made in tourism are spent locally. The few outsiders who exploit tourism businesses on the island make no difference in this respect. Their businesses are small scale, they employ locals and the leakages are minimal.

In the Trawangan case all the objectives defined within local participation strategies have been achieved. The locals have participated in planning, as well as decision-making and implementation. They are at the centre of the development of their tourism product, a product that matches their present capabilities. In areas where tourism is developing rapidly and where new requirements determine the direction and pace of the developments, local entrepreneurs should be given the opportunity to zero in on the new market. Government or NGO support in the area of credit facilities, education and management are then a prerequisite.

References

Azië (1995) 'Indonesië: meer dan Bali alleen', *Azië*, maart 1995: 62–3.

Bali Post (1995) 'Bara Pariwisata Gili Trawangan Berkobar' 20, 21, 22 April.

Bras, C. H. (1991) 'De Diola als Attractie – Het geïntegreerd ruraal toeristisch project in de Basse-Casamance Senegal', unpublished MA thesis, Cultural Anthropology, University of Amsterdam.

Bras, C. H. (1994) 'Toerisme: de ontdekking van het echte Afrika. Het dagelijks leven van de Diola in de Basse-Casamance (Senegal) als toeristisch attractie', *Vrijetijd en Samenleving* 12, 1/2: 15–29.

Breda, A. van (1997) 'Kleinschalig Toerisme Lombok onder Druk' *IFM*, maart 1997: 16–17.

Brohman, John (1996) 'New Directions in Tourism for Third World Development', *Annals of Tourism Research* 23, 1: 48–70.

Butler, R. W. (1980) 'The concept of a tourist-area cycle of evolution and implications for management', *The Canadian Geographer* 24: 5–12.

Cederroth, Sven (1981) *The Spell of the Ancestors and the Power of Mekkah. A Sasak Community on Lombok*. Gothenburg Studies in Social Anthropology 3. Acta Universitatis Gothoburgensis.

Cohen, E. (1982) 'Marginal Paradises: Bungalow tourism on the islands of Southern Thailand', *Annals of Tourism Research* 9 (2).

Cohen, E. (1996) *Thai Tourism. Hill Tribes, Islands and Open-Ended Prostitution*. Studies in Contemporary Thailand no. 4. Bangkok: White Lotus Co., Ltd.

Dinas Pariwisata Prop.DATI I NTB (1995) 'Selected Tourism Data 1988–1995'. Nusa Tenggara Province Mataram, November 1995.

Harrison, David (ed.) (1992) *Tourism and the Less Developed Countries*. London: Belhaven Press.

Hussey, A. (1989) 'Tourism in a Balinese village', *Geographical Review* 79, 3: 311–25.

Inside Indonesia (1993) 'Social Conditions and the Economy. The battle for Gili Trawangan' March 1993: 25.

de Kadt, E. (1979) *Tourism – passport to development? Perspectives on the social and cultural effects of tourism in developing countries*. New York: Oxford University Press for the World Bank and UNESCO.

Kamsma, M. J. (1996) 'Grof geweld in het Paradijs'. *Volkskrant* 23 November 1996.

Lübben, Christel (1995) *Internationaler Tourismus als Faktor der Regionalentwicklung in Indonesien: untersucht am Beispiel der Insel Lombok*. Berlin: Dietrich Reimer Verlag.

Mabbeth, H. (1987) *In Praise of Kuta. From slave port to fishing village to the most popular resort in Bali*. Wellington: January Books.

McCarthy, J. (1994) *Are Sweet Dreams Made of This? Tourism in Bali and Eastern Indonesia*. Indonesia Resources and Information Program (IRIP), Australia.

Mucipto, I. (1994) 'Development for whom? The tourism industry in Lombok, Indonesia', unpublished article.

Murphy, P. E. (1985) *Tourism: A Community Approach*, London: Methuen.

Murphy, P. E. (1994) 'Tourism and sustainable development' in Theobald, William F. (ed.). *Global Tourism. The next Decade*, Oxford: Butterworth-Heinemann Ltd.

Saglio, C. (1979) 'Tourism for Discovery: A Project in Lower Casamance, Senegal' in E. de Kadt (ed.), *Tourism – Passport to Development? Perspectives on the social and cultural effects of tourism in developing countries*. New York: Oxford University Press for the World Bank and UNESCO.

Sammeng, A. S. (1995) 'Tourism as a Development Strategy. International Tourism, Development and Policy-Making', plenary session at the 1995 Indonesian–Swiss Forum on Culture and International Tourism. Universitas Gadjah Mada, Yogyakarta, Indonesia, August 1995.

Schaardenburgh, A. van (1995) 'Local Participation in Tourism Development. A study in Cahuita, Costa Rica', unpublished MA thesis. Tilburg University.

Schlechten, M. (1988) *Tourisme balnéaire ou tourisme rural intégré? Deux modèles de développement sénégalais*. Editions Universitaires Fribourg Suisse.

Shaw, G. and Allan M. Williams (1994) *Critical Issues in Tourism. A Geographical Perspective*. Blackwell Publishers, England.

Simmons, David G. (1994) 'Community Participation in tourism planning', *Tourism Management* 15(2): 98–108.

Smith, Valene L. (1994) 'Privatization in the Third World: small-scale tourism enterprises' in Theobald, William F. (ed.) *Global Tourism. The next Decade*. Butterworth-Heinemann Ltd.

Smith, Valene L. and William R. Eadington (1992) *Tourism Alternatives. Potentials and Problems in the Development of Tourism*. Philadelphia: University of Pennsylvania Press.

Suara Nusa (1995) 'Mengapa terjadi tragedi Trawangan yang menumpahkan air mata?' 26–4.

Telfer, David J. and Geoffrey Wall (1996) 'Linkages between Tourism and Food Production', *Annals of Tourism Research* 23, 3: 635–53.

Vickers, A. (1989) *Bali: A Paradise Created*. California: Periplus Editions Inc.

Wall, Geoffrey (1995) 'People outside the Plans'. Plenary II Human Resource Development in Special Interest Tourism. The 1995 Indonesian-Swiss Forum on Culture and International Tourism. Universitas Gadjah Mada, Yogyakarta, Indonesia, 24 August 1995.

Wall, Geoffrey and Veronica Long (1996) 'Balinese homestays: an indigenous response to tourism opportunities' in Richard Butler and Tom Hinch (eds) *Tourism and Indigenous Peoples*, London: International Thomson Business Press.

World Tourism Organisation (1987) *Tourism Development Planning Study for Nusa Tenggara. Tourism Development Plan Package A – Lombok*. United Nations Development Programme/Directorate General of Tourism Jakarta, Indonesia. World Tourism Organisation, Madrid, Spain.

12 Tourism in Friesland

A network approach

Janine Caalders

Introduction

Community development and bottom-up planning have become popular strategies for rural (tourism) development. The exploitation of the endogenous potential of regions is of central importance in these concepts. Ideally, however, concepts for rural development should take internal factors, external influences and contextual elements into account. It has been suggested that the concept of 'networks' can be useful in this respect (Caalders 1997).

In this chapter, the concept of networks for rural tourism development will be explored. First ideas on endogenous and exogenous development will be placed into perspective in a general introduction on rural tourism development. A brief overview of some relevant ideas from regional economic and planning literature and their applicability for rural tourism development follows. Finally, some recent planning efforts in the south-west of Friesland (the Netherlands) will be described and evaluated using these ideas as a guideline.

Rural tourism development

Developing tourism is popular among rural areas searching for ways to escape from the negative spiral of agricultural decline and depopulation (Butler *et al.* 1998). It is regarded as one of the most promising sectors for employment creation and regional economic development. In an era where nostalgia, the search for authenticity and 'untouched' nature, as well as the need for action space are important trends in tourist demand, rural areas are becoming increasingly popular tourist destinations. The tourist market is still growing, but competition among destinations is also increasing. More remote rural areas have to compete not only with each other, but also with cities and 'the city's countryside' (Bryant and Johnston 1992) for the favour of visitors.

In this process of increasing competition regions are pressed to continually adjust their product to the needs and wants of the market. In many cases, market-led developments in recreation and tourism have generated negative impacts for host communities, and eventually for tourism revenues as well. Hawaii and the Spanish coast are much-quoted examples in this respect. Here tourism development has completely transformed the existing regional physical and social structures: landscapes become dominated by large hotel buildings, local culture changes and traditional employment sectors lose the battle for resources (rising prices of land, water, food etc., migration of young employees). What remains is a standardised tourist place, whose attractions can easily be copied in other locations: a product that has lost its unique character and can compete almost only on price (Dietvorst 1996). In rural areas, there is also a danger of these kinds of developments taking place. Many rural areas are already focusing on tourism development and this '. . . may well be creating a situation in which many regions will have to continue to invest in tourism infrastructure, facilities and promotion not so much to gain competitive advantage, but just to survive' (Jenkins *et al.* 1998: 62).

Such negative impacts have led to a search for other concepts for tourism development through which these negative impacts can be avoided. These include for example soft tourism, sustainable tourism, green tourism and community development. These approaches tend to emphasise endogenous, locally based and small-scale developments. The basic argument is that this type of approach will lead to a more sustainable development of tourism and that it can help preserve local identity against the pressure of external (tourism) influences.

This alternative way of tourism development is gaining momentum. There is a danger, however, of stereotyping two development paths: the 'good', small-scale endogenous and community-involved development against the 'bad' large-scale, exogenous, market-led development. Many mixtures of these two types exist, and besides this, the possible drawbacks and weak elements of the alternative approaches should not be ignored. Too much introspection or endogeneity can lead to an 'entrophic death' (Camagni 1991: 140) by slowly strangling innovation. A focus on bottom-up development may lead to an inward view and tourism development that is based only on nostalgia. It may cause the 'musealisation' of the countryside and a loss of dynamism.

Exogenous influences should not be regarded as threats, but as a means for innovation. A clear view of the world outside the region is needed in order to be able to attune the regional product to the market, an aspect often lacking in rural tourism (Jenkins *et al.* 1998). What is needed, therefore, is an approach to rural development that can unite both dynamics and development, as well as a preservation of identity and sustainable development. In order to create a framework for such an approach, insights from regional economics and regional planning disciplines can be useful. We will elaborate on these ideas below. Subsequently we will evaluate how these insights can be applied in the context of rural tourism development.

Some theories on regional development and innovation

In regional economic planning the emphasis has long been on attracting large foreign industries capable of creating regional 'growth poles' and serving as a pull factor for other economic activities. Usually this meant the settlement of branch-plants of large multinational enterprises, operating like 'cathedrals in the desert'. These types of developments exposed regions to global economic trends, making them vulnerable to forces they were unable to influence. The 'foot-loose' industries stayed in the region only as long as there were competitive advantages (Verhoef and Boekema 1986; Camagni 1995).

Such negative effects led to a search for alternative economic strategies. It was suggested that economic activities should have a local base in order to avoid the negative impacts of 'foot-loose' industry (e.g. Verhoef and Boekema 1986). When economic activities are locally embedded, they cannot simply be transferred to another region and therefore they provide a more sustainable basis for further development. There are other positive effects too. Changes that are driven locally or that are adapted to local circumstances are less disruptive of existing structures. New developments should link up with already existing 'carriers of development' in a region (Elsasser 1987). An endogenous approach can thus provide more prolonged and (socially) more desirable development. The importance of exogenous links and influences should, however, not be marginalised.

Camagni has developed a model that stresses the importance of both endogenous and exogenous linkages. He states that innovations are more likely in certain types of environment. These environments are labelled 'innovative milieux' (Camagni 1995; Camagni 1991; see also Fromholt-Eisebith 1995). This concept is based on the observation that the local environment performs some important functions for firms. Locality or proximity is of central importance in this respect, because of:

1. the presence of local resources of human capital, that are mobile within the area, but 'quasi-immobile' outside the area; this creates an effect of 'collective learning' and contributes to the creation of a local image;
2. the presence of mostly informal contacts among local actors, which creates a certain 'atmosphere';
3. the presence of synergy effects that stem from a common background (cultural, political and psychological), creating 'tacit codes of conduct', helping to 'uncode complex messages' and to form common representations and beliefs (Camagni 1991: 133–4).

A milieu achieves an innovative character mainly through linkages between different types of *local* actors. However, openness (i.e. linkages to actors external to the milieu) is also essential. These external links are important amongst others for retaining creativity and innovation.

> Autarchy in a cultural and technological sense and a sole reliance upon local entrepreneurial capabilities are definitely mistakes in the long run. This is due to the limited reaction capability and competitiveness of any small area in the face of massive international evolutionary processes. Therefore, co-operation with external institutions, firms or public agencies and research centres is crucial for the continuous recreation of local competitiveness and innovation capability [. . .].
>
> (Camagni 1995: 324)

Though *informal* relations are very important, especially as catalysts for innovation, after a while it becomes desirable for firms to have a more formal or institutionalised regulation of linkages:

> [. . .] endogenous and exponentially growing locational costs, which may be considered as the opportunity cost of utilisation of the 'milieu', and evident limits in the static or dynamic performance of the 'milieu' itself, push towards the creation of a new organisational and behavioural model, a new 'operator' enhancing the control capability of the firm upon its turbulent environment.
>
> (Camagni 1991: 135).

This new operator means that more formal and explicit linkages should be created. The reduction of uncertainty is the most important motive for firms to seek co-operation and regulation through such formal linkages. For the region as a whole, formalisation is important for retaining an innovative milieu.

Some specific types of actors within networks are of crucial importance. Actors who act as a 'bridge' between different types of economic and socio-cultural activities can play a central role in the innovation process (Fromhold-Eisebith 1995; Kamann 1989; for an application of this concept in relation to physical planning see Dietvorst and Hetsen 1996). These actors are positioned on the intersection between different local networks, or between local and external networks. They can provide a network with information on market trends, developments in other regions or developments in relevant sectors. When bridge actors are specifically given the task to translate information from different external sources for a firm or sector, they have been labelled 'gatekeepers' by some authors (see Kamann and Strijker 1991). A bridge position does, however, not necessarily bring advantages; if the actor is marginalised in both networks, then none of the parties will profit.

Ideas on innovative milieux have been inspired by the fast-growing 'new' industrial areas in the 1980s and the question of why some areas were able to react to the possibilities offered by the processes of global economic restructuring more successfully than others. It would be most interesting, however, to try to see to what extent the creation of innovative milieux in 'lagging' regions can be stimulated.

Planning for innovation through network management

For the stimulation of innovative milieux, two types of approaches are relevant. In the first place the approach of Camagni, who discerns some guidelines for regional economic development, focusing on policy efforts of the regional government or development agencies. On the other hand, ideas on the design of planning *processes* and their implementation in the regional context are of interest. These approaches are examined below.

In line with his argument on innovative milieux, Camagni identifies five elements for regional planning that can stimulate the creation of these milieux in lagging regions:

1. the integration of different policies, in order to attune sectoral policies and create the necessary preconditions for a balanced development;
2. the utilisation of external production units that can provide jobs and function as catalysts;
3. policies should be directed towards precise elements of the production sector and a policy of pure assistance should be avoided;
4. efforts should be concentrated spatially in the most promising parts of the area;
5. the stimulation of co-operation agreements between local and external firms and institutions.

(Camagni 1995: 338).

These elements focus mainly on the content of regional policy and have their roots in regional economic development theory. In recent policy and planning theories, emphasis is mainly put on the design of the planning process as such. These ideas have been developed as a reaction to approaches to government planning, where the aim was to implement 'blueprints' of the desired future situation. The most important part of planning in this tradition was the creation of the actual planning document, which is based on expert (and politicians') knowledge. The implementation of the plan in societal and political processes was not an issue (Hidding 1995). In complex planning environments, where the commitment of various societal groups is needed in order to achieve the goals that have been set in centrally developed plans, this approach has continually failed to deliver the desired results. Network approaches to planning therefore focus on the process of planning and the interaction with the planning context, rather than on the desired final results.

In network approaches, society is regarded as a web of (formal and informal) interrelations between different types of actors: individuals, societal (interest) groups, government bodies, commercial concerns and so on. Each of these actors has their own values, goals and interests. They try to achieve their goals through using instruments to try to steer other actors. In network-management approaches, this mutual dependence of all parties is not regarded as negative, but as a fact that should be acknowledged (de Bruijn and ten Heuvelhof 1991: 27). This means that those actors aspiring to steer societal processes should adopt a specific attitude in order to be successful. An important prerequisite is that the complexity of networks is accepted and all important actors should be involved in the decision-making process. Problems arise when actors that are in a position to block this process are left out of decision-making. Thus top-down government planning is bound to fail when the government is dependent on other parties (with different aspirations) to achieve the goals set. A centralist approach may be regarded as more vigorous, but in fact leads to less effective policy.

A common critique of network management as a policy instrument is that it is time-consuming and complex. It is important to realise that the reason for this lies not in the network approach but in the complexity of the problems being addressed. Complexity is inherent in the fact that the actors involved may have very different perceptions and goals. On the one hand, the urge to score in the short term should be sacrificed to the building up of sustainable relationships with other actors in the network. These relationships should be based on mutual trust. On the other hand, needless consumption of time and unnecessary complexity should be avoided. If processes take too long, the commitment of actors to co-operate may be lost. An important role is reserved for the co-ordination of network processes. A sensitivity to steer between different opinions and well-timed action are required. This requires professional management.

In practice, the position of governments or government bodies in network situations is a difficult one. When bargaining over goals and means of regional development, the public can get a feeling that the 'public good' is traded off in the bargaining process. Especially when several planning networks are operating in a region at the same time, this can have negative effects on its credibility and legitimisation. There thus exists a tension between effectivity of planning processes and the need for a consistent planning content. This points to an additional element deserving attention: the design of the planning process. In the following section, the various elements mentioned will be translated into principles for rural tourism development.

Principles for rural tourism development

Theories on regional development and innovation make it clear that both linkages within the area, as well as linkages with firms and institutions outside the region are important. When focusing on rural tourism, we can more specifically deduce three types of linkages that are essential for an innovative development. These are linkages within the tourism sector (marketing and product development), between the tourism sector and other regional actors (integration) and with the outside world (innovation and market knowledge) (cf. Caalders 1997). How can the formation of such linkages be stimulated? In order to answer this question, the guidelines mentioned in the previous paragraph should be applied to tourism development. It is important to take into account, however, the specific nature of tourism as an economic sector and its impact on the region as a whole. Aspects that are relevant in the context of this chapter include the following:

- The tourist product differs from most other products in that it is consumed in the region itself. In the case of rural tourism, this often means that the region as a whole is considered an attraction by the visitors.
- The tourist product is partly constituted by the dreams and fantasies of its customers (Seaton and Bennett 1996). This means that *coding* is an essential part of the product (Dietvorst and Ashworth 1995). Moreover, a region can have a very different image for various tourist groups. These groups will have different perceptions of the tourist product and different holiday experiences.
- Rural tourism is often characterised by a large number of entrepreneurs who jointly produce the regional product.
- The rural tourism sector generally consists of relatively small enterprises.
- There is a tension between successful tourism development (which generally means more visitors) and the attractiveness of rural regions (which includes quietness). This tension should be dealt with by tourism planners. Retaining the unique character of a region should be a central principle of tourism plans if success in the long run is desired (Dietvorst 1996; Caalders 1997).
- The landscape is an important part of the tourist product. This means that all activities that have an impact on the landscape also have an impact on the attractiveness of the region for tourists.
- Tourism is a fragile industry, susceptible to external forces beyond the control of its suppliers (Seaton and Bennett 1996).

When linking these characteristics of rural tourism to the planning guidelines mentioned in the previous paragraph, the following principles for rural tourism planning can be deduced.

Integration of different policies

Policy integration is needed between sectors (sectoral), but also between different local authorities in a region (spatial). Spatial integration is needed in order to create complementary rather than competing infrastructure and services. Sectoral integration is necessary in order to prevent innovations that are stimulated in one sector being hampered through unintended or unexpected effects in other sectors.

In the context of tourism development, integration might be even more important than for rural development in general. This is related to the fact that many activities in other sectors have a direct impact on what can be regarded as the tourist product. In agriculture for example, enlargement of scale, changes in the European Common Agricultural Policy (CAP), the subsidisation of certain sectors rather than others, or the fall or rise of world market prices, will all have visual effects in the landscape. But also housing or industrial development have such effects. Investments in infrastructure can be of vital importance for the tourist sector. Also a less tenacious element like the image of the region (that is influenced not only by tourist promotion, but much more by political events and so on) is of crucial importance for tourist development. This type of aspect is, however, difficult to manage.

Utilisation of external production units

Tourism investments by external actors can be useful catalysts for tourism development, but only if they can be integrated in a more comprehensive regional strategy. This means that initiatives should preferably link up with the (desired) regional image and identity and that the coherence of the total regional product should be strengthened by these investments. Other types of investments should be handled with great care, especially those that bring about substantial physical transformations. As investments in the physical infrastructure, such as hotel buildings, roads and holiday parks, are projected to have a life cycle of several decades, their attraction as tourist products should be measured on this time-scale as well. If (large-scale) external investments do not link up with time-and-place specific characteristics, the (comparative) attraction will not last for long.

On an organisational level, efforts should be taken to promote co-operation between local and external entrepreneurs, for example in the field of promotion.

Precise allocation of intervention policies and avoiding pure assistance

Aid should be given mainly to those types of projects that can create a momentum for the whole sector rather than furthering the interests of one enterprise. In general, they should either be directed towards projects that stimulate entrepreneurial activities, or towards innovative activities that can be passed on to the region. It is desirable that projects are co-financed by those actors that benefit. Attention should be paid to the viability of the projects in the long run and the institutionalisation or formalisation of linkages in order to create a durable effect on the region. Projects can be directed towards investments in physical infrastructure, but also in human capital, towards the stimulation of co-operation and so on.

Focusing on the most promising parts of the area

Focusing on the most promising parts helps to allocate investments and efforts where they will have most effect. Not only physical aspects are of importance here, but also linking up with already existing initiatives and activities. Dynamics that are already present in the area can thus be pushed a little further.

In general, the principle of zoning is very useful for tourism planning. This provides a means both to develop the most promising parts of the area, as well as protecting the more vulnerable parts. It can be useful when different tourist types are interested in different aspects of the regional product. It can also help to keep tourist numbers within certain limits, or at least to keep the impact spatially concentrated. In practice, however, it is very difficult for popular areas to keep visitor numbers down without taking strongly restrictive measures.

Stimulation of co-operation agreements

Not only is the stimulation of co-operation between local and external firms important, but also co-operation between local firms. The fact that the tourist product consists of the joint products of many entrepreneurs and the fact that the rural tourism sector consists mainly of small firms makes co-operation in the field of marketing and promotion all the more important. The offering of holiday packages or the co-operation of several small entrepreneurs can help to approach markets that cannot be served by individual firms. A professional 'gatekeeper' that can provide the sector with relevant information, for example, on market trends will benefit the whole sector.

Design of the planning process

Involvement of all relevant parties in the decision-making process is a principle that goes beyond the integration of policies. It has the effect of involving new parties in the plan-making and decision-making process. This is important not only because it is presumed to be a more democratic way of planning, but also because it is a more effective way in the long run. New types of opportunities or development options can come up in this way. It can help identify relevant activities or dynamics already present and help avoid negative impacts or conflicts. Tourism development not only affects the tourism sector, but also has an impact on the region as a whole. Many authors have stressed that those who benefit from tourism development are often other than those who experience the negative impacts (e.g. Pearce 1994).

Professional management is not only needed for effective promotion and marketing of the regional tourist product, but is also important for the planning process. This is true especially in situations where conflicts among various interest groups are likely, or when many groups are involved. If processes do not proceed as desired, precise interventions or stimulation of groups might be a good instrument. Local governments or development agencies should be aware of the different goals of local actors and the instruments they can adopt for achieving their goals.

In order to illustrate this conceptual framework, we will describe the progression of some policy efforts in the province of Friesland, in the north of the Netherlands. The south-west of this province has been designated as 'Valuable Cultural Landscape' (*Waardevol Cultuurlandschap*: WCL). In our evaluation we will try to establish to what extent these planning efforts have been successful in the light of the principles mentioned.

WCL policy and the Frisian WCL region

WCL policy in the Netherlands

In 1994, 11 regions in the Netherlands were designated Valuable Cultural Landscapes (WCL) (Ministerie van Landbouw 1994). The WCL policy aims to create a new dynamic for the countryside, while at the same time retaining valuable aspects of nature and landscape and the economy. This means that agriculture should remain an important economic sector in these areas. WCL areas are characterised as follows:

> Valuable Cultural Landscapes are regions with substantial nature and landscape values and an important geological meaning, that are attractive from a tourist point of view. Agriculture – and on some occasions forestry – is an important economic carrier in these areas. Functions are usually strongly interwoven; interdependence and influence can lead to tensions between agriculture, nature and landscape. The specific qualities and values are under pressure or affected negatively, which threatens to cause a loss of the special character of these areas and thus a loss of their attractiveness.
>
> (Ministerie van Landbouw 1994)

The main aims of the policy are:

- to stimulate a process of change aimed at preservation and strengthening of the specific qualities of the area coupled to sustainable agriculture. This should take place through a project-wise bottom-up approach. Projects are judged by the 'regional perspective': a policy statement that should be created jointly by the various regional actors.
- Agriculture should remain an important economic sector in the area. A minimum of 60 per cent of WCL money should be used for agriculture-related projects. If income from agriculture diminishes, additional sources of incomes for farmers should be sought.

The Frisian WCL region

The WCL in Friesland is situated in the south-west of the province (see Figure 12.1). It is designated 'Objective 5b' and has received some finance through

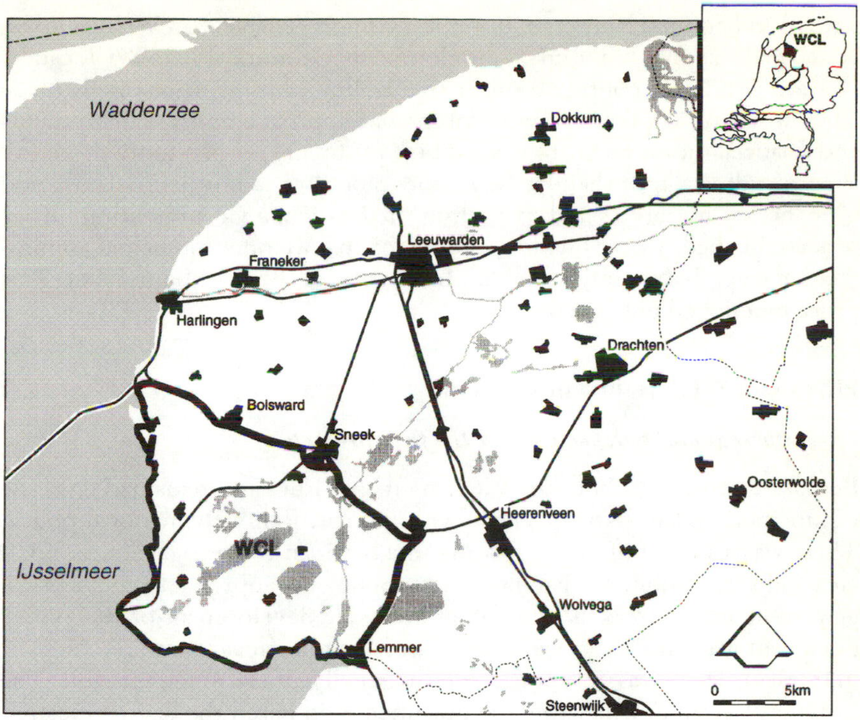

Figure 12.1 Location of the Frisian WCL area.

the EU LEADER-programme. It covers an area that has a varied landscape, with lakes, woodlands and picturesque villages and towns. It stands out from the rest of the province because of its relief, formed during the penultimate ice age. The municipality of Gaasterlân-Sleat, the heart of the WCL region, has a particularly long history as a tourist destination. At present, the area is most renowned for its sailing and yacht basins.

The municipality of Gaasterlân-Sleat is traditionally the most important tourist destination in the area. Tourist accommodation here suffers from the dialectics of progress, in the sense that most enterprises are rather small and their quality is not in accordance with contemporary tourist demand. Because of their small-scale character room for investment is limited. Large-scale investments, mainly in bungalow parks and yacht basins, increased during the 1980s and the beginning of the 1990s and existing tourist enterprises were upgraded. The municipality has long tried to frustrate the development of agri-tourism and other similar small-scale accommodation and campsites. In the other municipalities these issues did not play such an important role in tourism policy.

Before the region received its WCL status in 1994, a different landscape policy applied for the area, named 'National Landscape'. This policy had some goals that are similar to WCL, notably the preservation of the small-scale natural and cultural landscape. It was less comprehensive than WCL as it was concerned mainly with tourism development (Grontmij 1989). WCL can be characterised as a continuation of this policy, though its scope is more encompassing, its aim is a more integrated approach and the instruments and financial means make the desired bottom-up philosophy more powerful. For co-ordination of the National Landscape Programme, stimulation and guidance of initiatives and marketing of the area, a Co-ordinating Office was established. Co-ordination took place mainly on a municipal administrative level, between the different local policies (Brandenburg 1996). This office functioned until 1996.

History of WCL policy in Friesland

1994: the regional perspective and the first projects

Because uncertainty about the WCL status for Friesland remained until the last moment, a late start was made with the creation of the regional policy. There was a period of only a few months between the recognition as WCL area and the deadline for handing in the policy for approval to the national government. Whereas in some other areas the development of this policy document was a truly regional activity, in which all relevant parties took part, in Friesland there was not enough time to get all parties around the table. The provincial government wrote the document and based the text on existing (democratically effected) provincial and national policy for the area. Policy for nature, recreation and tourism, landscape environment and agriculture were integrated in this *WCL Regional Perspective*. Furthermore, the WCL policy was integrated with the Objective 5b policy of the European Union. Many projects have been financed from both funds.

It is important to notice that local entrepreneurs, farmers and population were not directly involved in WCL policy but were represented by interest groups (agriculture, tourism, landscape and nature) and municipalities. This is not the same for all WCL areas, for example in Limburg there has been much more direct community involvement, through a large number of sectoral and municipal committees. The organisational structure in Friesland was deliberately kept very simple, however, because it was assumed that a heavy organisational structure and a large number of meetings would not function well in this area.

For recreation and tourism the provincial policy was taken as a guideline. The aims of this policy included:

- growth of tourism employment in the province by 2000–2500 jobs between 1991 and the year 2000; this should take place through 'further development of a sustainable tourism product', by giving the protection of nature, culture, landscape and the environment a central position;
- stimulate co-operation between entrepreneurs;

- joint promotion of Frisian regions at the provincial level;
- stimulating new initiatives to strengthen the coherence of the regional product, particularly in south-west Friesland and to increase integration of tourism, agriculture and nature (Provincie Friesland 1991).

There was very little time to decide how to spend the budget for 1994. This meant that parties that were most closely involved and best organised were the ones that got projects financed. The existing relationships between the project co-ordinator of the 'National Landscape' (and thus the municipal governments) and the provincial government was most influential in this respect. Projects were mainly concerned with the improvement of tourist-recreation infrastructure (recreational cycle paths, canoeing routes). Co-operation between tourist entrepreneurs was stimulated, which resulted amongst other things in a collective presentation of some of the bigger firms involved in water sports at 'Boot' in Düsseldorf (the world's largest fair for water sports). The success of this action encouraged co-operation between the local tourist offices in the area.

1995: persistence of existing communication structures

From 1995 onwards, communication with other partners in the area on WCL policy took off. In the beginning mainly the tourist-recreation sector and a landscape organisation were involved in this process. Communication with farmers was difficult and proceeded slowly. This difference could be explained from the existing relationships between the different parties involved.

- The previously existing organisational structure of the 'National Landscape' (especially as the Projects Office is still active) facilitated communication with partners already involved in this programme. Because the National Landscape policy focused mainly on recreation and tourism as an economic sector, these sectoral interests were also well represented in the WCL. However, the involvement was mainly at the level of organisations, while tourist entrepreneurs were generally not aware of the existence of WCL policy.
- Farmers in the area were not very interested in developing non-agricultural activities or changing their farming strategy, in contrast to some of the other WCL areas. In the WCL area of Waterland in the province of Noord Holland, for example, farmers had been interested in developing pluriactivity for quite some time. Differences in the attitude of farmers in both areas can, among other things, be attributed to differences in farm structure. In south-west Friesland, farms are relatively large and prosperous. Changing attitudes towards alternative developments is a slow process.

- Relations between the provincial government and the agricultural sector were somewhat strained by plans to create an Ecological Main Structure in the Netherlands (a plan also developed by the Ministry of Agriculture, Fisheries and Nature Conservation and also implemented by the provincial government). This policy included plans to transform 1000 hectares of agricultural land in the municipality of Gaasterlân-Sleat into nature reserves in the long run. Farmers' organisations were closely involved in the negotiating process and had agreed on the plans. When farmers in the area realised the content of the plans, a small war broke out. Despite the fact that these plans had no direct link with WCL, they affected the process of WCL tremendously as farmers had become suspicious of all plans coming from the Ministry of Agriculture and the provincial government.

This explains why the goal of spending 60 per cent of the WCL budget on agriculture-related projects in 1995 was not reached. In fact hardly any agricultural projects were initiated. Integration of agriculture and tourism, one of the goals of WCL policy, was not stimulated through explicit projects.

1996: centre for agricultural support

It was then decided that something substantial should be done in order to involve the agricultural sector in the WCL policy. A centre for agricultural support related to WCL policy was created, consisting of one person working to initiate agriculture-related projects and to involve farmers in the WCL policy. The centre was largely subsidised by WCL. Projects should: (i) strengthen the position of agriculture and/or (ii) stimulate sustainable agricultural production (combining agriculture and nature protection) and/or (iii) create additional income for farmers. Since the creation of the centre, farmers have become more involved in WCL projects. This was further helped by the fact that at the beginning of 1996 the conflict around nature conservation was more or less solved and farmers became somewhat more inclined to co-operate.

In 1996, projects were mainly of the first type: strengthening the position of agriculture. Examples of such projects are courses for farmers aimed at implementation of (compulsory) environmental programmes, audit of farm enterprise results and optimising the use of fertiliser. Furthermore, some farmers started to develop pluriactivity: processing of agricultural products at the farm (e.g. cheese production) or agritourism. These activities were employed mainly on an individual basis (Kranendonk 1997). The centre explicitly tried to integrate these projects with other regional activities and to create linkages among them, for example in the case of a farm producing home-made fresh cheese. The centre intervened so that they could sell their product not only at home but also through a regional chain of supermarkets. Local bakeries used the cheese to make cakes and pastry. These cakes were sold as a regional product in the bakeries. This last experiment has, however, recently been cancelled, because this fresh cheese proved less practical than the powder normally used. The structure of the cakes was not quite as good according to the bakeries.

1997: further integration

In 1997, more initiatives for creating linkages between the separate projects were undertaken. An important initiative was the foundation of a platform for farmers involved in agritourism or home production. This platform helped create linkages between the different agritourism products being developed; joint marketing strategy, holiday packages, and so on. A good example of this would be the special tourist brochures that have been issued. In these brochures cycle routes in the area are described, including visits to several farms. Most of these farms produce products that visitors can buy, and they sell some general regional products as well. The brochures were available from the WCL centre, but also from the local tourist offices. Another positive aspect of the platform was that it could function as a spokesman for the sector towards the government. This was important because the regular tourist organisations did not represent the agritourism sector. Finally, it could continue initiating projects and generating dynamism and renewal, a function that was until then performed by the WCL centre.

Another important change was the fact that the National Landscape Project Office was closed. Efforts were made to hand the function of this Office over to the National Landscape Foundation that existed alongside the Project Office. This Foundation, in which the various municipalities are represented, should continue to function as a platform for regular communication. In practice, integration of recreation and tourism policies is not an issue for municipalities at the moment.

1998

It is too early yet to give an overview of developments in 1998, but there are some interesting plans to create more integration between different sectors. The most promising one is the attempt to create a link between locally produced agricultural products ('products with a story') and the regional hotel and restaurants. Though the idea itself is not completely new (it has been done in other regions), the plan definitely has potential.

Evaluation

Below, we will evaluate how these developments in Friesland can be understood in the light of the principles that were identified for rural tourism development.

Integration of different policies

The WCL can be considered a successful example of policy integration, because all relevant regional plans were taken into account in the Regional Perspective. However, some critical remarks can be made as well.

1. The integration of farmers into the WCL policy was hampered by the fact that plans for the Ecological Main Structure were running parallel to WCL. Though both plans came from the same Ministry (though from different departments of that Ministry), the consequences of this coincidence were not thought through well enough.
2. The integration of policies from different municipalities remains a difficult issue. While the National Landscape Project Office was still operational, communication at this level was greatly improved. The moment this office was closed, however, communication became very limited again. An effort to retain a platform for municipal contact on a regular basis has not been very successful.
3. The local tourist offices in the area have integrated part of their promotional activities. This is not in accordance with the provincial goal of centralisation and co-ordination of tourism promotion at provincial level.

Utilisation of external production units

The utilisation of external production units is not an explicit goal of any of the actors involved, neither in terms of the attraction of external capital into the area, nor when regarded as the creation of formal linkages with external production units. As far as the latter is concerned, it is not clear to what extent formal linkages between entrepreneurs inside and outside the area are being created. Many local entrepreneurs do not originate from the area, however, which means a certain external input is present. The creation of new large-scale attractions or accommodation has long been stimulated by the provincial and municipal governments. At the moment, however, the policy is to allow large-scale developments only if they are innovative and if they add value to the regional product. This may mean that external investors will become less interested in the area.

As far as WCL policy is concerned, visits to other WCL areas are made and the centre for agricultural support definitely has a clear idea of developments taking place in other areas.

Precise allocation of intervention policies; avoiding pure assistance

The idea behind the WCL is that project finance should benefit not just one entrepreneur but have an impact for the sector or the area as a whole. Most of the projects financed are concerned with creating preconditions for tourism development or improving the regional tourist infrastructure. Individual firms have received funds as well, however, mainly in the case of agritourism. This type of subsidy was legitimated by stating that this type of aid can generate a broader support for WCL within the sector.

It is now increasingly realised that priority should be given to projects that not only create dynamism, but which can subsequently sustain dynamism on their own. The example of the Project Office, whose achievements of co-ordination evaporated when it ended its activities, shows that this is not something that comes naturally.

Focusing on the most promising parts of the area

In the Regional Perspective, choices of the types of developments desirable in different parts of the area have been included. The Perspective is based upon existing provincial plans, in which spatial zoning has been adapted. Goals are not only to focus on most promising parts of the area, but also to spread developments more evenly. Agritourism, for example, is stimulated mainly outside the already touristic municipalities of Gaasterlân-Sleat (partly also due to restrictive municipal policy).

Stimulation of co-operation agreements

Co-operation between entrepreneurs has been stimulated successfully in the agritourism sector. Though further improvement is still desirable, it is in the first place the co-operation *between* tourism and agritourism that is most difficult. The tourist sector and tourist organisations are very negative towards agritourism developments. On the other hand, some tourist organisations (mainly the tourist offices) claim that they have not been fully incorporated into WCL policy.

The best way to reduce the friction between both sectors would be bottom up, through establishing joint projects that (financially) benefit both farmer and tourist entrepreneur. This offers a basis for changing the policy of tourist organisations from within. It also will provide a stimulus for new, innovative product development in recreation and tourism.

Design of the planning process: involvement and management

Involvement of those sectors most relevant for rural economic and spatial development was an explicit goal of WCL. This means that a large proportion of the local population is generally not involved in or aware of the existence of WCL. As mentioned earlier the involvement of relevant sectors in Friesland has mainly taken shape by involving a number of interest groups and municipalities. The rank and file of these parties was, however, not aware of the possibilities that WCL had to offer until the programme was already well on its way. The late start with developing the regional perspective further added to this problem. Farmers especially were little involved in WCL projects in the beginning. The opening of a centre in order to overcome this barrier and initiate agricultural WCL projects has been successful.

The centre offers a good example of the role of a bridge actor, in this case between farmers and politicians:

- The centre functions independently of governments or interest groups; this is important in order to have enough room for manoeuvre.
- Despite this independence, there exists a strong relationship with the provincial administration and policy, which smoothes the way for bureaucratic settlement of, for example, project funding.
- Contacts with the target group of local farmers are also very good, which is necessary to generate support and enthusiasm for plans.

It also functions as a gatekeeper because it possesses knowledge of national developments and market trends that is indispensable for judging the viability of intended projects.

The reason that the centre could function as a bridge actor was because it gained a sufficiently important position in both the agricultural and the WCL policy network. The fact that project funding was available in Friesland, as well as the attention paid to speeding up bureaucratic processes, have both contributed to this.

Conclusion

WCL Friesland provides an example of an attempt to create more community involvement in regional development. The main instrument used to create this involvement was a financial one: projects fitting in the proposed strategy were eligible for subsidies. Tourism development was regarded as one of the elements in a broader, integrated strategy and more interaction between tourism and other regional sectors was an important goal.

In practice, finance alone proved an insufficient stimulus to involve the local population. This may partly have been due to the fact that only interest groups and municipalities were attending WCL meetings, while tourist entrepreneurs, farmers and other inhabitants were only indirectly involved.

The success of the WCL centre shows the role that bridge actors and gatekeepers can play in involving local actors in regional development. Farmers have been more implicated and have benefited from what WCL has to offer. Because bridge actors are able to create linkages between different parties, they can serve as a catalyst for regional development. In the ideal case, they can create opportunities to combine both innovation and community involvement.

Acknowledgement

The research on which this chapter is based was carried out as part of a PhD project on rural tourism development in Europe. This project is financed by the Dutch Council for Scientific Research (NWO).

References

Brandenburg, W. (1996) 'Verdere samenwerking tussen de gemeenten in Zuidwest Friesland na 1 januari 1997'. Unpublished memorandum. Stichting Ontwikkeling Nationaal Landschap Zuidwest Friesland, Sneek.

Bruijn, J. A. de and Heuvelhof, E. F. ten (1991) *Sturingsinstrumenten voor de overheid. Over complexe netwerken en een tweede generatie sturingsinstrumenten*, Houten: Stenfert Kroese.

Bryant, C. R. and J. R. R. Johnston (1992) *Agriculture in the city's countryside*, Belhaven: London.

Butler, R., Hall, C. M. and Jenkins, J. (eds) (1998) *Tourism and recreation in rural areas*, Chichester: John Wiley.

Caalders, J. (1997) 'Managing the transition from agriculture to tourism: an analysis of tourism networks in Auvergne', *Managing Leisure* 2, 3: 127–42.

Camagni, R. (1991) 'Local "milieu", uncertainty and innovation networks: towards a new dynamic theory of economic space', in R. Camagni (ed.) *Innovation networks: a spatial perspective*, London: Belhaven Press.

Camagni, R. (1995) 'The concept of *innovative milieu* and its relevance for public policies', *Papers in Regional Science* 74, 4: 317–40.

Dietvorst, A. G. J. (1996) 'Over uniek en standaard in het toeristisch produkt', paper presented at the congress Amsterdam unlimited, Diemen, October 24, Wageningen: Centre for Recreation and Tourism Studies.

Dietvorst, A. G. J. and Ashworth, G. J. (1995) 'Tourism transformations: an introduction' in G. J. Ashworth and A. G. J. Dietvorst (eds) *Tourism and spatial transformations*, Oxon, UK: CAB International.

Dietvorst A. G. J. and Hetsen, H. (1996) 'Landelijke gebieden en economische ontwikkeling: een netwerkbenadering', *Stedebouw en Ruimtelijke Ordening* 77, 2: 39–45.

Elsasser, H. (1987) 'Regionalismus und endogene Entwicklung in der Schweiz', *Agrarische Rundschau* 6: 3–6.

Fromholt-Eisebith, M. (1995) 'Das "kreative Milieu" als Motor regionalwirtschaftlicher Entwicklung. Forschungstrends und Erfassungsmöglichkeiten', *Geografisches Zeitschrift* 83, 1: 30–47.

Grontmij (1989) *Produktontwikkelingsplan Nationaal Landschap Zuidwest Friesland, Hoofdrapport*, Drachten: Grontmij.

Hidding (1995) 'Veranderingen in het planningsdenken. Wageningse bijdragen in perspectief', *Planning in perspectief*, Wageningen: Vakgroep Ruimtelijke Planvorming.

Jenkins, J., Hall, C. M. and Troughton, M. (1998) 'The restructuring of rural economies: rural tourism and recreation as a government response' in R. Butler, C. M. Hall and J. Jenkins (eds) *Tourism and recreation in rural areas*, Chichester: John Wiley.

Kamann, D.-J. (1989) 'Actoren binnen netwerken' in F. W. M. Boekema and D.-J. Kamann (eds) *Sociaal-economische netwerken*, Groningen: Wolters-Noordhoff.

Kamann, D.-J. and Strijker, D. (1991) 'The network approach: concepts and applications' in R. Camagni (ed.) *Innovation networks: a spatial perspective*, London, Belhaven Press.

Kranendonk, R. P. (1997) *Resultaten monitoring WCL Zuidwest-Friesland 1994–1996*, Wageningen: DLO-Staring Centrum.

Ministerie van Landbouw (1994) *Structuurschema Groene Ruimte,* 's-Gravenhage: SDU.

Pearce, P. L. (1994) 'Tourist-resident impacts: examples, explanations and emerging solutions' in W. F. Theobald (ed.) *Global Tourism: the next decade*, Oxford: Butterworth Heinmann.

Provincie Friesland (1991) *Stategisch groeibeleid voor recreatie en toerisme*, Leeuwarden: Provincie Friesland.

Seaton, A. V. and Bennett, M. M. (1996) *Marketing tourism products. Concepts, issues, cases*, London: International Thomson Business Press.

Verhoef, L. H. J. and Boekema, F. W. M. (1986) *Lokale initiatieven; naar een nieuwe conceptie van regionale ontwikkelingen in theorie en praktijk*, Van zee tot land. Rapporten en mededelingen inzake de droogmaking, ontginning en sociaal-economische opbouw van de IJsselmeerpolders, 54, Lelystad: Rijksdienst voor de IJsselmeerpolders.

13 Understanding community tourism entrepreneurism

Some evidence from Texas

Khoon Y. Koh

Introduction

For over two decades now, tourism continues to be advocated as one industry that socio-economically at-risk communities could pursue to help improve their community health. Like any industry, when successfully developed, tourism earns real income, creates jobs, produces tax revenues, stimulates infrastructural improvement and beautification projects, and encourages community resource conservation and preservation. This, in turn, improves the community's attractiveness for in-migration of businesses and people, thereby strengthening the community tax base and enhancing the quality of life. More enticingly, tourism has been described as a 'smokeless industry' that almost all communities could viably develop: a low-tech industry that does not require high capital investment or highly skilled workers; and almost every community possesses some type of tourist attraction, namely, natural, ethnocultural and/or historical. For communities that are really ill-endowed, artificial attractions could be created and special events could be staged (Inskeep 1991; Blank 1989; Gunn 1988; McNulty, Jacobson and Penne 1985).

Hence, it is no surprise that the number of communities in both developed and under-developed countries are enthusiastically embracing tourism development as evidenced by the growing number of tourism promotion and development offices founded worldwide, and tourism programmes established in educational institutions. Although many success stories have been reported (measured by the number of tourism enterprise births in a community), there are also a great number of communities that continue to experience developmental stagnation despite significant investments of money, effort and time. Why such contrasting scenarios? That is, why are some communities able to enjoy a relatively high level of tourism enterprise birth while others seem to be less fortunate?

Review of the literature

Since tourism enterprises do not come into existence by themselves, but are products of human decisions, understanding community tourism entrepreneurism (the tourism enterprise creation process) seems to be a logical step towards answering the question posed in the preceding paragraph. Unfortunately, as noted by Simms (1981: 51), 'The tourism literature covers entrepreneurship in a quite spotty fashion. The presence of entrepreneurial activity in tourism settings is taken largely as a given.' More than a decade later, Shaw and Williams (1994: 120) repeated the same observation, '... within the considerable literature on tourism's economic potential, little attention has been paid to the role of entrepreneurial activity and, in particular, to how tourism enterprises operate in different economies.' Indeed, a comprehensive survey of the tourism literature indicated that the subject has only been rendered a cursory treatment. A review of the primary tourism research journals (1975–1995), namely, the *Annals of Tourism Research*, *Journal of Travel Research*, *Tourism Management*, *The Tourist Review*, and the *Journal of Tourism Studies*, indicated that only a few authors have recognised the role of entrepreneurship in community tourism development (Echtner 1995; Shaw and Williams 1994; Barr 1990; Kaspar 1989; and Kibedi 1979). An extensive review of the available tourism textbooks also yielded little enlightenment. In fact, the subject index of the *Encyclopedia of Hospitality and Tourism* (Khan, Olsen and Var 1993) does not even contain an entry for the word, 'entrepreneur'! In view of this paucity of knowledge in the tourism literature, an external review was carried out. An expansive review of the external literature suggested that spatial variation in levels of entrepreneurism may be explained by three basic schools of thought: 'people school', 'environmental school' and 'integrative school'.

According to the 'people school', differences in enterprise birth rates between communities is a function of differential supply of entrepreneurs. The greater the supply, the greater the level of enterprise birth rate, and vice versa (Krueger and Brazeal 1994; Krueger and Carsrud 1993; Asiedu 1993; Davidsson and Delmar 1992; Landry *et al.* 1992; Robinson *et al.* 1991; Gilad and Levine 1986; Ahmed 1985; Shapero and Sokol 1982). Differential supply itself is attributed to different socio-cultural factors (such as child rearing and educational practices, socio-economic reward system, historic conditioning and power structure) and personal factors (such as need to achieve, risk assumption and desire for autonomy).

The 'environmental school', on the other hand, argued that in every community and in every time period, there exists a latent supply of entrepreneurs. However, the extent to which latent entrepreneurs manifest themselves into active and serial entrepreneurs is a response to the attractiveness of the relevant investment environments, namely, economic, legal, socio-cultural, industrial and logistical. If nascent entrepreneurs perceive the relevant investment environments for a given industry as favourable (such as growing demand, low entry barriers, easy access to supplies, capital and labour) more would be encouraged to found and vice-versa (Gynawali and Fogel 1994; Baumol 1990; Allen and Hayward 1990; Mokry 1988; Drucker 1985; Chilton 1984; Rosen 1982).

The 'integrative school', however, contended that entrepreneurism is a complex and dynamic process involving the interplay of a variety of personal and environmental factors. As such, adopting an either/or orientation to explaining differences in levels of entrepreneurism between regions is an unsatisfactory approach. Bird (1992: 11) explained, 'New ventures are not coerced into being nor are they the random or passive product of environmental conditions. Ventures get started and develop through initial stages based largely on the vision, goals and motivations of individuals. New organisations are the direct outcome of these individuals' intentions and consequent actions, moderated or influenced by environmental conditions.' Van de Ven (1993: 211) expressed the same sentiment, 'The study of entrepreneurship is deficient if it focuses exclusively on the characteristics and behaviors of individual entrepreneurs, on the one hand, and if it treats the social, economic and political factors influencing entrepreneurship as external demographic statistics, on the other hand.' Perhaps the statement by Bygrave (1994: 4) best summed this school of thought: 'Entrepreneurship is a human endeavor and like all human endeavors, it is greatly determined by personal attributes and the environment.'

The integrative school of thought seems more plausible as tourism enterprises certainly do not come into existence by themselves but are creations of tourism entrepreneurs – people who harbour a favourable attitude towards the industry; believe there are opportunities for new tourism enterprises; have the desire to own and operate a tourism enterprise; feel confident in their ability to be enterprising; and are willing to deal with the risks and uncertainties associated with tourism entrepreneurship. However, even highly entrepreneurial people are unlikely to found unless they also deem the industry's relevant investment environments, namely, economical, legal, logistical, physical and socio-cultural climate as favouable. Therefore, it seems more logical that the difference in levels of community tourism entrepreneurism is more likely to be a function of two interactive factors, the differences in supply of tourism entrepreneurs and perceived attractiveness of the investment climate for launching tourism enterprises. Koh (1996) called the former a community's P-factor (populace's propensity to enterprise) and the latter a community's Q-factor (quality of the industry's investment climate as perceived by nascent tourism entrepreneurs). Accordingly, the more favourable the two factors in a community, the higher the level of tourism entrepreneurism that can be expected, and vice-versa.

According to Koh's conceptualisation, the interaction of the two factors produces four basic types of communities. Quadrant 1 communities are those with very favourable P- and Q-factors. Thus, they are theorised to be characterised by a moderate to high level of tourism enterprise births as both resident and alien tourism entrepreneurs are attracted to found in the community. In contrast, Quadrant 3 communities are those whose P- and Q-factors are very unfavourable. As such, they are theorised to experience little or no tourism enterprise births. Quadrant 2 and 4 communities,

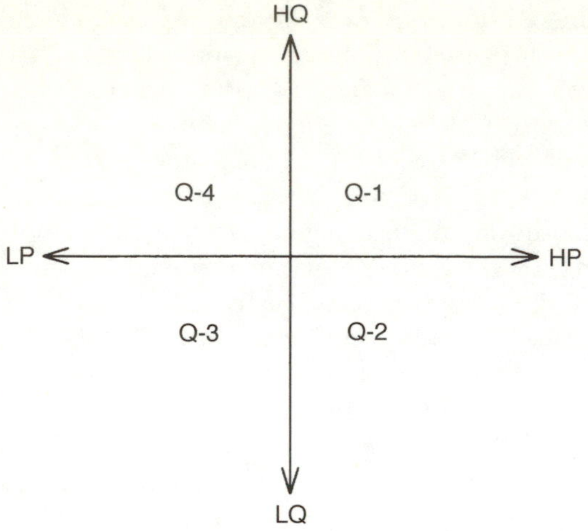

Figure 13.1 Conceptual model of community tourism entrepreneurism.

however, are expected to experience a low to moderate level of tourism enterprise births as only one of their factors is favourable (Figure 13.1).

A review of empirical studies on spatial variations in enterprise birth rates seems to provide support for Koh's 2-factor model (Shane 1994; Anderson 1993; Siu and Martin 1992; Bull and Winter 1991; Dubini 1988; Greenberger and Sexton 1988; Wilken 1979) with the strongest statement coming from Reynolds (1994: 440), whose study of autonomous firm births in the USA between 1986 and 1990 concluded, 'only when a region is seen in a favourable context and the right kind of people are available to take the initiative will new businesses be created'. Moreover, as per the criteria of a good model of entrepreneurism as discussed by Low and MacMillan (1988), Koh's 2-factor model seems to have effectively and efficiently integrated the two basic factors for entrepreneurism to occur as argued by integrative theorists; could explain variations in levels of entrepreneurism within and between communities over time; could be empirically tested by using cross-sectional or longitudinal studies; is parsimonious, and possesses normative values (that is, the model does not hold a community's level of tourism entrepreneurism as a product of some uncontrollable forces nor an unalterable state, rather, as a temporal situation which could be modified by human intervention).

Purpose of the study

The purpose of this study was to ascertain the extent of concordance between the model's theorised and observed scenarios. Specifically, the study sought to measure the P- and Q-factor scores of two small Texan communities, to determine their quadrant positions, and to evaluate the congruency between their quadrant's theorised and field scenarios.

Usefulness of the study

Tourism entrepreneurism is the engine that drives community tourism development. The greater the level of tourism entrepreneurism, the greater the likelihood that tourism attractions and support enterprises would be created, and consequently, the greater the socio-economic benefits associated with tourism development. Thus, communities that seek to increase their level of tourism development would appreciate information that could help them achieve their goals. In this vein, if the theorised scenarios of the 2-factor model were supported by field data, then communities could more confidently take appropriate actions to enhance their P- and Q-factors.

In terms of research implications, since there is no known empirical study of the dynamics of tourism entrepreneurism in small communities, the findings of this study would serve as baseline data for future studies. Moreover, as Babbie (1992) explained, it is through testing and reconceptualisation that theories are built, and a field advances its body of knowledge. In this regard, it is hoped that this study would rouse some tourism scholars into partaking in further exploration of the topic and testing of the model.

Study design

To increase the confidence that differences in levels of community tourism development could be more attributable to differences in P- and Q-factor scores, two comparable Texan communities (Athens and Mount Pleasant) were selected based on the following variables: geography, history, demography, and expressed interest in tourism development. Each community's P- and Q-factors were measured via two questionnaires which were developed with items adapted from the published literature, and refined using established social science research procedures. The final P-questionnaire contained 29 items (Cronbach's alpha = 0.82) while the Q-questionnaire contained 25 items (Cronbach's alpha = 0.84). All items in each instrument were measured with a 7-point interval scale with 1 = very unfavourable to 7 = very favourable.

Residents constituted the P-sample while tourism entrepreneurs comprised the Q-sample. To obtain the P-sample, each community was divided into four quadrants, and from each quadrant, two residential streets were randomly picked. All households located along each street were then sampled. All adult occupants were invited to participate in the study. As for the Q-sample, a sampling frame was created for each community by extracting all known tourism enterprises listed in each community's Yellow Pages (the commercial section of a community telephone directory) and compiled into an unarranged list. Following Krejcie and Morgan's (1970) sample size table, eighty-six tourism enterprises were randomly drawn for sampling. All samplings were conducted using personal delivery method. Participants were given five days to return their completed questionnaires either by mailing them back in prepaid envelopes or having them collected in person. To increase response rate, one telephone reminder was made on the third day, and a site visit was made to each household and establishment on the fifth day. The study was conducted in the spring of 1995.

Two assumptions were made in this study. One, both P- and Q-samples were assumed to be comparable to past populations of interest. Two, no significant changes occurred in each community's tourism industry's relevant environments during the past five years. To further determine the latter, each community's primary newspaper was systematically perused for events that might have directly or indirectly stimulated/discouraged tourism entrepreneurial activities during the period 1990–94. Except for the proposed building of the Texas Fishing Hall of Fame which would include a freshwater fish hatchery in Athens (expected to be completed in late 1996), there was no other known event that was deemed to have had a significant effect on the two communities' Q-factors during the period 1990 to 1994.

Results

The final returns for Athens were 103 P-responses (27.8 per cent) and 44 Q-responses (51.2 per cent) while the returns for Mount Pleasant were 156 (42.2 per cent) and 62 (72.1 per cent) respectively. Analysis of P-samples' representativeness using Chi-square and Kolmogorov–Smirnov goodness-of-fit statistics indicated that the P-sample for Athens was surprisingly more representative of its community workforce than the Mount Pleasant sample even though the response rate for the Athens sample was lower. As for the Q-samples, both samples were found to be typical of their community's tourism enterprise composition and owner's profile. Tables 13.1–13.4 summarise the P- and Q-factor scores for the two communities while Figure 13.2 shows their quadrant positions on the 2-factor interactive model.

The observed P- and Q-factor scores indicated that Athens and Mt Pleasant should be categorised as quadrant 2 communities. According to the 2-factor interactive model, quadrant 2 communities are characterised by a low to moderate level of tourism enterprise births; most of the tourism enterprises at time of birth are small scale (defined as employing less than fifty workers) and predominantly launched by indigenes (because alien tourism entrepreneurs are unlikely to be attracted to the community since the Q-factor is unfavourable). In this study, a 'moderate' tourism enterprise birth rate was operationalised as having a mean birth rate that lies within 5 per cent of the state's mean tourism enterprise birth rate for the same period. Hence, a 'low' tourism enterprise birth rate was operationalised as having a mean birth rate below the 'moderate' range while a 'high' tourism enterprise birth rate was operationalised as above the 'moderate' range.

To evaluate the congruency between the theorised and observed scenarios, each community's tourism enterprise birth rates between 1990 and 1994 were compiled from each community's Yellow Pages, and compared against the

Table 13.1 Athens' P-factor ($N = 103$)

Variable	*Items*	*Alpha*	*Mean*	*SD*	*SE*	*MOE*	*95 per cent CI*
Attitude towards tourism	6	0.88	4.96	0.34	0.03	0.07	4.89–5.03
Attitude towards entrepreneurism	6	0.79	5.58	0.12	0.01	0.02	5.56–5.60
Perceived opportunity	5	0.86	3.92	0.49	0.05	0.09	3.83–4.01
Perceived ability to enterprise	6	0.85	5.56	0.39	0.01	0.02	5.54–5.58
Willingness to enterprise	6	0.82	5.16	0.29	0.03	0.06	5.10–5.22
Composite score	29	0.82	5.08	0.65	0.06	0.12	4.96–5.20

Notes
SD = standard deviation *SE* = standard error
MOE = margin of error *CI* = confidence interval

Table 13.2 Athens' Q-factor ($N = 44$)

Variable	*Items*	*Alpha*	*Mean*	*SD*	*SE*	*MOE*	*95 per cent CI*
Social environment	5	0.82	4.38	0.21	0.03	0.06	4.32–4.44
Physical environment	5	0.78	5.13	0.71	0.11	0.21	4.92–5.34
Economic environment	5	0.76	4.47	0.83	0.12	0.24	4.23–4.71
Legal environment	5	0.73	4.84	0.26	0.04	0.08	4.76–4.92
Logistical environment	5	0.82	4.67	0.14	0.02	0.04	4.63–4.71
Composite score	25	0.89	4.70	0.54	0.08	0.16	4.54–4.86

Notes
SD = standard deviation *SE* = standard error
MOE = margin of error *CI* = confidence interval

Table 13.3 Mount Pleasant's P-factor ($N = 156$)

Variable	*Items*	*Alpha*	*Mean*	*SD*	*SE*	*MOE*	*95 per cent CI*
Attitude towards tourism	6	0.93	5.65	0.12	0.01	0.02	5.63–5.67
Attitude towards entrepreneurism	6	0.77	5.82	0.15	0.01	0.02	5.80–5.84
Perceived opportunity	5	0.84	4.31	0.36	0.03	0.06	4.25–5.37
Perceived ability to enterprise	6	0.82	5.30	0.19	0.02	0.03	5.27–5.33
Willingness to enterprise	6	0.84	5.19	0.19	0.01	0.03	5.16–5.22
Composite score	29	0.86	5.29	0.55	0.04	0.08	5.21–5.37

Notes
SD = standard deviation *SE* = standard error
MOE = margin of error *CI* = confidence interval

Table 13.4 Mount Pleasant's Q-factor ($N = 62$)

Variable	*Items*	*Alpha*	*Mean*	*SD*	*SE*	*MOE*	*95 per cent CI*
Social environment	5	0.70	5.35	0.54	0.07	0.13	5.22–5.48
Physical environment	5	0.68	4.94	0.75	0.10	0.19	4.75–5.13
Economic environment	5	0.72	4.51	0.64	0.08	0.16	4.35–4.67
Legal environment	5	0.62	5.00	0.13	0.02	0.03	4.97–5.03
Logistical environment	5	0.74	4.42	0.49	0.06	0.12	4.30–4.54
Composite score	25	0.81	4.85	0.61	0.08	0.15	4.70–5.00

Notes
SD = standard deviation   *SE* = standard error
MOE = margin of error   *CI* = confidence interval

state's birth rates, compiled from the Texas Business Directory for the same period. These historical reviews indicated that the state's annual tourism enterprise birth rates for the five-year period, ranged from 1.48 per cent to 2.47 per cent with a mean birth rate of 1.93 per cent. As for Athens, the annual tourism enterprise birth rate during the same period ranged from 1.04 per cent to 2.08 per cent with a mean birth rate of 1.87 per cent (Table 13.5) while that of Mount Pleasant ranged from 1.02 per cent to 2.04 per cent with a mean birth rate of 1.63 per cent (Table 13.6).

Since the state's mean tourism enterprise birth rate for the study period was 1.93 per cent, the 'moderate' range was determined to be between 1.88 per cent and 1.98 per cent (as per the operationalised decision criteria). In

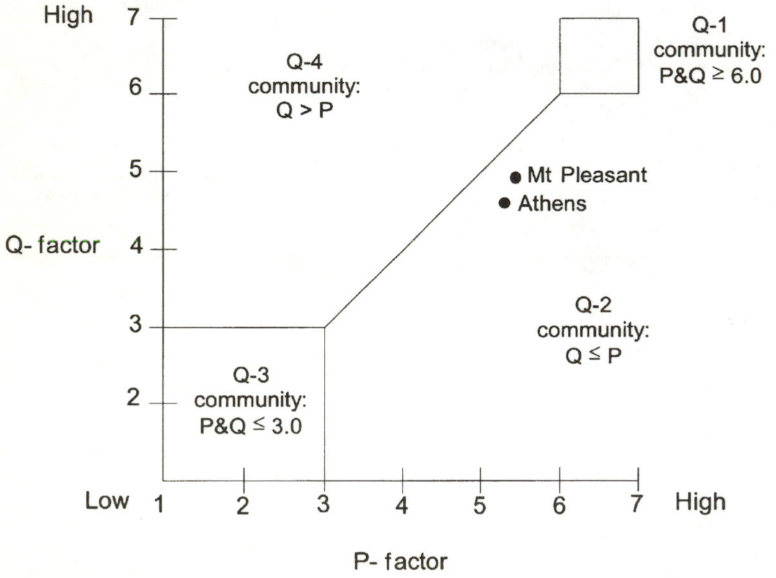

Figure 13.2 Quadrant positions of Athens and Mount Pleasant.

Table 13.5 Tourism enterprise births in Athens

Tourism sectors	*1990*	*1991*	*1992*	*1993*	*1994*
Attractions/recreation	0	1	0	0	0
Lodging	0	1	0	0	1
Food & beverage	0	0	1	1	1
Travel facilitation	1	0	0	0	0
Travel retail	0	0	1	1	0
Total births	1	2	2	2	2
Percentage change	1.04 %	2.08 %	2.08 %	2.08 %	2.08 %

Notes
Total number of tourism enterprises in base year (1989) = 96
Mean tourism enterprise birth rate (1990–1994) = 1.87 %
Per capita enterprise ratio = 1:1230.8

Table 13.6 Tourism enterprise births in Mount Pleasant

Tourism sectors	*1990*	*1991*	*1992*	*1993*	*1994*
Attractions/recreation	0	0	0	0	0
Lodging	0	0	0	1	0
Food & beverage	1	1	1	0	1
Travel facilitation	0	0	0	1	0
Travel retail	0	1	0	0	1
Total births	1	2	1	2	2
Percentage change	1.02 %	2.04 %	1.02 %	2.04 %	2.04 %

Notes
Total number of tourism enterprises in base year (1989) = 98
Mean tourism enterprise birth rate (1990–1994) = 1.63 %
Per capita enterprise ratio = 1: 1536.4

accordance with these operational boundaries, both Athens' and Mount Pleasant's mean tourism enterprise birth rates of 1.87 per cent and 1.63 per cent were thus considered to be 'low'. To determine the size and resident status of owners at the time of their tourism enterprise's birth, all owners of each neonate tourism enterprise were contacted via telephone. Tables 13.7 and 13.8 show the two study communities' findings on these two variables.

Evidently, all the tourism enterprises in both communities were small scaled and primarily founded by indigenes at birth. Thus, it could be concluded that indeed, both communities exhibited evidence of a quadrant 2 community as theorised by the 2-factor model.

Conclusion

The purpose of this study was to ascertain the extent of concordance between the theorised scenarios as propounded by Koh's 2-factor model of tourism entrepreneurism and the observed field scenarios. The rationale was that if

Table 13.7 Tourism enterprise size and ownership at time of birth in Athens

Tourism sectors	*Births*	*Employees*	*Owner's status*
Attractions/recreation	1	3	Local
Lodging	2	2,8	Local, alien
Food & beverage	3	2,12,4	Local, local, local
Travel facilitation	1	2	Local
Travel retail	2	1	Local, alien
Total births	9	34	7 locals and 2 aliens

Table 13.8 Tourism enterprise size and ownership at time of birth in Mount Pleasant

Tourism sectors	*Births*	*Employees*	*Owner's status*
Attractions/recreation	0	0	Not applicable
Lodging	1	9	Alien
Food & beverage	4	2,3,6,2	Local, local, local, local
Travel facilitation	1	3	Local
Travel retail	2	3	Local, local
Total births	8	28	7 locals and 1 alien

field scenarios support the theorised scenarios, then communities that seek to increase their levels of tourism development could more confidently proceed to enhance their P- and Q-factors as suggested by the model. Although one study does not establish its validity nor refute its assertion, the results of this study seemed to offer some empirical support for the model, at least, the theorised scenario of a quadrant 2 community.

One practical implication arising from this study is that quadrant 2 communities, such as Athens and Mount Pleasant, could now use the instrument developed in this study to diagnose their Q-domains for remedial action if they seek to increase their level of tourism development. Another implication is that communities need to implement a reliable management information system to track the number of tourism enterprise births and deaths in their community each year. Without such information (as experienced in this study), any well-intentioned intervention programmes aimed at stimulating development of the industry may not be effective or efficient, and worse, could be debilitating.

As for research implications, since only one quadrant of the model was tested, future researchers should seek to test the remaining quadrants. Also, in view of the assumption of a static population and the lack of experimental control in this study, longitudinal studies would be a superior study method. Finally, the two survey questionnaires developed in this study (P-factor and Q-factor questionnaires) should be further refined to improve their psychometric properties, and perhaps integrated into a single instrument for easier administration.

Regarding tourism development, Pearce (1991: 282) wrote, 'Any general increase in our understanding of the nature of tourism development will only come about if research efforts are cumulative, if general issues are identified and pursued and explicit links are made between different areas of research and the methods employed in them.' In this light, it is hoped that this study would be a contribution to a better understanding of community tourism development.

References

Ahmed, S. U. (1985). 'NAch, Risk Taking Propensity, Locus of Control, and Entrepreneurship'. *Personality and Individual Differences*, 6: 781–2.

Allen, D. N. and Hayward, D. J. (1990). 'The Role of New Venture Formation/Entrepreneurship in Regional Economic Development: A Review'. *Economic Development Quarterly*, 4, 1: 55–63.

Anderson, A. R. (1993). 'Understanding Entrepreneurship'. *Frontiers of Entrepreneurship Research*, 13: 246–8.

Asiedu, S. T. (1993). 'Some Socio-cultural Factors Retarding Entrepreneurial Activity in Sub-Saharan Africa'. *Journal of Business Venturing*, 8, 2: 91–8.

Babbie, E. (1992). *The Practice of Social Research* (6th edn). Belmont, CA: Wadsworth Publishing Co.

Barr, T. (1990). 'From Quirky Islanders to Entrepreneurial Magnates: The Transition of the Whitsunday's'. *Journal of Tourism Studies*, 1, 2: 26–32.

Baumol, W. J. (1990). 'Entrepreneurship: Productive, Unproductive, and Destructive'. *Journal of Political Economy*, 98, 5: 893–921

Bird, B. J. (1992). 'The Operation of Intentions in Time: The Emergence of the New Venture'. *Entrepreneurship Theory and Practice*, 17, 1: 11–20.

Blank, U. (1989). *The Community Tourism Industry Imperative: The Necessity, The Opportunities, Its Potentials*. State College, PA: Venture Publishing.

Bull, I. and F. Winter (1991). 'Community Differences in Business Births and Growths'. *Journal of Business Venturing*, 6, 1: 29–43.

Bygrave, W. D. (1994). *The Portable MBA in Entrepreneurship*. New York, NY: John Wiley & Sons.

Chilton, K. W. (1984). 'Regulation and the Entrepreneurial Environment' in C. A. Kent (ed.), *The Environment for Entrepreneurship*. Lexington, MA: Lexington Books, pp. 91–115.

Davidsson, P. and Delmar, F. (1992). 'Cultural Values and Entrepreneurship'. *Frontiers of Entrepreneurship Research*, 12: 444–58

Drucker, P. E. (1985). *Innovation and Entrepreneurship: Practice & Principles*. New York, NY: Harper and Row.

Dubini, P. (1988). 'The Influence of Motivations and Environment on Business Start-Ups: Some Hints for Public Policies'. *Journal of Business Venturing*, 4, 1: 11–26.

Echtner, C. M. (1995). 'Entrepreneurial Training in Developing Countries'. *Annals of Tourism Research*, 22, 1: 119–24.

Gilad, B. and Levine, P. (1986). 'A Behavioral Model of Entrepreneurial Supply'. *Journal of Small Business Management*, 24, 4: 45–53.

Greenberger, D. B. and Sexton, D. L. (1988). 'An Interactive Model of New Venture Initiation'. *Journal of Small Business Management*, 26, 3: 1–7

Gunn, C. A. (1988). *Tourism Planning* (2nd edn). New York, NY: Taylor & Francis.

Gynawali, D. R. and Fogel, D. S. (1994). 'Environments for Entrepreneurship Development: Key Dimensions and Research Implications'. *Entrepreneurship Theory and Practice*, 18, 4: 43–62.

Inskeep, E. (1991). *Tourism Planning.* New York, NY: Van Nostrand Reinhold.

Kaspar, C. (1989). 'The Significance of Enterprise Culture for Tourism Enterprises'. *The Tourist Review*, 44, 3: 2–4.

Khan, M. A., Olsen, M. D. and Var, T. (1993). *Encyclopedia of Hospitality and Tourism.* New York, NY: Van Nostrand Reinhold.

Kibedi, G.B. (1979). 'Development of Tourism Entrepreneurs in Canada'. *The Tourist Review*, 34, 2: 9–11.

Koh, K. Y. (1996). 'Understanding Differing Levels of Community Tourism Entrepreneurship: A Conceptual Model.' Proceedings of the Society of Travel and Tourism Educators Conference, Ottawa, Canada, 1996, pp. 195–9.

Krejcie, R. V. and Morgan, D. W. (1970). 'Determining Sample Size for Research Activities'. *Educational and Psychological Measurement*, 30: 607–10.

Krueger, N. F. and Brazeal, D. V. (1994). 'Entrepreneurial Potential and Potential Entrepreneurs'. *Entrepreneurship Theory and Practice*, 18, 3: 91–104.

Krueger, N. F. and Carsrud, A. L. (1993). 'Entrepreneurial Intentions: Applying the Theory of Planned Behavior'. *Entrepreneurship and Regional Development*, 5, 4: 315–30.

Landry, R., Allard, R., McMillan, B. and Essiembre, C. (1992). 'A Macroscopic Model of the Social and Psychological Determinants of Entrepreneurial Intent'. *Frontiers of Entrepreneurship Research*, 12: 591–606.

Low, M. B. and MacMillan, I. C. (1988). 'Entrepreneurship: Past Research and Future Challenges'. *Journal of Management*, 14, 2: 139–61.

McNulty, R. H., Jacobson, D. R. and Penne, R. L. (1985). *The Economics of Amenity: Community Futures and Quality of Life*. Washington, DC: Partners for Livable Places.

Mokry, B. W. (1988). *Entrepreneurship and Public Policy*. Westport, CT: Quorum Books.

Pearce, D. (1991). *Tourist Development* (2nd edn). Essex, UK: Longman.

Reynolds, P. D. (1994). 'Autonomous Firm Dynamics and Economic Growth in the US: 1986–1990'. *Regional Studies*, 28, 4: 429–42.

Robinson, P. B., Stimpson, D. V., Huefner, J. C. and Hunt, H. K. (1991). 'An Attitude Approach to the Prediction of Entrepreneurship'. *Entrepreneurship Theory and Practice*, 15, 4: 13–31.

Rosen, S. (1982). *Economics and Entrepreneurs*. Lexington, MA: Lexington Books.

Shane, S. (1994). 'Why Do Rates of Entrepreneurship Vary Over Time?' Proceedings of the Academy of Management, Dallas, Texas, pp. 90–4.

Shapero, A. and Sokol, L. (1982). 'The Social Dimensions of Entrepreneurship' in C. A. Kent, D. L. Sexton and K. H. Vesper (eds), *The Encyclopedia of Entrepreneurship*. Englewood Cliffs, NJ: Prentice-Hall, Inc., pp. 72–90.

Shaw, G. and Williams, A. M. (1994). *Critical Issues in Tourism: A Geographical Perspective*. Oxford, UK: Blackwell Publishers.

Simms, D. M. (1981). 'Tourism, Entrepreneurs, and Change in Southwest Ireland'. Unpublished Doctoral Dissertation, State University of New York, Albany, NY.

Siu, W. S. and Martin, R. G. (1992). 'Successful Entrepreneurship in Hong Kong'. *Long Range Planning*, 25, 6: 87–93.

Van de Ven, A. H. (1993). 'The Development of an Infrastructure for Entrepreneurship'. *Journal of Business Venturing*, 8, 3: 211–30.

Wilken, P. H. (1979). *Entrepreneurship: A Comparative and Historical Study*. Norwood, NJ: Ablex Publishings.

Part 4

Rural communities and tourism development

14 Can sustainable tourism positively influence rural regions?

Jan van der Straaten

Introduction

It goes without saying that the negative impact of many modern tourist activities on nature and the environment influences national and international policy. In addition to the negative influences of current activities, there is also the issue of how tourist activities can positively influence the economic development of traditionally agricultural regions which have experienced sharp declines in their agricultural activities in the last decades, though they are attractive rural sites, such as Tuscany, inland Spain and Central France. The crucial question is to what extent communities in these areas are able to organise economic life in such a way that sustainable tourism can be facilitated and promoted to the benefit of the local stakeholders. Bätzing (1991 and 1996) states that tourism development should be based on ecological and landscape limitations. Rural or regional development has to be used as a framework for all problems relating to nature conservation and tourism.

The concept of sustainable tourism has become the catchword in the industry in the 1990s, particularly in the Alps. It has been argued that tourism development has to take the protection of landscape and nature as a prerequisite for every type of development. However, the concept of sustainable tourism is, in itself, still undefined. Therefore, it is necessary to operationalise the concept, though in practice this goal is far from easy, as these preconditions are not clear in many concrete cases as evidenced by Laws and Swarbrooke (1996), and Evans and Henry (1996).

Therefore, there is a need for operationalisation of the sustainable tourism concept. Furthermore, attention should be paid to the broader context of sustainable tourism. Can it be seen as a special case of sustainable development, as articulated by the World Commission on Environment and Development (1987)? Another relevant question is to what extent sustainable tourism as a concept can solve the environmental problems as well as the regional problems of peripheral agricultural areas? We can answer these questions only by looking at agricultural development in Europe in more detail, since nature tourism and agricultural activities are closely connected. Additionally, we should pay attention to the supply and demand of nature in rural areas; this is necessary since tourism and agriculture compete with each other regarding the use of space. Finally, we come to the crucial question of which groups, e.g. farmers, tourists, local communities or modern industries are the losers or winners of the current competition for space in Europe.

Polluting sectors

Traditional neoclassical economics claims that an optimal allocation of production factors can be realised by letting the market work. Environmental problems are defined in this framework as negative external effects which disturb the optimal functioning of the otherwise well-functioning market process. The authorities should define these external effects and calculate the costs which are shifted away to non-market parties. These social costs outside the normal market process have to be paid by the polluter: the Polluter Pays Principle finds its roots in these ideas.

It became clear, in the course of time, that there are many barriers to implementing these measures as defined in the theoretical framework (Dietz and Van der Straaten 1992). Legislative procedures were necessary to curb pollution, but even these environmental laws were not able to alter production to reduce environmental problems in number and scope. On the contrary, Western societies were increasingly confronted with serious environmental problems.

Societal groups sought other concepts and ideas aimed at reducing these problems to an acceptable level. The World Commission on Environment and Development (1987) created the concept of sustainable development as a new paradigm for modern societies. This concept has been accepted as a guiding principle for all economic activities worldwide.

Tourism sector and sustainable tourism

The reaction of the tourism sector was quite weak after publication of the report by the World Commission on Environment and Development. This was surely influenced by the idea that the tourist industry does not belong to the polluting production sectors, which are in that view concentrated in industrial activities such as oil refineries, heating, intensive agriculture, heavy industry, etc. Tourism can be considered in that view as an environmentally friendly sector. This attitude changed rapidly; at the beginning of the 1990s, sustainable tourism became a well-accepted concept in the tourism sector.

There are several reasons why the industry adopted the concept in that period.

1. Nobody could foresee the wide acceptance of the concept of sustainable development. In fact, the International Union for the Conservation of Nature (IUCN 1980) propagated the concept in its World Conservation Strategy as early as 1980. Nevertheless, it was not accepted as a general

principle, so there was no reason to expect it to become popular seven years later. However, there is a significant difference between the two concepts. The World Commission on Environment and Development consolidated the issue of the development of Third World countries and environmental problems which made the latter concept acceptable to rich and poor countries. In particular, governments and international authorities were among the first to advocate the sustainability concept. They felt it could help to solve environmental and development problems.

The polluting sectors were in a difficult position as they became aware of potential problems from new massive attempts by the government to reduce environmental degradation and disruption. For example, the European Commission called its Fifth Environmental Action Plan *Towards Sustainability* (1992). Polluters predicted that authorities would make the environment a serious policy aim. Polluting sectors would not gain, in that situation, from a reluctant attitude towards environmental claims, but they could win with a pro-environmental attitude. By adopting the ill-defined concept of sustainable tourism, the sector could provide this gain, while, additionally, accepting the concept would not put them in a difficult position *vis-à-vis* the authorities due to the weakness of the concept itself.

2. Furthermore, many tourists changed their general attitude towards environmentally friendly activities. A substantial number reacted positively to the environment label, so that the demand side of the market influenced the attitude of the industry. In particular, a general concept like sustainable tourism met these goals.
3. In the environmental debate, it is impossible to prioritise all the issues. By definition, there are only a few concrete environmental problems which can penetrate people's perceptions. The result of this is that the environmental debate can be strongly influenced by any problem which attracts enough attention at a particular moment. At the beginning of the 1990s, the enhanced greenhouse effect was a priority on the agenda. The greenhouse effect is mainly caused by the emissions of carbon dioxide resulting from the burning of fossil fuels. The tourism sector is strongly dependent on planes, cars or trains as a mean of transport. All of them use fossil fuels. This put the industry in a difficult position. If the authorities called for a reduction in the use of fossil fuels as an element of sustainable development, a reduction in traffic might result. It would be good for the sector to develop the concept of sustainable tourism by excluding the use of fossil fuel as a topic in the debate. However, this could only be done if the tourism sector itself used and manipulated the concept. Therefore, it was in the industry's interest to accept the loosely defined concept and to define it to its advantage.
4. Until the 1980s, there were no problems on the demand side of the tourist market. The demand for tourist services increased every year

> resulting in a booming industry, particularly in mass tourism. However, certain segments of the market became saturated, especially as a result of tourists exploring the new opportunities of far and exotic destinations. Certain types of mass tourism developed in the late 1960s and 1970s without being aware of nature, and the environment was confronted with a new type of tourist for whom sun, sand and sea were no longer sufficient for a holiday on the Mediterranean. In particular, many parts of the Eastern coast of Spain were confronted for the first time with fewer tourists.
>
> There was no choice for the authorities in these regions. Most of them recognised the seriousness of the situation. If they continued as before, they would surely lose many customers who preferred, generally speaking, a higher level of services and a more environmentally friendly situation. The sustainable tourism concept became, in the eyes of these authorities, a solution to the problems of the sector. With the concept of sustainable tourism, the old-fashioned resorts along the Spanish coast could be saved. Upgrading the resort including the environment became, from a marketing point of view, a good option. The concept of sustainable tourism fits in these plans perfectly.

From the previous discussion, it can be concluded that the concept of sustainable tourism was increasingly accepted by the tourist industry, mainly due to certain aspects of the sector itself. Additionally, the vagueness of the concept made it a good marketing instrument. If it had not been available, it would have been developed by the tourism sector itself!

The confusion in the use of the concept grew when it became associated with the need for rural development. The concept of rural development is mainly related to the German idea of 'sanftes Tourismus', already propagated at the end of the 1970s as a means to reduce mass tourism development, particularly in the Alps (CIPRA 1985; Krippendorf 1986; Bätzing 1991). In many publications, the concept of sustainable tourism is used in connection with the need for sound rural development. Before we answer the question whether this new combination of sustainable tourism and concepts such as rural development and sustainable development make sense, we need to address the general development of European rural areas in recent decades.

Rural development in Europe

Many environmental groups claim that mass tourism in the Alps will put more pressure on fragile ecosystems in the Alps (Danz 1989; Lorch 1995; Hasslacher 1992). From these statements it is often generally concluded that the Alps are overloaded with tourist activities. Bätzing (1993) made it clear, however, that many parts of the Alps are, in reality, underdeveloped. Traditional agriculture is hardly possible any more in the remote mountain valleys with a short growing season, which creates barriers against intensification of agriculture. This results in an unfavourable competitive position for mountain farmers compared with farmers from the warm fertile plains which are, indeed, nearer to the centres of consumption. From this research it can be concluded that the Alps are deviating from a generally accepted idea, in fact, not overrun with tourists.

After visiting the Alps, people are convinced that this region is overdeveloped, because the great majority of tourists, by definition, only visit resorts. They do not go to the remote valleys without modern motorways and comfortable hotels everywhere. These tourists believe that all parts of the Alps are similar to the section they see. In reality, the Alps are geographically polarised, meaning that overdevelopment and underdevelopment can be found in the same region.

If this is the case in such a 'well developed' area as the Alps, we might ask about the situation in other more remote parts of Europe such as Central France, inland Spain, Portugal, and the Scandinavian countries. Here the development is not only influenced by agricultural activities, but also by a strong geographical polarisation of economic activities.

The development in the agricultural sector can be sketched by a continuing intensification of agricultural practices in the fertile plains not far from the market in agricultural products, the modern conurbations. These areas can realise a lower cost price than nearly all farmers located in remote parts of Europe, particularly in hilly and mountainous areas. These farmers can only compete to a very limited degree. In most cases, they abandon their land, which is a form of extensification, or they intensify on those segments of their soils which are relatively fertile and flat. This process leads to many environmental problems such as the high use of fertilisers, resulting in water pollution, and pesticides. The intensification process and, in particular, abandoning the land causes other serious environmental problems such as erosion and a decrease in the surface of grassland with a high ecological value. One should not overlook the fact that the breakdown of the community system in these regions had a dramatically negative effect on the functioning of the ecosystem. The traditional agricultural system, in which community activities played a significant role, was placed in a very difficult competitive position. This resulted in negative environmental effects, as the value of these ecosystems is strongly connected with these traditional agricultural and community activities.

The European Union, fully aware of the social aspect of these problems, constructed numerous instruments aimed at redeveloping the countryside. Nowadays, more than 50 per cent of the European Union's budget is transferred as an income guarantee to one sector, namely agriculture. Farmers are losing support in the political arena for this immense income support contributed by the taxpayers. Therefore, many attempts have been and will be made to reduce this percentage. Furthermore, paradoxically this money is often used in such a way that it exacerbates the problem. If farmers want to generate a higher level of income, they can only do so by raising the labour and capital intensity of their activities. By doing so, they intensify unemployment in the region. This is a circular process, which will undoubtedly solve neither the social, nor the ecological problems of the European countryside.

Furthermore, there is the process of geographical polarisation leading to a certain pattern of competition in the use of land. We can explain this development using the example of the regions around Grenoble in France. Grenoble has always been the capital of the Dauphiné which now includes the Départements of The Drôme, Isère, and the Hautes Alpes. Grenoble was and is an important centre for that region. Recently, with the expansion of the high-speed train system in France, the centre of Grenoble can be reached from central Paris in three hours and 10 minutes. Obviously, opportunities for the economic development of Grenoble have increased dramatically. Additionally, it has the same type of connection with Lyon, and to the north to Chambéry, Annecy and Geneva. Needless to say, in the greater Grenoble region the population density has substantially increased. But where do these people come from? They migrate from the remote areas of the country. Generally speaking, axes of transport and of communication result in a concentration of economic activities near these axes, but in particular, in the existing centres themselves. So far, modern communication systems such as e-mail, Internet and fax have not changed that pattern.

The people of Grenoble have profited from this development. They are able to increase labour and capital intensity of the economic activities taking place in these favourably located regions. Their income is higher than that of the EU-supported farmers in the neighbourhood. Therefore, they are able to claim nature and space in the surroundings. For example, there is a road project under construction to connect Grenoble via the A51 highway with the highway system of Sisteron in the South. This will extend Grenoble's links to the Mediterranean which puts them in a highly competitive position. Construction of this road through the Drac valley is accompanied by a huge destruction of unspoiled mountain valleys. This polarised development of space in which the concentrations of population are connected with each other mainly by means of car transport without paying attention to the 'empty' space between, will shift the high burden of social costs to regions with a low density of population and economic activities, as has been clearly demonstrated by the research of Brückl and Molt (1996).

Grenoble is located in the valley of the Isère surrounded by many beautiful mountains. For recreation, tourists can visit many of the well-known ski resorts such as Val d'Isère, Les Deux Alps and Alp d'Huez. Additionally, people from the Grenoble region buy holiday houses in the surrounding mountains. The effects are varied. In some cases, these activities 'save' these small mountain villages by generating income from the spending of these people. In many other cases, large numbers of new apartments are built with all the detrimental effects on the social structure and nature of these areas.

The essential point is that a complete reconstruction of these regions is based on the impetus of market processes in a competitive market. So far, no social or environmental policy has been able to nullify these developments. Grenoble is not a special example; we see this type of development nearly everywhere in Europe. The crucial question is to what extent the sustainable tourism concept connected with the idea of rural development can change this situation. What role can be played by the rural communities in these regions?

Some case studies

The underdevelopment of remote agricultural regions in Europe due to falling employment in the agrarian sector is not easy to solve. The combination of the concept of sustainable tourism with traditional agriculture is often suggested as a solution. Some case studies can provide insight into how viable that concept is.

Grande Traversata delle Alpi (GTA)

The valleys of the Italian Western Alps suffer particularly from stagnation in the agrarian sector. Some of these valleys had a quite developed industry in the past based on ores mined from the mountains. These ore mines attracted small-scale industries and provided work for the locals who no longer worked in the agrarian sector. Over time, however, the quality of the ore decreased and the mines were closed (*Grande Traversata delle Alpi* 1981). The employment situation deteriorated. The authorities of the Regions Piemonte and Lombardia initiated small-scale tourist development for these regions: they created a long-distance footpath which started north of Ventimiglia in the south and ended south of the Matterhorn in the north (*Grande Traversata delle Alpi* 1982).

The structure of this long-distance footpath is based on existing paths and tracks which connected the villages in one valley with those in the adjacent one. The trail was marked by special signs, and booklets (in Italian) were written to facilitate the use of the mountain paths. No new mountain huts were built; existing buildings which were no longer in use were reconstructed as a type of mountain hut. Small restaurants and shops provided sufficient services for the mountain walkers. The idea was to generate income for the villages in the mountain valleys, giving an extra impetus to the local economy.

The trail is very attractive, as the alpine scenery is of high quality. It is not really difficult since the old tracks and mountain paths have been transformed. The total number of visitors, however, is quite low. Though one crosses the most scenic parts of the Italian Alps, walkers are, in most cases, alone on these trails. Mountain lovers from abroad were afraid that the trail would no longer be maintained due to the low number of visitors, so they publicised the path in various publications (Bätzing 1986; Neubronner 1992; Van der Straaten and Verhagen 1996). A quite similar long-distance footpath, the GR 5, is located on the French side of the Western Alps. It is very popular and the number of visitors is many times higher than on the parallel Italian footpath.

Why is there such a difference? Of course, the tradition of walking, particularly in the mountains, is different in Italy from that in France. Many foreign visitors, however, walk on the GR 5, so this cannot be the only explanation. The most significant difference is that the French path is very well known and guidebooks in French, English and Dutch have been published and can be bought in every specialised bookshop in Europe. The initiative in France was taken by the French ramblers organisation which is perfectly equipped to publicise and support this type of tourism. The situation is completely different in Italy. In Italy, the authorities of the Piemonte and Lombardia regions took the initiative; they are not equipped to promote this type of tourism. The result is that many walkers are not aware of the Italian path and, additionally, the quality of the huts is not always as it should be due to the neglect of local or regional authorities. Local communities should be more involved in the maintenance of the trail, as this will increase the community's interest in the trail and its economic impact on the villages along the trail.

Bonneval in Maurienne[1]

Bonneval is a small mountain village in the upper valley of the Maurienne, located at an altitude of approximately 1800 metres. The total population of the village is 200 inhabitants. It is traditionally dependent on agriculture. When neighbouring villages such as Val d'Isère developed mass tourism after the Second World War, the municipality of the village decided not to follow that type of development. They did not want to lose their independence and were afraid that with that type of development, capital from outside the village would dictate the economic development of the village. That decision made them more dependent on other economic activities. There is a small-scale tourist infrastructure of 1,300 beds, whereas the hotel owners intend to sell traditional agricultural products to the tourists. Therefore, a combination of low-impact tourism and traditional agriculture is the basis for the village.

The municipality sold the right to generate hydro-electric power from their own mountain river to the French utilities company, which generated a new source of income and improved the financial position of the municipality. However, in 1990, the electricity company stopped paying for the power which reduced the income of the municipality. The total infrastructure of the village could only be paid for when enough tourists visited the village. In the period when the electricity company paid, there was a need for 1,300 tourist beds to balance the village's budget. This number of beds was available in the village. In the new situation the budget of the village could only be balanced if the number of beds was expanded to 5,000. This development would, however, completely change the character of the village.

The village is in a difficult position. There are several options, such as co-operating with other nearby villages which have developed mass tourism. However, this type of development would compromise the original character of the village. Nevertheless, it is argued that without a broader base the village cannot survive. The total cost of the infrastructure cannot, in fact, be supported by this combination of low-impact tourism and traditional agriculture.

Parc Naturel Régional du Vercors (PNR Vercors)[2]

This concept is different from a national park, whose first aim is to protect landscape and nature. The aim of the PNR is to protect nature and to stimulate rural development. Of course, this leads to conflicts. Nevertheless, France has developed quite a number of these regional parks.

The Vercors is a limestone tableland, south of Grenoble in the French Western Alps. The main elevation of the plateau is 1000 metres, while in the east there is a mountain ridge over 2,300 metres high. Traditionally, forestry and sheep grazing are the main economic activities. It goes without saying that these activities cannot guarantee a high level of employment. Tourism is the only option. Since the mountain scenery is of outstanding beauty, such a development is possible. Additionally, in winter the plateau is an excellent area for cross-country skiing.

The model of the PNR offered new opportunities to develop small-scale tourism in combination with traditional agriculture. In fact, such a development can only be recognised to a limited extent. Sheep grazing on the Hauts Plateaux is supported by the PNR authorities; however, the influence on tourism is quite weak, mainly due to the remoteness of grazing locations.

The tourism concept in the region is different. Some places such as Villard-de-Lans and Lans-en-Vercors have chosen a typical mass tourism option with huge apartments completely in conflict with any concept of landscape protection. Other places, such as Autrans, Meaudre, Vassieux, La Chapelle, and Gresse-en-Vercors gave priority to small-scale development based on summer walking and cross-country skiing in winter. These places have been able to maintain their services such as schools, post offices, shops, etc. Landscape and nature protection in these villages have a much higher profile.

However, one cannot overlook the difficult position of many of these villages. Gresse-en-Vercors, for example, had 900 inhabitants in 1910. More than 100 farmers and their families had to live from the agricultural land around the village. Now there are four farmers in the village who have to make a living from the same surface area. Since 1910 the number of inhabitants has dropped steadily to 185 ten years ago. The small-scale development in the village provided new jobs and the population is now up to 220. These villages are absolutely dependent on favourable developments in this type of tourism. Over 90 per cent of the tourists in the region are French from the Grenoble and Lyon regions.

Conclusion

Some general conclusions can be drawn based on the discussion earlier in the chapter and the previous case studies.

- The Italian case study demonstrated that, without effective marketing, the concept of sustainable tourism cannot be successful. Marketing is often a weak point in small-scale tourism. Which segments of the market are the most appropriate and how can they be reached?
- The combination of traditional agriculture and low-impact tourism is not successful *per se*. Traditional agriculture provides more jobs, but makes the agricultural products more expensive. Only with well-developed marketing, are tourists willing to pay more for these products. The Bonneval example demonstrated that scale is always an important factor. Costs have to be covered by a certain number of tourists.
- Economic development on a macro level is one of the biggest problems in the discussion of rural development and traditional agriculture. General market transactions have compromised these regions. The products of intensive agriculture are cheaper only because this method of production is not confronted with all the social and environmental costs it generates, such as water pollution, the use of pesticides, the destruction of the landscape, air pollution, and the decrease in biodiversity. Generally speaking, these social costs become society's burden. It is not the polluter who pays, but the taxpayer. Nor are the full costs of spatial polarisation paid by those segments of society who profit from the development. On the contrary, the regions suffering from this type of development have to pay the costs in this model.
- This brings us to the conclusion that particularly the European Common Agriculture Policy should be drastically changed. In the current model income support is given to farmers' without heeding the effects of modern agriculture on nature and the environment. The reform of the CAP should be used to create new procedures aiming to combine farmers' support with the level of environmentally friendly agriculture.
- The French PNR concept has positive influences on the development of regions. Overdevelopment, e.g. in certain parts of the Vercors, should be prevented. On the other hand, we should not overlook the limitations of the concept. If all attractive areas in Europe are given this status, not all of them will have enough tourists. Many of these regions are too far away from the densely populated areas. Supply of attractive landscapes often does not meet the demand; this is especially true in spatial polarised development models which are currently the normal practice.
- All these developments are strongly related to policies. Market development is the cause of the problems described here. The market cannot, by definition, solve these problems. But how can a strong policy be developed in a climate where the market should function as freely as

possible? A strong policy is necessary to counteract the negative effect of the polarisation of the market on nature, the environment, the community and rural development. The concept of sustainable tourism can only increase the potential of regions which are more or less favourably located; the model cannot, however, solve the problems of other regions.

- Special attention has to be given to the role of the regional communities. Plans of rural development in which the communities are not involved from the beginning, will have a difficult start. A bottom-up approach, as has been the case in Bonneval, can create a stronger connection with the development concept than the top-down approach of the GTA in the Italian region, where the initiative has been taken by regional authorities without sufficiently incorporating the communities along the trail.

Notes

1. This section is mainly based on Meyer-Küng (1991).
2. This section is based on field studies carried out between 1993 and 1997.

References

Bätzing, W. (1986) *Die GTA. Der große Weitwanderweg durch die piemontesischen Alpen. Teil 1: der Norden*. Friedberg: Verlag Buchhandlung in der Ludwigstraße.

Bätzing, W. (1991) *Die Alpen. Entstehung und Gefährdung einer europäischen Kulturlandschaft*, München: Verlag C. H. Beck.

Bätzing, W. (1993) *Der Sozio-ökonomische Strukturwandel des Alpenraumes im 20. Jahrhundert*, Bern: Geographica Bernensia, P 26, Geographises Institut der Universität Bern.

Bätzing, W. (ed.) (1996) *Landwirtschaft im Alpenraum unverzichtbar, aber zukunftlos?* Berlin/Wien: Blackwell Wissenschaftsverlag.

Brückl, S. and W. Molt (1996) *Kostenwahrheit; Verkehrsinfrastruktur und wirtschaftliche Entwicklung,* Augsburg: Süddeutsche Institut für nachhaltiges Wirtschaften und Oeko-Logistik.

CIPRA (1985) *Sanfter Tourismus – Schlagwort oder Chance für den Alpenraum?*, Vaduz: CIPRA Schrift, Band 1, CIPRA, Vaduz.

Danz, W. (ed.) (1989) *Umweltpolitik im Alpenraum – Ergebnisse der Internationalen Konferenz vom 24–25 Juni 1988 in Lindau*. München: CIPRA Schrift Band 5.

Dietz, F. J. and J. van der Straaten (1992) 'Rethinking Environmental Economics: Missing Links between Economic Theory and Environmental Policy', *Journal of Economic Issues*, 26, 1: 27–51.

European Commission (1992) *The Fifth Environmental Action Plan: Towards Sustainability*. Brussels: European Union.

Evans, D. M. and I. P. Henry (1996) 'Sustainable Tourism Projects: a Trans Border case Study of Samava National Park and Associated Landscape Protected Area, Czech Republic', in B. Bramwell *et al.* (eds) *Sustainable Tourism Management: Principles and Practice*, Tilburg: Tilburg University Press: 201–16.

Grande Traversata delle Alpi (1981) Torino: Centro Documentazione Alpina.

Grande Traversata delle Alpi (1982) Ivrea: Priuli and Verlucca.
Hasslacher, P. (1992) *Alpine Rühezonen; Bestandsaufnahme und Zukunftsperspektiven.* Vaduz: CIPRA, Kleine Schriften 4/92, CIPRA, Vaduz.
International Union for the Conservation of Nature (1980) *The World Conservation Strategy*, Gland: IUCN.
Krippendorf, J. (1986) *Alpsegen, Alptraum – für eine Tourismusentwicklung im Einklang mit Mensch und Natur*, Bern.
Laws, D. and J. Swarbrooke (1996) 'British Airways and the Challenge of Environmental Management and Sustainable Practices' in B. Bramwell *et al.* (eds) *Sustainable Tourism Management: Principles and Practice*, Tilburg: Tilburg University Press: 171–200.
Lorch, J. (1995) *Transportarten in den Alpen.* Vaduz: CIPRA Kleine Schriften 12/95, CIPRA, Vaduz.
Meyer-Küng, H. (1991) 'Ein Experiment im Gefahr', *Berge*, December 1991, 232–6.
Neubronner, E. (1992) *Der Weg, vom Monte Rosa zum Mittelmeer*, München: Verlag J. Berg.
Straaten, J. van der and M. Verhagen (1996) Over het Wasbord. De Grande Traversata delle Alpi. *Op Lemen Voeten* 1996/2: 10–13.
World Commission on Environment and Development (1987) *Our Common Future*, Oxford University Press, Oxford.

15 Cultural tourism and the community in rural Ireland

Jayne Stocks

Introduction

This chapter is concerned with the development of cultural tourism in the Gaeltacht regions of Ireland. The Gaeltacht is the term given to those regions where, although the inhabitants all speak English, Irish is still spoken as the community language. The regions possess a rich heritage, a stong cultural identity and, in most cases, spectacular scenery. The question which stimulated this research was whether the tourism development in the Gaeltacht was characterised by the top-down approach of Irish tourism strategy of the late 1980s and early 1990s (Stocks 1996) or whether more sympathetic, consultative methods were applied. This chapter will examine the process of strategy development for tourism in the Gaeltacht, highlighting the main action areas emanating from this process, and considering a contemporary example of tourism development in these areas.

Before progressing further it is important to clarify the location and nature of the Gaeltacht regions. The Gaeltacht can be divided into seven specific regions, which can be seen in Figure 15.1. These are Donegal, Mayo, Galway, Kerry, Cork, Waterford and Meath. The areas of Gaeltacht are not necessarily contiguous and the regions are not specific to any county boundaries nor do they cover any particular counties in their entirety. They are made even more difficult to distinguish due to the fact that they do not appear on the standard, or more commonly used, maps of Ireland. It can be seen that the majority of these regions are peripheral. Historically, the difficulty of access to the more peripheral regions combined with poor agricultural land and less favourable climate, amongst other factors, led to financial poverty. Connaught, which includes the Mayo and Galway Gaeltacht, is an example of such a region. The term 'to Hell or Connaught' was coined at the time of Cromwell when the only choice given to many of the Irish population was to leave the areas taken over by the British and either be killed or allowed to go to Connaught, which was considered to be a similar fate. However, it can be argued that this historical legacy has allowed the strength of culture, in terms of language, music, story-telling and dance to survive.

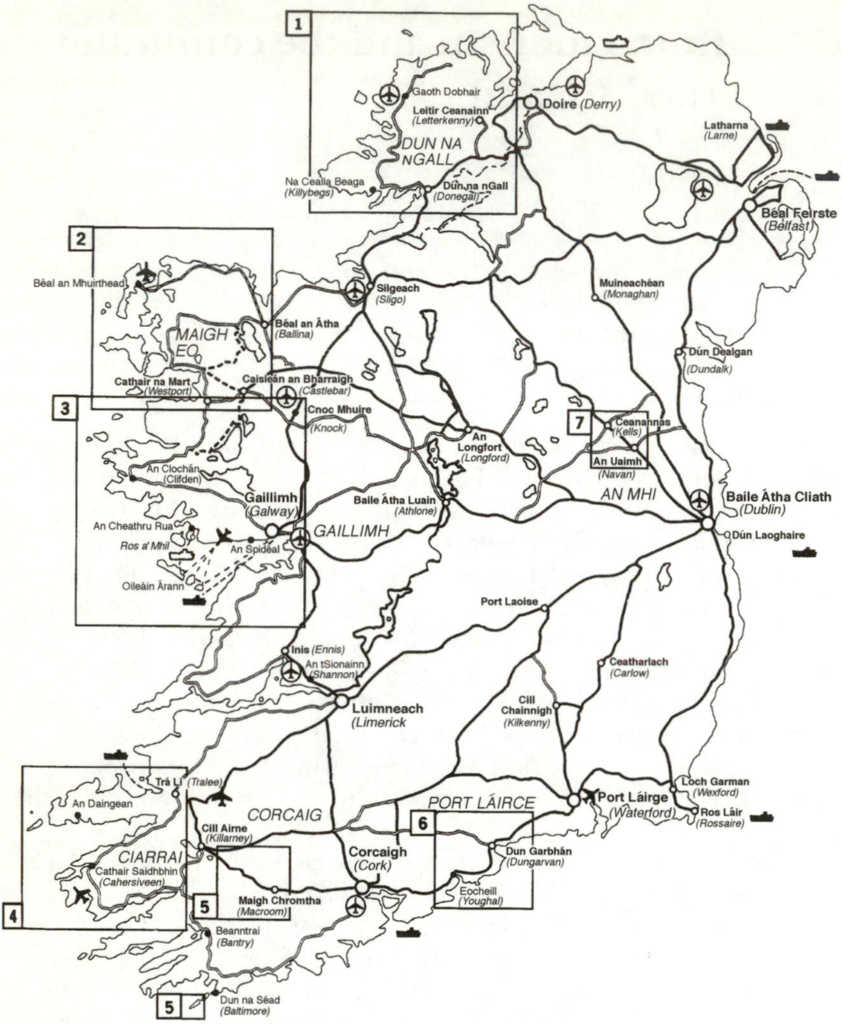

Figure 15.1 The Gaeltacht regions in Ireland.

The Gaeltacht areas are seen to present an opportunity to maintain naturally the continuation of Irish tradition and language and to provide an environment where other people can learn Irish (Udaras na Gaeltachta 1995).

The following quote gives a poetic but interesting description of the areas. 'The Gaeltacht is many things – a physical space, a state of mind, a symbol, the residuary legatee of an ancient civilisation, a metaphor for a central strand of national identity. But for the residents, it is also a space in which to earn a living, in a place and time where such is difficult to achieve' (MFG 1994: 1). The fine balance between the importance of the cultural heritage of the areas and the need of the host populations to survive, and hopefully prosper, is highlighted in this quote. The same document acknowledges that 'The very achievement of even a modest prosperity demands a level of interaction with the "outside world" which threatens the existence of linguistic and cultural identity' (MFG 1994: 1).

The 1991 census recorded just over 81,000 people living in these regions (Udaras na Gaeltachta 1995) with their traditional forms of employment being in farming and fishing. However, more recently there has been growth, supported by government initiatives, in textiles, engineering, electronics, aquaculture, media and tourism. It is important to note that tourism is only one of a number of industries which is moving these regions into the twenty-first century. Too often tourism is seen as the major instigator of all perceived negative impacts brought into a region. It is also important to consider whether any negative impacts created by tourism are deemed to be negative by outsiders, academics, environmentalists and the like, or by the host communities themselves.

Tourism strategy in the Gaeltacht

In the Republic of Ireland the Department of Tourism and Trade formulate tourism policy and Bord Failte (the Irish Tourism Board) and Shannon Development have overall responsibility for the development and implementation of specific strategies. There is thus another tier of regional tourism organisations (RTOs) which has responsibility for planning at sub-national level but it does not correspond to the Gaeltacht regions. However, there are government agencies which have specific concerns for tourism development, within a larger remit, in the Gaeltacht. The two most prominent of these, whose strategies are examined in this research, are Udaras na Gaeltachta and Meithal Forbartha na Gaeltachta (MFG).

Udaras has the task of developing the economy of the Gaeltacht so as to facilitate the preservation and the extension of the Irish language as the principal language of the regions (Udaras na Gaeltachta 1995). MFG has the responsibility for operating the LEADER Programme in rural development in the Gaeltacht. These two agencies consider more than just tourism, but tourism plays an important role in their portfolio.

Udaras na Gaeltachta's guiding tourism policy is to develop initiatives that will integrate visitors into the Gaeltacht, or at least educate shorter-stay visitors about the distinct nature of the regions. They also emphasise helping local communities to develop their own initiatives based on special features of particular areas (Udaras na Gaeltachta 1995). It is recognised that the tourist who chooses to visit the Gaeltacht may be drawn by the natural and built heritage, the unique cultural traditions and the slower way of life. But it is acknowledged that they also want comfortable accommodation, reasonable access and a certain level of entertainment, facilities and service at affordable prices.

To realise these tourist requirements, Udaras accept that help in the form of grants is not sufficient to build a sustainable tourism base. Additional support in the form of short- and long-term training programmes in tourism and related subjects are required in order that the community can properly manage and have realistic expectations for their developments.

Community approach to tourism

MFG commissioned a report to identify a more specific tourism strategy for the Gaeltacht. This was completed independently by the Tourism Research Unit at the University College Dublin in association with University College Galway. The resulting document entitled *From the Bottom Up* established that 'Priorities are to be determined by the communities of the areas they served, they are to be involved in carrying out the actions which flow from these priorities and they are to be the primary beneficiaries of same' (MFG 1994: 2).

The study did not limit itself to one consultative method but employed a series of techniques which would appear to maximise the community involvement in the process. Initially they conducted a series of workshops for the general public which also included a wide range of informed individuals so that overambitious, inappropriate or nonviable ideas could be discussed as well as the complexities involved with possible developments. The consultants held public 'wrap up' review sessions on draft reports before they proceeded to a further stage and written comment was invited to include members of the public who had not attended the meetings. Visits were also made to those members of the community already active in various capacities in the tourism business.

This process of involvement and consultation is very different from that at the national level. Research into the development of the *Operational Programme for Tourism 1994–1999* indicated that asking for public feedback was impossible, and therefore not undertaken, due to the quantity and very individual nature of expected response (Aylwood 1994). Whilst this may be criticised it is reasonable to consider that it is logistically easier to organise and administer a more consultative, 'bottom-up' process with a small percentage of the population facing similar issues than with the scale and complexity of national strategies.

The results of this consultative approach do not attempt to stand alone as idealistic strategies which could never be achieved. They have been integrated with those strategies emanating from the national process of Bord Failte's plan, 1994–1999 *Developing Sustainable Tourism*, and the five regional plans. This should assist with acceptance of the strategies at various levels and their implementation, particularly when this involves various departments outside tourism. It is important to realise that the broader framework, including issues of access, sewage disposal, quality of drinking water, loss of welfare and marketing were brought up by the local communities and taken into consideration in the development of this strategy.

Strategies by action

It is outside the remit of this chapter to consider in detail each of the strategies proposed, which are examined by action, by policy instrument and by geographical region. However, in order to understand the direction tourism is following in the Gaeltacht the strategies by action are reviewed. The strategy information contained in this section is taken from *From the Bottom Up* (MFG 1994: 8–11).

'Define sustainability and manage for it!' (MFG 1994: 8). This appears to be a mammoth area for one action point in a far larger document. If the answer was here someone could be very rich. However, the key points which are raised are the desire for quality, and increased revenue, rather than quantity; the need to limit numbers where carrying capacity is being exceeded and the acknowledgement that making infrastructural improvements, such as road widening, will have implications for visitor numbers and product quality.

The move towards activity-based accommodation, particularly sold as a package, is seen to be a way forward. The intention to keep as much tourism benefit as possible within the community is reinforced by a proposal to establish a small accommodation providers' co-operative to link up with touring agents to provide these packages. It is also suggested that activity combined with language, heritage or culture could also be offered.

In relation to marketing it is believed that the Gaeltacht should be marketed as a 'place apart' with entry and directional signs to indicate that the tourist is in a Gaeltacht area. Cultural associations and links with other European areas such as Brittany should be expanded and specific Gaeltacht tourism marketing material should be developed. All these suggestions would increase the tourist's awareness of the Gaeltacht, particularly those in search of a slightly different tourism product.

The importance of maintaining and developing the use of the Irish language has already been mentioned. This is actively addressed in the proposed 'carrot and stick' strategy ideas. Included in this section is the strategy to help with planning permissions and grant aid to those operators willing to make a commitment to the use of Irish as a working language, and to monitor the ongoing use of this. To assist with this it is suggested that backup and special support should be available and training in the use of Irish as a language of work and customer service should also be given.

It is hoped to develop local empowerment by encouraging networking within the Gaeltacht and to extend the 'block grant' scheme so that real devolution can take place rather than relying on central government approval on a project by project basis. This will require an organisation to monitor progress and inspire action to ensure the distinct needs of these regions are recognised.

The provision of finance needs to be addressed in areas where banking is often limited to mobile or part-time services. Financial credibility between potential entrepreneurs and bank managers could then be built up and banks could be encouraged to lend in areas from which their deposits derive.

Finally, it is felt that more could be achieved in relation to the environmental attributes possessed in these areas. Education could assist in the understanding of the damaging effects of litter, pollution and abandoned cars. 'Flagship' environmental features, such as scenic parks and wildlife refuges could be developed and a code of practice to characterise buildings and refurbishment in styles which are indigenous to the area should be established.

The strategy document does not pretend that all members of the community agreed with all the actions, or that issues felt to be important by the authors were deemed significant by the local people. However the report maintains: 'The real authors are the residents of the Gaeltacht, who gave generously of their time, attention and accumulated wisdom and expertise to decide what is important, to identify opportunities and constraints, and to suggest effective ways of dealing with them' (MFG 1994: 2).

These strategies do not cover all the proposals made in the document, but they do give direction to the nature of further development in these areas. The emphasis is clearly with the move towards quality rather than quantity, formulating a distinct marketable image for these areas, encouragement of the use of the Irish language, involvement and empowerment of the local people and utilising, but not exploiting, the natural environment. The positive nature of these strategic proposals is that they are feasible, compliment the strategies identified in Bord Failte's *Developing Sustainable Tourism* plan and are what the local communities, not just the planners, desire.

This chapter will now consider a specific destination in the Gaeltacht where evidence of community-based tourism development can be found.

Glencolumbkille (Gleanne Cholme Cille)

Primary research, undertaken in Glencolumbkille in the Donegal Gaeltacht (see Figure 15.2), illustrates the present tourism situation in one particular community.

The potential of tourism for revitalising this area was acknowledged long before various government departments took such a proactive role. The village priest, Father McDyer, saw the potential for tourism development but stressed that it must be done in an environmentally and culturally sensitive manner. He was the driving force behind the development of a folk museum which aimed, with a group of purpose-built cottages, to depict 300 years of domestic life in South West Donegal. In 1967, when this museum was being built, the Dublin bureaucrats were not so supportive and planning permission was not received until the day after the museum opened (FOG 1997). The folk museum is still community owned and run, attracts approximately 30,000 visitors per annum and acts as an outlet for locally made produce. Shortly afterwards, in 1968, a 'village' of traditional houses was built by a local co-operative. These houses are available for rent by tourists in the summer

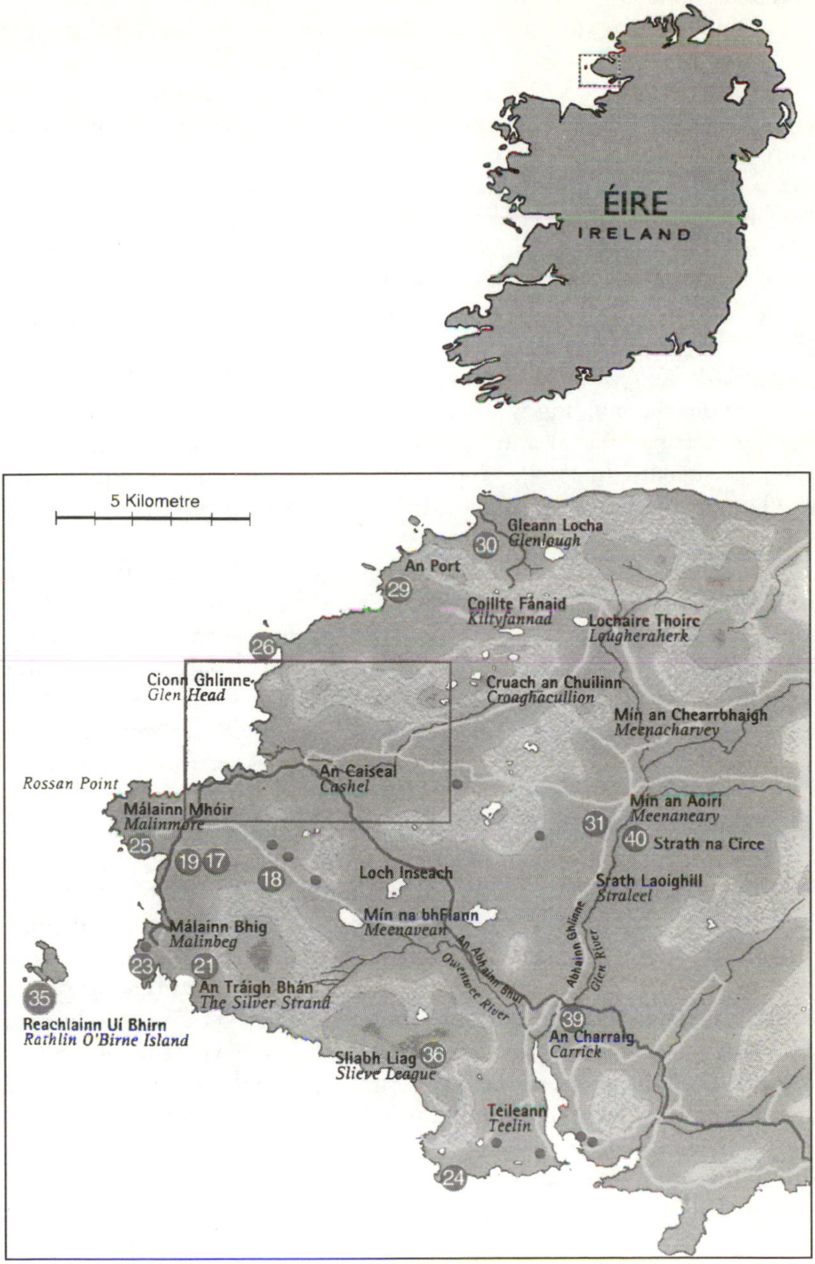

Figure 15.2 The Donegal Gaeltacht region.

months. It is amusing to consider that community development is viewed by many as a *new* way forward when classic examples such as these are thirty years old.

More recently, in 1983, Oideas Gael was founded in Glencolumbkille. This is an organisation which aims to promote the Irish language and foster Irish culture by offering language courses to adults as well as bilingual courses in subjects such as archaeology, painting and set dancing. The idea is that the young and old can come to their centre, bringing people together from Irish and other Celtic backgrounds and participate in everyday Gaeltacht life (MFG 1994: 145). The success of Oideas Gael created the need for a specifically designed building with classrooms, library, hall, kitchen and restaurant. The £150,000 required was made available from the International Fund for Ireland, Udaras na Gaeltachta and the promoters themselves. A local contractor did the building work, a local family leases the restaurant and the hall is used for national and international conferences as well as local community events. In 1994, 700 places were filled giving the equivalent of 4,900 bed nights to owners of local accommodation (O'Gallachoir 1995: 93–4).

Weaving is a traditional industry in this area but had fallen into decline since the late 1940s. Udaras na Gaeltachta gave a grant to a group, known as Tapeis Gael, to train six young people in this traditional craft. They now have a thriving business and, as well as encouraging tourists to their studio, sell and display their work depicting the culture and heritage of the area locally, throughout Europe and in the United States. This obviously provides income to local families and the fact that many of the materials used are sourced locally adds to the local economy (O'Gallachoir 1995: 95).

Observational research, and comparison with old photographs, indicates that the village has not changed dramatically since the advent of tourism development. Only one craft shop was evident in the village itself which sold generally quite tasteful souvenirs, a high proportion of local goods, many books on the Irish language and culture, literature in Irish and bilingual guides to the area. Discussion with local people indicated that, at present, tourism was only of significant importance for two or three months of the year. However, it was also mentioned that the money made through tourism initiatives made the difference between some of the community staying in the area, which they wanted to, or migrating due to financial problems.

In the case of Glencolumbkille the community, or bottom-up, approach to tourism development appears to have been in action for the last thirty years. This approach has been strengthened, rather than diminished, with the influence and grant aid from bodies such as Udaras na Gaeltachta which are actively supporting community projects. Research indicates that the cultural heritage of the area has been enhanced rather than compromised and the problem of migration of young people, which is a serious problem in many of the more remote Gaeltacht areas, has been alleviated to an extent with the economic benefits associated with the tourism.

Conclusion

It cannot be claimed that the method of strategy development outlined in this chapter or the resulting strategies are perfect. In fact it would be dangerous or foolish to propose that perfection has been or could be achieved in relation to any tourism strategies. However, the evidence presented indicates that the development of the strategies, the resulting strategies and the example cited are positive moves forward in the development of tourism in the Gaeltacht. There appear to be no misconceptions that heritage tourism could be the panacea to resolve the long-standing difficulties in the Gaeltacht regions. Rather it is seen as one of a number of initiatives to assist with the problems of depopulation, economic decline and loss of cultural identity.

Expanding tourism on an *ad hoc* basis could work to destroy the Irish language if fostering Irish as a component of the tourism product is not actively promoted. The strengthening of the language in the Gaeltacht is not perceived as a fossilisation of the culture, a move towards disunity with the rest of Ireland or a simple marketing ploy. The fact is that Irish is an integral part of the Irish culture which has a place in modern society. The active involvement of the local people at every stage in the development process considers the direction they, the people who know the places best, want their community to go.

This paper would assert that this process is grounded in realities and could help to retain the integrity of local cultures and heritage through the sympathetic development of 'bottom-up' tourism. It is possible that this process could be adapted to similar regions within Europe and beyond to counteract the problems of commodification, exploitation, fossilisation, commercialism or plain destruction of cultural heritage through inappropriate tourism strategy.

References

Aylwood, A. (1994) *Personal interview*, Dublin, Principal, Department of Tourism and Trade, Government of Ireland.

Bord Failte (1994) *Developing Sustainable Tourism – Tourism Development Plan 1994–1999*, Dublin, Bord Failte.

FOG (Foilsithe ag Oideas Gael) (1997) *Gleann Cholm Cille, Seoid oidhreacht na hEireann*, Donegal, Oideas Gael.

MFG (Meitheal Forbartha na Gaeltachta) (1994) *From the Bottom Up – A Tourism Strategy for the Gaeltacht*, Dublin, Tourism Research Unit, University College Dublin, University College Galway.

O'Gallachoir, C. (1995) *Culture as an Economic Resource with Special Reference to the Donegal Gaeltacht*, Unpublished Masters Dissertation, University College Galway.

Stocks, J. (1996) *A Comparative Analysis of the Tourism Development Strategies of The North and South of Ireland*, Unpublished Masters Dissertation, University of Derby.

Udaras na Gaeltachta (1995) *Tuarascail agus Cuntais*, Co. Donegal.

16 Agritourism – a path to community development?

The case of Bangunkerto, Indonesia

David J. Telfer

Introduction

In the context of globalisation there is increasing dissatisfaction with current top-down models of development necessitating self-reliance at the community level. Trends in tourism literature have mirrored trends in development literature, with both placing increased emphasis on involvement at the local level. In order to evaluate the potential of agritourism generating community development, the initiatives taken by the residents of the village of Bangunkerto, Indonesia are examined. Located in the highly productive agricultural lands of central Java, Indonesia, the community is utilising local resources to attract tourists and promote development. Altering their production to the more profitable salak crop, the villagers have established a plantation centre for tourists. In order to get an understanding of how their lives have changed since the introduction of tourism and the adoption of salak as the dominant crop, thirty-seven area farmers were interviewed. This chapter will initially examine the changes in development theory and consider how these trends have been reflected in tourism research and planning. The chapter will then attempt to answer the question through the investigation of the case study, is agritourism a path to community development?

Community development

Development theory has been shifting to include the role of community development as the result of the inappropriateness of the meta-narratives for those at the local level. There has been a call for increased local involvement in the development process (Clark 1984; Haq 1988; Pretty 1994; Smith and Blanc 1997) focused on local inputs and local participation. Indigenous theories of development have been put forward to incorporate local conditions and knowledge systems (Chipeta 1981) and the process of involving local populations and empowering them has been the focus of grassroots approaches such as Participatory Rural Appraisal (PRA) (Chambers 1994). There is an increased recognition of the role of women in local development (Norem *et al.* 1989), and the South Commission (1990) suggests that development should be self-directed and focused on self-reliance. Mitchell (1997) points out that key aspects of sustainable development include empowerment of local people, self-reliance and social justice. Nozick (1993, cited in Joppe 1996) developed the following list of principles of sustainable community development:

1. economic self reliance,
2. ecological sustainability,
3. community control,
4. meeting individual needs and
5. building a community culture.

Nozick's five points will be revisited at the end of the chapter in order to evaluate the case study.

Pretty (1994) developed a typology of how people participate in development programmes. Participation ranges from passive participation where people are told what development project is proceeding to self-mobilisation where people take initiatives independent of external institutions. Pretty argues that if development is to be sustainable, then at least the fifth level of participation or functional participation must be achieved which includes the forming of groups by local people to meet predetermined objectives related to the development project. On a more cautionary note, Li (1996) suggests that advocates of community-based resource management have represented these communities as sites of consensus and sustainability; however, it is important to recognise that communities may not be homogeneous. Critics of indigenous development paradigms cite problems of consensus building, barriers to participation, lack of accountability, weak institutions, and lack of integration with international funding sources (Wiarda 1983; Brinkerhoff and Ingle 1989).

Use of tourism as a community development initiative

Similar to development theory, tourism research and planning have placed increased emphasis on the role of community development. Researchers have begun to examine indigenous tourism development recognising that indigenous communities are not only impacted by tourism but they also respond to it (Long and Wall 1993; Wall 1995). Similar to the trends in development paradigms of dissatisfaction with existing development philosophies, many tourism analysts have become disillusioned with mass tourism in favour of 'alternative tourism'. Brohman (1996) argues that while the term has been abused, there are a number of recurring themes to alternative tourism which can be utilised to define the concept. Alternative tourism strategies stress the following: 'small scale, locally-owned developments, community participation, and cultural and environmental sustainability' (Brohman 1996: 65). Brohman (1996) cautions that despite these similarities, it is important to take into consideration the changing conditions and interests of individual countries before any strategies are adopted.

Recent studies have examined the role of indigenous entrepreneurial activity in response to tourism (Shaw and Williams 1990; Wahnschafft 1982; Stott 1996). Long and Wall (1993) studied small-scale lodging establishments in Bali and in Tufi, Papua New Guinea; Ranck (1987) found that small-scale guest houses based on local ownership and management were a viable industry. They used local labour and construction materials and few imported foods. Archer (1978) stated that domestic tourism may be a better generator of local income than international tourism as it relies more on local sources. Authors in the field of tourism planning also now stress the need for local community involvement and co-operation in the planning process (Murphy 1985, 1988; Simmons 1994; Gunn 1994). McIntosh *et al.* (1995: 342) indicate that tourism development should contain elements of community involvement including: raising the living standards of local people, developing facilities for visitors and residents, and ensuring the types of development are consistent with the cultural, social and economic philosophy of the government and the people of the host area. Advocating an ecological approach to tourism planning, Murphy (1985) stresses that each community should identify tourism development goals to the extent that they satisfy local needs. Long (1993) argues that if affected populations are not included in the planning of the development then even the most planned mitigation programme is likely to be altered in the implementation phase by the host population. Similarly, Brohman (1996) argues that community-based tourism development should strengthen institutions which enhance local participation and promote the economic, social and cultural well-being of the popular majority.

Hall (1994) cautions, however, that issues of power must be considered along with the inability of some groups to participate in decision making. While community involvement is promoted, there are often institutional obstacles in developing countries which may be difficult to overcome (Sofield 1993). Reed (1997) examined power relations in community-based tourism planning in Squamish, Canada and found that power relations are endemic features in emerging tourism areas. It is difficult for independent agencies to convene differences in power across stakeholders and she argues that power relations may alter the outcome of collaborative tourism planning and may even preclude collaborative action. Joppe (1996) calls for additional research to determine how much local residents benefit from tourism in community tourism development.

The purpose of this chapter is to further investigate this area through a case where agritourism has been adopted and to evaluate if it is promoting community development. Agritourism is becoming part of the wider trend of expanding rural tourism which is increasing in popularity in many countries. The relationship between tourism and agriculture is complex and can range from conflict to coexistence (Telfer and Wall 1996; Telfer 1996). Rural tourism has been increasing in demand in Europe with many regions opting for rural tourism development (Lowyck and Wanhill 1992; Caalders 1997). Cavaco (1995) has indicated that demographic shifts have resulted in the demand for tourism in the countryside. Countries such as Switzerland, Austria, Sweden and Germany have had strong links between farming and tourism (Oppermann 1996; Cavaco 1995). In Portugal the number of working agritourism facilities increased from twenty in 1989 to sixty-seven by February 1994 and the development of rural tourism has also made it 'possible to realise the economic value of specific, quality based production of foodstuffs' (Cavaco 1995: 145). Sharpley and Sharpley (1997: 31) argue that 'within the European context, there is an identifiable relationship between promotion of rural tourism, agricultural policies and broad, regional development policies demonstrating an increasing recognition of the countryside as a resource for tourism.'

In a North American context, both Joppe (1996) and Go *et al.* (1992) have examined the case of Alberta where in the late 1980s approximately 400 community tourism action plans were developed. Hjalager (1996) warns of the difficulties in establishing rural tourism and that while it may bring innovation in terms of product, it does not necessarily achieve financial success. 'The village as a community is withering away' (Hjalager 1996: 109) and community level interorganisations needed for marketing are slow to develop. Hjalager (1996) developed a list of elements which would assist in moving rural tourism towards making a significant contribution for community development in rural districts which include: efficient marketing, quality monitoring, joint organisation of attractions, efficient linking of food production with tourism in order to use tourism to showcase products and capital provision. Finally, tourism and agricultural policies can be used to assist grassroots involvement.

Agritourism case study

The results of a 1995 study released by the Indonesian Business Data Centre predicted that tourism will become the third largest non-oil foreign exchange earner after textile products and wood by the end of the century (*The Jakarta Post* 1995). Prior to the current economic crisis, earnings from tourism were expected to reach US$8.9 billion a year from 6.5 million foreign visitors by the end of the century. Arrivals exceeded four million for the first time in 1994 (*The Jakarta Post* 1995). The National Tourism Strategy (Directorate General of Tourism 1992) for Indonesia has as one of its recommendations that special interest tourism including agritourism should be developed.

Located on Java, one of the more agriculturally productive islands in Indonesia, the 27 hectare agritourism site is situated in the Kampung of Gadung within the village of Bangunkerto. The village of Bangunkerto is 20 km to the north of the city of Yogyakarta in the Kabupaten (district) of

Sleman in the Special Province of Yogyakarta. The project began in 1980 when community farmers started to switch to a higher quality salak fruit called salak pondoh. Salak, also known as snake fruit is small in size and brown in colour. The term snake fruit is applied due to its distinctive sharp, scaly thin brown skin. Inside are two to three sections of hard white flesh surrounding large pits. Before switching to salak, the residents grew a wide variety of fruit crops including rambutan, mango, coconut and bananas. While these crops are still grown, they have been relegated to corner sections. With the shift to salak, the average income has reportedly increased by three to five times with the best months for production occurring from November to January. The price for the fruit can range from 2000 Rp in peak season to 7000 Rp in low season. The cost to grow salak is, however, quite high as the plants require a great deal of attention, taking two years before any fruit is produced.

The project started in association with university students who were in the area doing community service work. In 1983, local residents developed the initial idea that agritourism could be established based on salak and in 1984 the Provincial Agricultural Department assisted in the effort by giving the villagers 4000 salak plants. A proposal was put forward to the local district government and in 1990 the agritourism site was selected on 27 ha of land in the village of Bangukerto. The agritourism facilities were constructed in 1992.

The purpose of the project is not only to expose tourists to the natural environment but also to stimulate awareness and demand for the product. By creating a promotion medium for agricultural production it is hoped that the market will expand not only nationally but also internationally thereby increasing the level of income for local farmers. The following seven objectives have been put forward for the project:

1. increase agricultural production
2. increase income to farmers
3. develop the area in an environmentally sound manner
4. generate a promotional medium for agricultural production
5. expand the market
6. increase foreign exchange earnings and
7. promote human development.

While the attraction is initially targeting domestic tourists, the long-range plans include an expansion of facilities with the hope of attracting international tourists.

Site description

The area has been divided into three zones. In the inner zone lies the agritourism centre built on communal government land and managed by local residents. There are sixteen households who are responsible for managing the twenty-seven hectare site. These households work in association with the Tourism Youth Association who work as guides at the site. The agritourism site is surrounded by a second or a support zone which is comprised of households whose primary crop is salak. This support zone also acts as a tourist attraction as visitors are allowed to drive or walk through the surrounding villages. Farmers in the second zone also grow a variety of other crops which the tourists can view. The third zone which is even greater in size is identified as a second support zone where contact with tourists is reduced and the main activity is farming salak. The hope is that there will be a trickle down effect from zone one to zone three. As awareness of the crop increases partly through tourism, there is potential that demand will increase, generating additional revenue for the community. The total area involved either directly or indirectly takes in three sub-districts or twelve villages.

Visitors arrive at the agritourism site and parking is provided off the main road next to a small guide centre (see Figure 16.1). A student guide meets the visitors and shows them the displays in the centre of the different

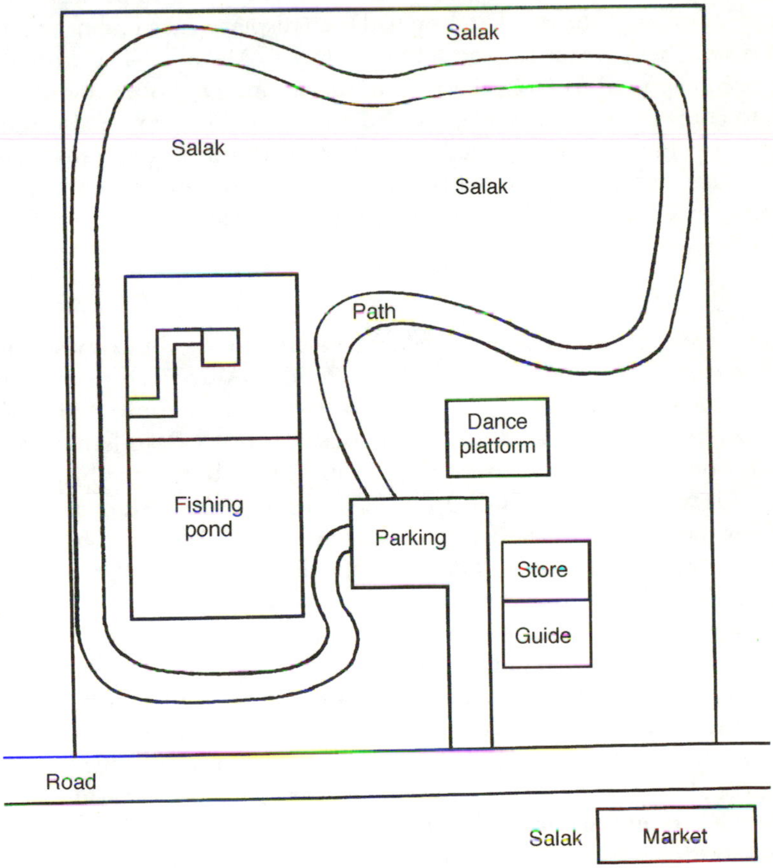

Figure 16.1 Salak tourism plantation.

varieties of salak illustrating how the fruit is grown and harvested. Next to the guide centre is a small store which is run on a co-operative basis which mainly sells drinks and snacks. Traditional dances are performed on the dance platform located beside the parking area. Visitors can take a guided tour or are free to wander along the path through the plantation viewing traditional methods of cultivation. On the far side of the plantation is a large fishing pond with a viewing or fishing platform built out over the water. All of the facilities were made by the local residents including the stone path which was laid by hand. Future plans include the construction of a homestay and a swimming pool. Across the road, locals have responded to the new opportunities and have set up a small market to sell salak to the tourists. Tourists are also allowed to walk or drive through the village in zone two beyond the agritourism site. In the area around the site there are some 2000 to 3000 households.

In addition to the tourist side, the project committee also looks after the agricultural component of the business. The villagers sell small salak trees and the leader of the committee was in the process of trying to promote the crop in Indonesia and abroad. The crop had been displayed at a fair in Jakarta and plans were being made to send some product to Malaysia. The crop has already been exported to Hong Kong and the committee hopes to expand exports to countries such as Singapore and Australia. The project also runs with the co-operation of the Agricultural Department and the local university (Gadjah Mada) which has offered training programmes on topics such as production, marketing, management and project evaluation.

Visitors

Tour Indo, a tour agent based in Yogyakarta, includes the agritourism site on their itinerary and advertises the site as 'Salak Pondoh Agrovillage' as part of their 'Special Interest Tours'. The agritourism site is combined with Borobudur temple on a five-hour tour or with both Prambanan and Borobudur on an eight-hour tour. After visiting Borobudur the brochure states 'On the way back, visit AGRO Village at the slope of Mt Merapi, where you'll enjoy the really fresh air around the Plantation of Salak Pondoh (Snake Skin Fruit), while having lunch.' The cost of the Borobudur and Agro Village tour was US$47. Tour groups can also arrange to go to one of the villagers' homes where salak and tea are served to the guests.

From September 1994 to May 1995 the monthly visitor totals at the site ranged from a high of 1076 in October 1994 to a low of fifty in February 1995. The majority of the guests were domestic tourists while foreign tourists only accounted for twenty to twenty-five visits per month. Differential pricing existed with locals paying 300 Rp and foreign tourists paying 2000 Rp. It has been found that individual tourists tend to stay half an hour while organised tourists tend to stay one and a half hours. The money raised from entrance fees is used for staff salaries and the creation of new buildings. In addition, the money raised is also used to help pay people to harvest the salak from the site.

Community aspects

One of the important aspects of the agritourism project is to foster local involvement, co-operation and community development. The site was constructed on community land and built by local residents. On one field visit, villagers were observed working together laying stones for road foundations gathered from the site of a nearby recent volcanic eruption so that tourists could get through on minibuses. Approximately forty people were gathering stones by the truckload from a nearby river which was almost totally covered by a recent eruption from Mt Merapi. When the researcher returned from investigating the site of the eruption, additional people in the community had joined the effort in the village with the road project. All of the stones were being sorted and laid by hand. Once the stones were in place, they were being covered with asphalt. A total of approximately two kilometres of new road was being built mainly by hand. Community spirit was also demonstrated by residents who owned corner lots. In order that the minibuses carrying the tourists could make their way down narrow windy streets, those with corner lots had taken down walls so that the curves in the road could be widened.

The youth also play a key role in the project and the dances which are performed at the site are put on by local students who belong to the Tourism Youth Association. To avoid conflict by hiring all of the staff from one village, the project organisers have hired student guides from the twelve surrounding villages. There are seven salary positions at the site which are distributed between the villages. One of the student guides interviewed on site had worked there for three years. Two other guides can also be hired to give tours in English.

Methods

In order to get a preliminary understanding of how the lives of the residents had changed since the introduction of tourism and the adoption of salak as the dominant crop, thirty-seven area farmers were interviewed through available sampling in June 1995. The respondents were interviewed either in their homes or as they exited from a village meeting. The purpose of this exploratory research was to investigate attitudes in the community as a result of the adoption of agritourism. Financial limitations and time restrictions prevented the implementation of a wider study with a larger sample. A survey with fifty-four questions was developed divided into the following categories: demographics, agriculture and tourism, agricultural practices and lastly marketing practices. Respondents were also asked for more general comments during the interviews. All respondents surveyed were from villages in zones one and two. Village officials and members from the agricultural department of Gadjah Mada university were also interviewed.

Interview results

Results indicate that the farmers are generally positive toward the initiative in the community and wish to see it develop further. In terms of the demographic attributes of the respondents, the majority (68 per cent) were from the village of Bangunkerto with the remaining distributed over six other villages. A broad distribution was obtained in age categories with the top two being ages 36 to 45 and ages 46 to 55 at 41 per cent and 30 per cent respectively; however, there was high rate of males interviewed at 87 per cent. The primary occupation was farmer at 81 per cent; however, almost all of the respondents also listed a secondary occupation covering a number of different positions such as market traders and teachers. Close to 90 per cent had lived in the community over twenty years and 60 per cent of the respondents had between four to six people living in their household. When asked how many members of the family worked in the salak fields, the highest response was three members at 41 per cent.

There is a heavy reliance on salak as 68 per cent of the respondents stated that over 76 per cent of their monthly income comes from salak. The highest grossing months for salak were reported to be December and January. While there is a high reliance on the crop, 46 per cent responded that the income from salak was sufficient to support the family. The main reason for income from salak not being sufficient was that the farm size was too small.

The second part of the survey dealt with questions which specifically addressed the relationship between agriculture and tourism and again, there appears to be strong support for the project by the majority of the respondents. Over 90 per cent felt that there was a good to excellent potential for the agritourism project. All respondents wanted more tourists to visit the site and have a positive feeling toward tourists. The respondents would like to see the programme expand by creating additional facilities including a swimming pool and a restaurant. When presented with a list of potential benefits from the agritourism project, the highest response rate was for better roads at 95 per cent. As outlined above, road improvements were observed as the interviews were conducted. In addition 89 per cent felt that the agritourism project would create more jobs in the community and 87 per cent felt that the project had increased the value of their land. Respondents also overwhelmingly rejected a list of potential problems resulting from the project with 92 per cent of the respondents stating there was no community disruption. Over 89 per cent rejected the idea that there was any increase in noise, traffic or litter.

In terms of contact with tourists, approximately 60 per cent encountered tourists at least two to three days per week. As outlined above, the tourists are encouraged to walk through the villages in zone two and all but two respondents indicated they wanted the tourists to walk through their villages. An interesting result was found when asked about full-time employment in the tourism sector. Approximately 65 per cent stated that they would not like a full-time job in tourism; however, 68 per cent stated that they would like their children to have a full-time position in tourism.

Some of the responses indicate the strong community connection which exists around the project. While few of the respondents work on (27 per cent) or helped plan (30 per cent) the agritourism site, 62 per cent stated that they helped in the construction of the facilities. In addition, 87 per cent of the respondents did not want a private company to take over the marketing of the agritourism site. The findings of the surveys also reflect the opinions of the committee leaders in that they do not want to sell out to a private company for marketing or production purposes. They want to maintain and control the production locally and keep a traditional atmosphere within the project. The majority (97 per cent) sell their crop to local collectors. One downside of keeping local control is that few travel agents outside the area know of its existence. If it was developed by a private company, marketing may increase and the number of tourists may increase.

A series of questions were asked to determine if the residents felt that the agritourism project leads to an increase in marketing and production of salak as well as generating an increase in earnings. The results indicated that at this time, while attitudes are positive towards tourism, the majority do not feel that it has resulted in an increase in market size for the product. Approximately 92 per cent indicated that the number of salak plants had increased in recent years; however, only 30 per cent felt production had increased as result of tourism. In addition, only 35 per cent stated that increased earnings were one of the benefits due to the project.

In addition to the general frequencies, cross-tabulations were calculated along with chi-squared statistics to determine if there were relationships between the variables which were statistically significant. Despite the range of ages in the respondents, no relationships with age were found to be statistically significant. In addition, due to the majority of respondents being male, this variable was not included in the analysis. The discussion which follows deals with relationships which were all found to be statistically significant.

A trend which developed repeatedly in the analysis of the cross-tabulations indicated that the more directly a respondent is involved with tourism, the more positive they are about the potential positive spin-offs of the salak project. In addition, those who are more directly involved with the project have a higher response for wanting an increased role in tourism. To illustrate this finding a few specific relationships are presented. At the outset of this analysis, it must be stated that overwhelmingly there is a positive attitude toward tourism and tourists regardless of the respondent's employment. However, if they are directly involved in planning or working on the agritourism site then they have strong views that it impacts positively on salak

Table 16.1 Occupation/has agritourism helped salak marketing?

Count total %	*Yes*	*No*	*Row total*
Farmer	6 16.2	24 64.9	30 81.1
Teacher	1 2.7		1 2.7
Village staff	6 16.2		6 16.2
Column total	13 35.1	24 64.9	37 100

Chi-squared – Pearsons: 15.938, DF: 2, significance: 0.00035

Table 16.2 Do you work on the agritourism site/has salak production increased due to tourism?

	Production			
Count total %	*Yes*	*No*	*No answer*	*Row total*
Work				
Yes	8 21.6	2 5.4		10 27.0
No	3 8.1	22 59.5	1 2.7	26 70.3
No answer		1 2.7		1 2.7
Column total	11 29.7	25 67.6	1 2.7	37 100.0

Chi-squared – Pearsons: 16.737, DF: 4, significance: 0.00217

Table 16.3 Have you been involved in planning the agritourism site/do you want a full-time job in tourism?

	Job		
Count total %	*Yes*	*No*	*Row total*
Planning			
Yes	8 21.6	3 8.1	11 29.7
No	5 13.5	21 56.8	26 70.3
Column total	13 35.1	24 64.9	37 100.0

Chi-squared – Pearsons: 9.707, DF: 1, significance: 0.00184

production and marketing. As illustrated in Table 16.1, the majority of farmers do not feel that agritourism has helped salak marketing while all the village staff feel that the project has resulted in an increase in marketing of the product. A similar situation is portrayed in Table 16.2. Those that work on the site believe the project has resulted in an increase in salak production while those who do not work on the site do not believe that there has been an increase in production. Finally as illustrated in Table 16.3, those that have been involved in the planning of the site want a full-time job working in tourism while those not involved in the planning have little desire to work full time in tourism.

Conclusions and implications of the study

In concluding the chapter it is important to reflect on the research undertaken in terms of the opening question on whether agritourism can be a path to community development. As a reference point, Nozick's (1993, cited in Joppe 1996) five principles of sustainable community development can be used as a preliminary measuring stick:

1. economic self reliance
2. ecological sustainability
3. community control
4. meeting individual needs
5. building a community culture.

Similar to the findings of Hjalager (1996), this project has not at this time created a lot of income for the community directly although they are hoping to expand the facilities and markets. Respondents indicated that the project will create additional jobs; however, the actual number of paying jobs at the site is very small. It is also important to recognise that the perception of those surveyed will be impacted upon by the level to which they are directly related to the tourism project. In this case it was found that those who worked closely with the project felt that it increased the production of salak along with generating additional marketing opportunities. While salak has increased the income of the local farmers, their farm size restricts their ability to operate on a larger scale. Many respondents have a second job indicating the project itself is not necessarily creating economic self-reliance. Although there has not been a large economic impact, the project has introduced a new tourism product in the community which could in the future further increase the multiplier effect.

In terms of community control and building community culture, this case study presents some compelling evidence. The initiative was taken at the community level in an attempt to utilise local resources to promote development. All aspects of the attraction are controlled at the local level and in doing so the operation has strengthened local identity. Cultural heritage has also been maintained through the performance of traditional dances at the site. Evidence from the questionnaire along with observations and informal discussions with residents of the area indicate a strong sense of community through the project. A large number of respondents helped in the construction of the site and many were observed laying the foundation for new roads. The jobs at the site have been specifically divided up among students from the surrounding villages and the local university has become involved in the training of the residents in salak production. There is a strong sense of place at the site and the respondents are very positive about the project. There is a desire to increase the number of international visitors.

In terms of ecological sustainability the community has maintained traditional agricultural practices and one of the objectives of the project is the conservation of the natural environment. Courses have been conducted on production with the local university, however, there are certain dangers of moving towards monoculture.

As outlined above it is important not to assume that rural communities are homogeneous. While this case is drawn from Indonesia where village life is strong, Li (1996: 508) points out 'most of Indonesia's rural areas, both on and off Java are complex mosaics of cultural groups and social classes, products of diverse agrarian histories and centuries of interaction with market and state.' Within this diversity this project does not necessarily meet the needs of every individual. Despite a majority viewpoint which is positive towards the project, a few of the respondents indicated that only the privileged were involved and benefited from the project perhaps reflecting on the power relationships in the environment referred to by Hall (1994) and Reed (1997).

The results of the study indicate that tourism can act as a generator of community development to a certain degree. Although it is important to understand this case within its cultural context it does indicate possible development structures. While village life in Indonesia is very cohesive and may have provided the framework for this project to develop, this case study has implications for small rural communities in other developing and developed nations utilising tourism in the rural environment. These rural regions may learn from the level of co-operation and community involvement from this example. If the local community is involved early on in the development of a project, those involved may have a stronger sense of control and an increased community awareness. Other regions may have to work harder at establishing strong community relationships especially if there is a lack of history of this activity. While this research is very preliminary and yielded some interesting results, a future research opportunity would include a return visit to broaden the sampling frame and investigate the changes that have occurred in the community since the last field visit. It would be interesting to investigate the power relationships which have evolved in the area as the project further develops.

Acknowledgements

The author would like to thank Dr Atsuko Hashimoto from Brock University for assistance with the data analysis. This research was conducted with support from scholarships from the Canada-ASEAN foundation and the University Consortium on the Environment along with a research award received by Dr Geoffrey Wall of the University of Waterloo from the Social Sciences and Humanities Research Council of Canada. The research was facilitated through linkages with the Bali Sustainable Development Project funded by the Canadian International Development Agency. A research permit was provided by Lembaga Ilum Pengetahuan Indonesia (LIPI), the Indonesian Institute of Science.

References

Archer, B. (1978) 'Domestic tourism as a development factor', *Annals of Tourism Research* 5, 1: 126–40.

Brinkerhoff, D. and Ingle, M. (1989) 'Integrating blueprint and process: a structured flexibility approach to development management', *Public Administration and Development* 9, 5: 487–503.

Brohman, J. (1996) 'New directions in tourism for Third World development', *Annals of Tourism Research* 23, 1: 48–70.

Caalders, J. (1997) 'Managing the transition from agriculture to tourism: analysis of tourism networks in Auvergne', *Managing Leisure* 2, 3: 127–42.

Cavaco, C. (1995) 'Rural tourism: the creation of new tourist spaces' in A. Montanari and A. Williams (eds) *European Tourism Regions, Spaces and Restructuring,* Chichester: John Wiley & Sons, pp. 127–49.

Chambers, R. (1994) 'The origins and practice of participatory rural appraisal', *World Development* 22, 7: 953–69.

Chipeta, C. (1981) *Indigenous Economics: A Cultural Approach.* Smithtown, New York: Exposition Press.

Clark, G. (1984) 'A theory of local autonomy', *Annals of the Association of American Geographers* 74, 2: 195–208.

Directorate General of Tourism (UNDP) (1992) *Tourism Sector Programming And Policy Development Final Report: Output 1 National Tourism Strategy.* Jakarta: Government of Indonesia.

Go, F., Milne, D. and Whittles, L. (1992) 'Communities as destinations: A marketing taxonomy for the effective implementation of the tourism action plan', *Journal of Travel Research* 30, 4: 31–7.

Gunn, C. (1994) *Tourism Planning: Basic Concepts Cases*, Washington: Taylor & Francis.

Hall, C. M. (1994) *Tourism and Politics: Policy, Power and Place*, Chichester, John Wiley & Sons.

Haq, M. (1988) 'People in development', *Development Journal of SID* 2, 3: 41–5.

Hjalager, A. (1996) 'Agricultural diversification into tourism: evidence of a European Community development programme', *Tourism Management* 17, 2: 103–11.

Joppe, M. (1996) 'Sustainable community tourism development revisited', *Tourism Management* 17, 7: 475–9.

Li, T. (1996) 'Images of community: discourse and strategy in property relations', *Development and Change* 27: 501–27.

Long, V.H. (1993) 'Techniques for socially sustainable tourism development: lessons from Mexico' in J. G. Nelson, R. W. Butler and G. Wall (eds), *Tourism and Sustainable Development: Monitoring, Planning, Managing*. Waterloo, Department of Geography, University of Waterloo, 201–18.

Long, V. H. and Wall, G. (1993) '*Balinese "Homestays": an indigenous response to tourism*. Paper, 13th International Congress of Anthropological and Ethnological Sciences, Mexico City.

Lowyck, E. and Wanhill, S. (1992) 'Regional development and tourism within the European Community' in C. P. Cooper and A. Lockwood (eds), *Progress in Tourism, Recreation and Hospitality Management* 4: 227–43, London: Belhaven Press.

McIntosh, R. W, Goeldner, C. R. and Ritchie, J. R. B. (1995) *Tourism: Principles Practices and Philosophies*, Toronto: John Wiley & Sons, 7th edn.

Mitchell, B. (1997) *Resource and Environmental Management*, Harlow: Longman.

Murphy, P. (1985) *Tourism: A Community Approach*, London: Methuen.

Murphy, P. (1988). 'Community driven tourism planning', *Tourism Management* 9, 2: 96–104.

Norem, R. H., Yoder, R. and Martin, Y. (1989) 'Indigenous Agricultural Knowledge and Gender Issues in Third World Agricultural Development' in D. M. Warren, L. J. Slikkerveer and S. O. Titilola (eds), *Indigenous Knowledge Systems: Implications for Agriculture and International Development*, Iowa State University, pp. 91–100.

Nozick, M. (1993) 'Five principles of sustainable community development' in E. Shragge (ed.), *Community Economic Development: In Search of Empowerment and Alteration*, Montreal: Black Rose Books, pp. 18–43.

Oppermann, M. (1996) 'Rural tourism in southern Germany', *Annals of Tourism Research* 23, 1: 86–102.

Pretty, J. (1994) 'Alternative systems of inquiry for a sustainable agriculture', *The Institute of Development Studies Bulletin* 25, 2: 37–48.

Ranck, S. (1987) 'An attempt at autonomous development: the case of the Tufi guest houses, Papua New Guinea' in S. Britton and W. C. Clark (eds), *Ambiguous Alternative Tourism in Small Developing Countries*, Suva, The University of the South Pacific, pp. 154–66.

Reed, M. (1997) 'Power Relations and Community-Based Tourism Planning', *Annals of Tourism Research* 24, 3: 566–91.

Sharpley, R. and Sharpley, J. (1997) *Rural Tourism: An Introduction*, London: International Thomson Business Press.

Shaw, G. and Williams, A. (1990) 'Tourism, economic development and the role of entrepreneurial activity', *Progress in Tourism, Recreation and Hospitality Management* 2: 67–81.

Simmons, D. G. (1994) 'Community participation in tourism planning', *Tourism Management* 15, 2: 98–108.

Smith, D. and Blanc, M. (1997), 'Grass-roots democracy and participation: a new analytical and practical approach', *Environment and Planning D: Society and Space* 15: 281–303.

Sofield, T. (1993) 'Indigenous Tourism Development', *Annals of Tourism Research* 20, 4: 12–24.

South Commission (1990) *The Challenge To The South: The Report of the South Commission,* Toronto: Oxford University Press.

Stott, M. A. (1996) 'Tourism development and the need for community action in Mykonos, Greece' in Briguglio *et al.* (eds), *Sustainable Tourism in Islands and Small States: Case Studies*, London: Pinter, pp. 281–305.

Telfer, D. J. (1996) 'Food Purchases in a Five Star Hotel: A Study of the Aquila Prambanan Hotel, Yogyakarta, Indonesia', *Tourism Economics* 2, 4: 321–37.

Telfer, D. J. and Wall, G. (1996) 'Linkages Between Tourism and Food Production', *Annals of Tourism Research* 23, 3: 635–53.

The Jakarta Post (1995) 'Tourism to become 3rd biggest foreign exchange earner'. July 3.

Wahnschafft, R. (1982) 'Formal and Informal Tourism Sectors: A Case Study in Pattaya, Thailand', *Annals of Tourism Research* 9, 3: 429–51.

Wall, G. (1995) *Change, impacts and opportunities: turning victims into victors.* Paper presented at the Sustainable Tourism Conference, December, Tilburg University, Netherlands.

Wiarda, H. (1983) 'Toward a nonethnocentric theory of development: alternative conceptions from the Third World', *Journal of Developing Areas* 17, reprinted in C. K. Wilber (ed.) (1988) *The Political Economy of Development and Underdevelopment,* Toronto: McGraw-Hill, pp. 59–82, 4th edn.

17 Community and rural development in Northern Portugal

Joachim Kappert

Introduction

Community and regional development in the northern region of Portugal is strongly connected to agriculture and rural tourism. The development of an organised rural community and of farm tourism throughout the region is regarded as a valuable contribution to local and municipal development. It is expected that the tourism industry will create a significant portion of future new job opportunities in the region so that the continued expansion of tourism is essential (Martins 1993).

This chapter examines the extent to which accelerated growth of the tourism industry, guided by public and private policies, will really benefit the region. The answer to this question depends in part on the kind, as well as number, of tourists the region attracts and the type, as well as the extent, of tourism infrastructure it allows and encourages. The nature of Northern Portugal's visitors and accommodation has changed as the industry has grown. Tourism expansion has been accompanied by changes in the composition of tourists and tourism infrastructure, which have major implications for those seeking to retain a larger share of benefits within their communities (Sampaio 1991). This chapter also attempts to explain the reasons for shifts in tourism and supplier characteristics; to identify their implications for retaining benefits in the northern region; and to suggest ways in which present planning efforts might be modified to benefit rural communities more fully in the future. The impact of tourism development cannot be divorced from the process by which it occurs.

Tourism and regional development

The structure of Northern Portugal's tourism is complex and therefore requires an honest and open dialogue between all the different parties involved, and in particular it requires the distinguished promotion of the Oporto, Minho, Douro, and Trás-os-Montes areas. Tourism could play a much more powerful role in stimulating rural development in Northern Portugal than it has in most regions. Achieving this objective will require more careful integration of tourism marketing and rural development strategies. What constitute 'high quality' tourists and accommodation may not be the same in rural and urban areas. What may be good for the region as a whole may not be good for its rural periphery. In the case of Northern Portugal, the expansion of tourism has not been planned as an integral part of the rural development process, despite the fact that most of the accommodation is located in non-metropolitan areas.

The territorial structures which sustain tourism are frequently the result of spontaneous initiatives which may pay scant attention to the principles of integrated regional planning. What has happened, however, is recognition of tourism as one of the important components in regional development. In this sense, the Municipal Development Plans (PDM) and other specific plans, have begun to present the tourism potential – real or imagined. Yet there is no real sense of how to develop tourism within a co-ordinated regional planning effort.

Tourism's role in national economic and social development has received increasing scrutiny over the last two decades. Those who had hoped that tourism would lead to rapid modernisation now have reason to be worried about its impact on natural resources and cultural values. Regional planning does not yet take into account the overlap between rural tourism and natural environment. The social, psychological, political and cultural effects of tourism and its infrastructure are seen as largely negative.

Much less is known about the role of tourism in stimulating regional, and especially rural, development. The integrated function of agriculture in rural development is not merely the production of food and seed, important though this is (Coelho 1992). Its functions, as generally recognised, also include populating the territory, guaranteeing the ecological and territorial stability and ultimately being a source of culture and leisure resource. Agriculture is the only way of maintaining the population of a territory. In Portugal, many of the problems apparent at the coast originate in a combination of coastal erosion and poorly integrated woodland settlement. Appropriate agriculture protects against heavy rain, prevents erosion and regulates the whole water system on which the urban and rural spaces depend. Agriculture is not only an economic question, but also a cultural one. As long as we pay only for food, the other aspects inevitably suffer.

A major error in Portuguese planning was the idea of matching rural productivity with countries like France and Germany by reducing agricultural activity to 10 per cent of the gross national product. First, Portugal's agriculture is very different to that of North Western Europe. It is based essentially on labour-intensive parcels and vineyards. Secondly, the part-time farmer, who exists in other European countries but who does not enter the statistics despite producing 30 per cent of all agricultural output, is not a feature of Portuguese rural life. Portugal reduced the agricultural population in all regions, without due regard to the consequences.

It is necessary to give dignity to agriculture, and to provide infrastructure to keep the people in their territory. A policy of solidarity which is not based exclusively on food production is required. Agriculture is an essential part of social development. Competing against other countries which are better positioned, with better soils and regular precipitation to produce a specific product, is a lost race. However, this does not necessarily mean that Portugal should not produce goods which are cheaper elsewhere. The cultivation of soils without the undue pressure of competition ensures the population in the territory and a minimum of security and independence. International competition is dangerous as it aims to provide larger, international markets. It is not a solution to wait for Brussels sprouts to come from Belgium – agriculture has to strengthen the local market. The rural areas in Northern Portugal desperately need an agropolitan connection, integrating agriculture with rural industry and diffusing innovations. Tourism can provide rural residents with a broader range of social and economic opportunities than presently exist.

Composition of Northern Portugal's tourism industry

The potential of Northern Portugal's tourism development is based on the exploitation of natural and human resources. The beaches and the sea, the uniqueness of the climate, the countryside, mountains and water are the natural attractions. The human potential is focused on the cultural and social varieties and the traditional hospitality towards visitors. These aspects may, however, become gradually compromised over time through increasing professionalism, becoming, inevitably, less genuine.

Tourism in Northern Portugal is characterised by a number of features (Sampaio 1991):

- the process of spontaneous expansion associated with polifunctional land-use
- tourism motives dominated by beach tourism
- poor job-creation capacity
- seasonality.

The improvement of some of these parameters is linked to the definition of what to present as a development model. It is difficult to combine urban development, beach tourism and the preservation of balanced natural environment. Therefore, the development model assumes that an increasing number of tourists will jeopardise natural resources. We have already become used to seeing Portuguese beaches without fishermen, but a field without farmers does not seem to have the same viability.

Regionalisation is one possible means of highlighting, politically and financially, the urgent actions needed, which require competence and responsibility from all agents involved. Nevertheless, Northern Portugal should be presented as a whole. This strategic perspective begins with collaboration. The success of tourism activity depends on the people involved. Tourism development in Northern Portugal is often confronted with an unfavourable economic conjuncture and official policies which tend to value the Algarve Coast, Madeira and Lisbon as prime destinations. Needless to say, this is something which does not help to encourage tourism incentives and activities in the north.

Regarding community planning, it should be noted that tourism events and facilities are in fact supported both by local and central public administration. Tourism projects enjoy a more beneficial and simpler tax regime than other urban and rural projects (Coelho 1992). However, subsidies are the most pernicious form of economic activity, something which is also true in tourism. In order to create wealth with tourism activities, dynamic entrepreneurs who are able to understand all its potentials are essential. On the other hand, the equally necessary infrastructure depends on funds coming for the most part from the European Union. To date, these two elements have not been well co-ordinated.

The difficulty in establishing links between community and rural development, and incentives to the tourism industry is especially evident in individual municipal planning efforts. These tend to be competitive rather than co-operative endeavours at producing a general strategic plan for leisure, tourism, work and housing. A tourism destination should offer a range of coherently developed products, emphasising the authenticity of place, in accord with expectations of potential consumers. This may include the promotion of aspects related to minorities, to farmers in the remote region of Trás-os-Montes or the gypsy community of Northern Portugal, as well as the promotion of typical local market fairs such as those in Ponte de Lima, Barcelos and Espinho.

The image of a country, a tourism region or a city is extremely important in terms of tourism marketing. The town residents want a city to live in, the tourists want a city to experience. The northern region of Portugal must promote corporate images, such as 'Port Wine', 'Vinho Verde', 'Douro Valley' or even 'F.C. Porto'.

Tourism in rural areas

What we have then is much potential and a developing infrastructure. However, there is still a long way to go to transform Northern Portugal into a tourist destination which will attract tourists who like to discover other peoples and cultures, and to transform it into a competitive destination for congresses and incentives. Sixty per cent of tourists in the north are domestic while 40 per cent are foreign. The overnight accommodation occupancy for foreigners is dominated by the Spanish (22 per cent) and French (14 per cent), followed by English (11 per cent), something which is not typical for the country as a whole (Martins 1993). The typical client of this region is from a higher income bracket and has a high disposable income, although their social and educational background mean that they are less consumption oriented.

Although most tourists travel around at least the Minho area, they stay there for shorter periods. Group travellers spend less time in the interior than independent travellers, irrespective of whether it is a first or subsequent trip. If continued tourism growth entails increasing reliance on foreign tourists and on group tours, then it is doubtful if average lengths of stay in areas like Trás-os-Montes will increase, whether or not more accommodation is built there.

Rural space as a 'new' area of leisure is a symbol of modernity, even if it does not directly reinforce agricultural earnings. It has become a background for new leisure activities, essentially because of stimulation coming from a wealthy upper social class, who have focused their wealth mainly on the commercial sector. Their activities have already convinced rural residents to change their way of life and to throw in their lot with the 'new' tourism industry. While lip service is paid to the concept of eco-tourism, in practice ecologically aware behaviour and activities are not in evidence. The rural establishments created by this process are the most visible consequences of this entrepreneurship, and are being described as an economic alternative to agriculture (Sampaio 1991).

Although the initial effect of tourism, especially in the Minho area, has been to stimulate agriculture by providing the market for meat, milk, and other farm produce, in the long run, however, tourism tends to harm agriculture. One reason is that the desire for building holiday homes and apartments in a municipality, which has only a small amount of useful land, causes land prices to rise drastically. This prevents farmers from purchasing land for agriculture because it is difficult to pay off such high purchase costs from agricultural profits. Moreover, landlords tend not to sell farmland to tenants because they wish to speculate on land values. The resulting difficulties are virtually insurmountable for organised regional planning.

A further difficulty arises from the international food market's requirement for standardised products determined largely by the globalised food industry. Except for certain fruit and vegetables from each region, most food is produced for national and international markets.

It is thus clear that tourism in Northern Portugal affects agriculture in at least two ways: it causes a labour shortage, and it brings about an enormous rise in prices. The influence of tourism development on agriculture is generally perceived as problematic at best, disastrous at worst. If it does not lead to the complete destruction of agriculture, it causes it to become dependent on subsidies. With the creation of the Metropolitan Area of Oporto, however, many farmers saw their land rezoned as 'urban' which, although liable to higher taxes and contributions as a consequence, can now be sold for construction, and has jumped in value.

If rural development is to be achieved through tourism, then marketing should be redirected to attract more repeat tourists, more Spanish tourists and more national visitors, and to convince them to spend more time in the interior of Portugal. Such a change does not conflict with the current view that visitors who spend the most money per day provide greater net public revenues than visitors who spend more money in total but less per day. To Northern Portugal's communities, what matters is not just how much is spent but also how much remains in local circulation. The best kind of tourist to attract from the perspective of an interior community is not necessarily the best kind to attract from an aggregate nation-wide point of view. Tourists make up heterogeneous social groups with homogeneous consumption characteristics. When they discover new destinations, they tend to search for the unusual, for that which is unfamiliar. The tourism industry, therefore, must constantly seek to diversify tourism products.

A tourism region is mutually dependent on other regions. Attempts to diversify tourism must bear in mind the danger of replacing traditional mutually beneficial patterns of behaviour with homogenised patterns at the behest of investors. So the tourism industry tries to attract innovating minorities with incentives like 'golf season', 'rural tourism', 'all-terrain vehicle trails'. By this means, society moulds alternative tourists who see themselves as 'travellers', but who are, in fact, intellectualised and élite consumers of 'travel packages'. Tourism begins to function as a symbolic consumption. Strategies must therefore be developed to attract more independent tourists to the interior, who will stay longer and spend more on local goods and services. It should be noted, however, that decentralisation and more rigorous marketing will not be enough. Local residents should be encouraged to co-operate in the decentralisation of decision-making (Healey 1997). Promoting travel to the interior of Northern Portugal will work in the long run only if tourists, expecting to enjoy the natural splendour, cultural attractions and the unique characteristics of Northern Portugal's communities, actually find them to be attractive. Promotion of Northern Portugal should stress the scenic and cultural differences from other holiday destinations in Portugal and in Spain, rather than constantly stressing a single emblematic element, such as the village church (Costa 1996).

Structure of Northern Portugal's accommodation

The quality of Portugal's tourism industry does not depend simply on good sites and beautiful architecture. 'Pousadas' are among the loveliest accommodation you can find on the Iberian Peninsula and Portuguese tourism policy is attempting to upgrade their image. But more information is needed to improve quality. There must also be a will to preserve the quality of small enterprises such as cafés, otherwise there is a risk of losing traditional local cuisine – something which is unique to every country and thus an object of tourist interest. The revival of century-old spas, especially in the rural interior, is equally a prime objective.

But tourism also demands good services and pleasant social environments. Rapid expansion of the tourism industry is frequently resisted by rural residents whose land and water are diverted from agriculture to tourism or speculative use. Still unclear, however, is the extent to which public intervention in future investment and management decisions can be directed to ensure that agricultural and tourism development is integrated, so that local communities share more fully in the benefits of this expansion.

The way diversification is implemented is of prime importance. Different localities can develop their own identity, stressing tradition, costumes, festivals, and local food specialities. This symbolic revolt against standardised normality results not in a conquest or reconquest of an identity, but in a collective re-examination by the population of their own identity, as manifested in the diversity of regions. Nevertheless, we have to keep in mind that predictions or 'scientific mythologies' can produce their own veracity if, as a result of their mobilising power and their overwhelming marketing, they succeed in dominating the collective beliefs.

Another key-element in the search for better quality is the natural environment. Its quality is the fundamental aspect in the quality/price relationship at the moment of deciding one's destination. It must be emphasised that Portugal has a long way to go to improve its coastal area and the urban image of the principal tourism destinations. Portugal must pay attention to the urgency of pollution-limiting measures, especially along the Algarve Coast, as in general for the country as a whole. Tourism promotion discourages tourists from visiting countries that do not yet practise sustainable tourism. Ignoring environmental measures and community development will turn out to be a dead-end for the industry if it continues to permit such conduct. It has become extremely urgent to provide more green spaces in cities and to improve their appearance. The management of natural reserves must be developed and the careful exploitation of Peneda-Gerês National Park belongs to strategic planning and the decisions cannot be left in the hands of a few investors.

Planning for integrated rural development

Tourism should constitute a complementary activity of small farms. It should be dominated by interest in helping the farmer's family and not the tourism industry. Rural tourism is seen as a means of raising farmers' expectations. Some farmers linked to rural tourism activities already obtain considerable economic benefit. In particular, vineyard owners have been able to exploit a prestigious image that can be promoted in tourism.

If, as previously suggested, more repeat tourists are to be attracted to Northern Portugal, then the regions must become more distinctive and differentiated from each other, as well as from Oporto. Lengths of stay would be extended if tourists had more attractions to visit and different activities to pursue. However, making communities more distinctive through historic preservation, thematic design, cultural pride, and product specialisation will not be easy to achieve.

Many old towns, caught in the transition from agriculture to suburban development, will continue to decline unless they can innovate and adjust. Unable to compete directly for new hotels, the towns which survive will be those that innovate and specialise, serving not just their rural hinterlands but larger markets. Towns should be encouraged to serve as production, distribution, and processing centres, for one or two specific agricultural, horticultural, aquacultural, silvicultural, or marine products for sale or use in restaurants and shops. Supported festivals featuring one of these towns each month, for example, could help reinforce that distinction without turning them into permanent tourism communities (Alberquerque 1997).

Present policies in Northern Portugal do not yet encourage co-ordinated regional planning in which the development of physical facilities, social programmes, and economic activities is well integrated. The planning efforts continue to be oriented toward functional planning, in which agriculture, tourism transportation, recreation, health, housing, conservation, and educational policies, programmes, and projects are all pursued separately. Integration is supposed to occur at the local level and through the municipal development plans. It is unlikely, however, that in event of conflict, municipal plans will take priority over nation-wide functional plans. Regional planning must be aimed at the co-ordination and integration, not at the separation, of activities.

In view of the unpredictability of tourist arrivals and the resulting erratic dependency on them, it seems of considerable importance to understand responses of farmers to tourism, to understand the particular conditions that make agriculture boom or decline, and to consider the viability of agriculture in event of a decline in tourism. Northern Portugal is neither a sun and sand destination, nor an exotic one; much is also still missing to make it a cultural destination. Cuisine is one of the most important factors in regional tourism. The regions try to maintain and rediscover their traditional dishes. Collaboration between tourism and the food and catering industries significantly improves the image of a region, turning food into a major element in the choice of destination. Regional specialities become prestigious. 'Food and drink' circuits are already part of many holiday programmes.

Agriculture in Northern Portugal can be profitable and relatively independent of tourist demand. Farmers are rational and innovative and over the last decade they have left agriculture only in limited numbers. It seems that tourism has promoted agriculture, directly or indirectly. However, this does not mean that Portuguese farmers do not have problems. On the contrary, in spite of profitability and a relative independence from tourism, the future of agriculture in Northern Portugal is doubtful for three specific reasons.

First, labour has become scarce and wages high – the two are interrelated and both are influenced by tourism. Wages in agriculture are much lower than in other sectors. It is therefore no wonder that it is difficult to find labour. Mechanisation has become a necessity as a result of this scarcity. Nowadays harvesting is done with the help of the extended family – mainly women in fact. It seems that because of the scarcity of labour, mutual assistance among farmers is again on the increase, something which is worth noting because of the often-reported decline in social relations and increasing isolation of the nuclear family as a negative impact of tourism.

Second, as everywhere, it is impossible for a farmer to obtain any concrete idea of supply and demand at the moment when he has to decide which crop to grow and in what quantity. The state interference in agriculture is low in Portugal – whether in the form of guaranteed prices, or in the form of loans or other help. The European Union's CAP is making more impositions on Portuguese agriculture than are necessary. Mixed farming could be a way of diminishing the price-risk, but it does not solve the problem. Uncertainty is inherent in agriculture; tourism is not the cause.

Third, the young generation does not want to work in agriculture. Again, however, this fact is not especially due to tourism. This is part of a much more general phenomenon – the decline in the prestige of agricultural work, a decline which seems to be much in evidence throughout the southern European countries.

The Portuguese inheritance system which granted equal shares to all sons and daughters seems to play a role in the successful continuation of agriculture as well. In any case, inheritance conflicts seem to have few consequences for willingness to work the land. Land in Northern Portugal is, nevertheless, broken up into numerous divisions and subdivisions, which severely complicates efficient spatial organisation.

The role of the *Cooperativa Agrícola* (agricultural co-operatives) is thus of crucial importance. It needs to perform various functions: as a bank, as an institution that tries out new crops and seeds, as a supplier of fertilisers and insecticides. It should encourage mixed farming, which was not, for many years, an important policy of the *Cooperativas*. As a result, most Northern Portugal's farmers are still mono-culturalists, dependent on the marketing of only one crop, and thus with no spread in their risks. Efforts toward diversification would prevent high dependency on tourist arrivals. A decline in tourism would certainly affect farmers' economic position because they would also get lower prices for their fruit and vegetables and their earnings from tourism-related activities would decline.

The consequences of tourism for agriculture are clearly not uniform. The argument of this chapter is that one needs to know much more about the characteristics of the land and of territorial organisation, the scale and timing of agriculture, and about wages in all sectors, before one can make any generalisations. But from the case of Northern Portugal, it is clear that the encounter is not always negative. Tourism is in fact essential in the commercialisation of agriculture, and an active market-diversification policy on the part of the *Cooperativa Agrícola* could prevent the farmers from becoming intolerably dependent on tourists for the sale of their produce.

References

Albuquerque, P. (1997) *Turismo, ambiente e desenvolvimento sustentável*, Aveiro: Universidade de Averio.

Coelho, M. (1992) *Administração Pública do Turismo em Portugal*. Lisbon: ISCSP.

Costa, C. (1996) *Towards the improvement of efficiency and effectiveness of tourism planning at the regional level: planning, organisation and networks. The case of Portugal*. Aveiro: Universidade de Aveiro.

Healey, P. (1997) *Collaborative Planning. Shaping Places in Fragmented Societies*. London: Macmillan.

Martins, L. P. S. (1993) *Lazer, Férias e Turismo no Noroeste de Portugal*. Porto: Afrontamento.

Sampaio, F. (1991) *O produto turístico do Alto Minho. Região Turística do Alto Minho*, Viana do Castelo: Região de Turismo de Alto Minho.

18 The market for rural tourism in North and Central Portugal

A benefit-segmentation approach

Elizabeth Kastenholz

Introduction

Tourism in rural areas is frequently studied from a perspective of planning and community development. Marketing research in this field is rather scarce, perhaps due to the small-scale nature and diversity of supply, insufficient available resources, or because of the diverse nature of the market, making data collection difficult. If market studies are rare, even rarer are psychographic segmentation approaches. The current study uses benefit segmentation in order to understand if and how tourists differ in what they look for in Portuguese rural areas. This should provide an important basis for marketing and tourism planning decisions.

Tourism plays an important role for the Portuguese economy, being responsible for about 8 per cent of its GNP. The country's main product 'sun and beach', mainly associated with the Algarve, is confronted with increasing competition from similar, i.e. substitutable destinations. Differentiation is difficult and competition very much based on price. Furthermore, the strong spatial and temporal concentration of demand has led to management, quality and environmental problems, making a more balanced spread of tourism desirable. On the other hand, Portugal possesses resources that might attract tourists also to other parts of the country, namely in 'rural areas' in the hinterland. On the demand side there is an increasing tendency towards split holidays together with a demand for a variety of tourist products, a trend towards 'authenticity', 'thematic holidays', and an increasing interest in 'nature', 'culture' and 'health', which may be satisfied by 'rural tourism'.

Definitions of rural tourism are diverse, ranging from the inclusion of 'the entire tourism activity in a rural area' (EC 1987, cited by Keane 1992: 44) to very specialised tourism products closely related to 'rurality',[1] such as 'agro-tourism' (Leite 1990). Definitions vary from country to country, as do manifestations of rural tourism. A comparison of the Portuguese situation with other European countries shows that the legally defined Portuguese product of rural accommodation (TER) is relatively recent, not as thematically developed as elsewhere and still of minor importance in most rural areas.[2] The TER product has an 'elite image' and caters to an upper market. Examples of other countries may indicate vast opportunities for broadening the market and for developing specialist products, based on outdoor activities, cultural and family offerings.

The still poor development and low profitability of Portuguese rural tourism may be overcome by a well-designed marketing strategy applicable at the destination level. An important first approach is the analysis of the market, its profile, needs and desires, the image tourists associate with the destination, its basic attractions, potentials and weaknesses. The identification of market segments that value different aspects of their holiday experience further permits an approximation between markets and products/destinations, i.e. the right strategy for the right segment in the right area.

The study

The major aim of the current study was the analysis of effective demand of summer tourists in the Portuguese countryside, namely in its North and Central regions. A questionnaire was developed to identify the profile of respondents, their tourist behaviour, attitudes and demographics. Identification of market segments was accomplished through factor-cluster analysis of importance ratings attributed to potential benefit items of a countryside holiday. Variables were found through a literature review, discussions with tourists and rural tourism promoters, as well as by the revision of a pre-test.

The population of the survey was defined as all tourists spending at least one night in the defined areas between the months of June and September 1996. Due to limitations of resources, the spatial and temporal scope of the study had to be limited. In the Central Region the counties of Mortágua, Santa Comba Dão, Carregal do Sal, Tondela and Tábua were chosen, mainly due to their rural nature and interest by ADICES, the regional agency for the EU LEADER programme. In the North, the county of Ponte de Lima and the village of Soajo were included, representative of diverse forms of rural tourism. Ponte de Lima can be considered the best-known 'rural tourism' destination in Portugal with a high concentration of rural and manor houses (TER). Soajo represents a unique, innovative offering of rural tourism in Portugal, namely 'village tourism' in typical, restored rustic houses, integrated in traditional communities.

Owners of lodging establishments and tourist offices distributed the questionnaire among clients and also local residents addressed friends and relatives staying with them. Despite being a convenient, non-probability, sample, it was attempted to make it as 'representative' as possible, including different accommodation forms, geographical areas and by providing the questionnaire in four languages (Portuguese, English, French, German). In total, 500 questionnaires were distributed, with 300 in the North (250 in Ponte de Lima, 50 in Soajo) and 200 in the Central Region; 200 useable responses were collected, with 71 per cent from the North and 29 per cent from the Central Region.

Demographic profile, tourist behaviour and attitudes of respondents

The most frequent nationality was Portuguese, followed by British, German, Dutch and French. The foreign market outnumbered the domestic, especially in the North. Most respondents lived in cities, confirming the often-proposed hypothesis that rural tourism is more attractive to urban dwellers. Cross-tabulation[3] shows that foreigners were more likely to live in low-density habitats than Portuguese (Pearson Chi Square: 16.295, sig: 0.000). Discovery of a different culture may be more significant for foreign tourists than change from an urban environment.

There were slightly more male respondents in the sample. A large majority were married; about 50 per cent had no children and 77 per cent no small children. Respondents were mostly between twenty and fifty years old, with foreigners tending to be older (Kruskal Wallis Test: H = 21.888, Asymp.sig. = 0.000). The respondents' level of education was perhaps surprisingly high, with 58 per cent having at least a university degree. Studies in other countries produced similar results (Crosby *et al.* 1993). Corresponding to this high level of education, 38 per cent of respondents occupied scientific, technical and liberal professions and 11 per cent upper management positions.

Generally preferred accommodations were hotels, followed by 'friends and family', country-houses, pensions and rented flats. However, in the countryside country-houses were preferred (67 per cent), especially by foreigners (83.5 per cent). Portuguese stayed naturally more with friends and family than foreigners did. They also visited 'village tourism', where the poor representation of foreigners may reveal a lack of knowledge of this new product. The average length of stay was 8.7 days, with 23 per cent staying one to three days, not suggesting a main holiday destination. Travel groups were mainly constituted of couples (58 per cent).

Most respondents took two holidays a year, with foreign tourists taking more holidays (Kruskal Wallis H = 17.895, Asymp.sig. = 0.000). Weekend breaks were also popular, with Portuguese taking more weekend breaks than foreigners (Kruskal Wallis H = 5.076, Assymp.sig. = 0.024). The countryside is popular for both longer holidays and weekends.

The holiday destinations most frequently mentioned were Portuguese regions, which is naturally biased by respondents' actual stay. Many foreign respondents repeat their visit to the Portuguese countryside and also Portuguese visit foreign rural tourism destinations. The most mentioned foreign rural holiday destination was France. The UK was mostly cited by its domestic market, which also happened for Germany. However, France, Italy and Spain seem to have a wider appeal. A large group of respondents (38 per cent) primarily spent their rural holidays in their home country.

However, 50 per cent preferred rural holidays abroad, particularly foreign respondents (Pearson Chi Square = 112.06655, sig. = 0.00000).

Twenty-seven per cent of respondents mentioned the role of personal recommendation for destination choice, another 14 per cent a catalogue and 12 per cent previous visit. Therefore, satisfied guests should be considered the most important 'marketing tool', recommending further their experience and being more likely to repeat it. Travel agencies played some role for the foreign market, suggesting the existence of a commercially integrated 'rural tourism product' offered by tour operators.

Respondents' average expenditure per person per day was about 17,500 escudos, with large variance among observations. Expenditure levels were higher in the North (Kruskal Wallis H = 10.936, Asymp.sig. = 0.001), and by foreigners (Kruskal Wallis H = 6.105, Asymp.sig. = 0.013). Most was spent on lodging and least amount on fun, eventually due to missing opportunities.

Asked via an open-ended question what they liked most about countryside holidays, respondents gave preference to 'peace and quiet', 'landscape and nature' and 'health/fresh air'. Also 'isolation/few people', 'authenticity/traditions' and 'hospitality' received favourable mentions, with 'walks' being the most frequently mentioned concrete activity. On the other hand, people disliked most noise and traffic conditions (bad street conditions, dangerous drivers and lack of signposting). Lack of infrastructures (services, public transport and cycling paths), poor accessibility, insects, lack of activities/entertainment, congestion, poor environment (including pollution, poor urban planning and fire) and commercialised countryside were all subjects of criticism. Generally, positive comments outnumbered the negative.

Looking at importance ratings of twenty-seven holiday features, 'peace and quiet' received the highest score, reflecting the above-mentioned positive comments. 'Unpolluted environment' was considered next important, but not indicated as a most liked aspect. Perhaps this item is considered an indispensable condition, without which people are dissatisfied, not implying a special attraction, though, like Herzberg's (1991) 'hygienic factors'. This may be confirmed by the indication of 'poor environment' as an aspect people 'disliked most'. One may consider those factors, which are freely elicited as most liked and also highly rated, as 'motivating' in Herzberg's terminology. Those only highly rated, but not freely elicited may correspond to 'hygienic' factors, resulting in the classification shown in Table 18.1.

There are items 'most liked' in the countryside, but not explicitly referred to in the importance rankings, e.g. 'walking' seems to be a motivation for tourism in the countryside in its own right (confirmed by its frequent mentioning in the context of desirable improvements). The item 'lodging' is related to the freely elicited items 'architecture' and 'hospitality' and thus partly motivational. The motivational versus hygienic character of items may be partially hybrid and vary among tourist types, making a clear

Table 18.1 Ranking of holiday factors

'Motivational/attractive' factors	*'Hygienic' factors*
1. Peace and quiet	1. Unpolluted environment
2. Landscape	2. Independence
3. Hospitality	3. Climate
4. Health	4. Lodging
5. Gastronomy	5. Price
6. Local history and culture	6. Ease of orientation
7. Architecture	7. Accessibility
8. Traditional way of life	8. Tourist information
9. New experience	9. Professional service

distinction between Herzberg's factors difficult. However, it may be an approach worthwhile deepening.

Corresponding to importance ratings are satisfaction ratings concerning respondents' actual stay. The items receiving the most positive satisfaction scores were:

1. Hospitality
2. Quality of lodging
3. Landscape
4. Peace and quiet
5. Climate; independence
6. Unpolluted environment
7. Healthy lifestyle
8. Gastronomy
9. Price
10. Professional service

Further, satisfaction deficiencies or surpluses may be suggested, since desires/expectations were implicitly confronted with perceptions of realities. Obviously, this requires some caution, as the study did not explicitly confront expectations with outcomes. The following major 'importance-satisfaction deficiencies' for items rated 'very important', but receiving a lower corresponding satisfaction score can be identified (subtracting the average importance level of each item from the average satisfaction score). Items with the highest average 'satisfaction-importance' deficit were:

1. 'Ease in finding locations'
2. Unpolluted environment
3. Tourist information
4. Getting to know local history and culture.

Extraordinary satisfaction levels were yielded by:

1. hospitality and
2. lodging.

Further, overall satisfaction levels per respondent may be computed by weighing satisfaction levels with corresponding degrees of importance and then summing. Even if such a 'compositional' measurement of satisfaction is debatable, the resulting, nearly normally distributed, overall satisfaction levels are later useful for segment comparison.

When finally asked how the countryside could become more attractive for a holiday, 'walking paths' and 'typical restaurants' were most frequently mentioned. Other activities looked for were swimming, horse riding, cycling and boating. Excursions/guided tours and cultural performances were also demanded. Further missed were vegetarian restaurants, especially by foreign tourists. A few, particularly Portuguese tourists, would like to be offered more places for socialising such as bars/pubs or discos. Many advocated 'no changes', fearing for destruction of original attractiveness through development. Information and signposts were missed, pollution should be reduced, more public transport offered and driving conditions improved. A more differentiated view on motivations and satisfactions will be provided on a market segment level.

Identification and profile of tourist segments

In order to identify benefit dimensions, a Principal Components Analysis was undertaken on the twenty-seven importance ratings. Thirteen responses were excluded for missing values, leaving 187 useable questionnaires. The first PCA run, including all variables, led to a seven-factor solution with a KMO of 0.80889 and 59.3 per cent of total variance explained. Unfortunately, two factors were not very meaningful, one only consisting of one variable, the other of two, but with low reliability (Cronbach alpha = 0.08). The existence of variables with low communalities (below 0.5) and factor loadings was also observed. These variables were analysed for deletion, using the criteria suggested by Hair *et al.* (1995).[4] Accordingly the following variables were deleted: 'opportunity for excursions', 'climate', 'novelty of experience', 'landscape', 'healthy lifestyle'. The resulting factor solution presented:

- respondents/variables ratio = 187/22 = 8.5
- 60.6 per cent total variance explained
- KMO = 0.82002
- Bartlett Test of Sphericity = 1286.0094 (sig. = 0.00000)
- generally low anti-image (negative partial) correlation (only 7.86 per cent > 0.2)
- correlation matrix with 24.5 per cent correlation > 0.3

- communalities: all > 0.5 except for 'opportunities for shopping' (Its corresponding factor was not used for cluster analysis, though.)
- MSA all > 0.695
- Reproduced Correlation Matrix: 39 per cent of residuals with absolute values > 0.05.

Factor extraction was undertaken according to Kaiser's criterion, yielding a six-factor solution, to which a Varimax Rotation was applied. Interpretation was relatively easy, as factor loadings of all relevant variables in the rotated factor matrix were above 0.5 and further clearly related to only one factor each (see Table 18.2).

Factor 1 explains most variance (25.9 per cent), expressing benefits related to independent travel, namely ease of orientation, independence/flexibility, information and price. The factor reflects a rather autonomous, rational, critical and demanding attitude, focusing particularly on 'hygienic' aspects of the trip. The factor may be called 'independent travel'.

Factor 2 is easy to name: 'Culture and tradition' is self-explanatory, this being a rather 'motivational' factor and further a 'pull motive', according to Iso-Ahola's (1987) terminology.

Factor 3 is also straightforward. Fun and social life, including family and sportive activities, is the focus of this factor, which may be summarised by 'social, active hedonism'.

Factor 4 is determined by an interest in 'handicraft', 'accessibility' and 'shopping' together with 'professional service', pointing at an interest in 'handicraft shopping'. However, this factor is less clear, with low factor loadings and Cronbach alpha.

Factor 5 is a clear component again, consisting of only two variables. 'Peace and quiet' and 'unpolluted environment' are clearly connected with high factor loadings. 'Calm and pure environment' reflects a strong 'pull motive' towards the countryside, but simultaneously a 'push' away from urban living conditions.

Finally, factor 6 refers to items of the destination's service, making people feel comfortable through the way basic needs are satisfied, namely 'gastronomy', 'lodging' and 'hospitality'.

Respondents' scores on the four clearest factors (1, 2, 3, 5) were used as composite variables for identifying tourist groups looking for similar benefits. Since the number of segments was unknown beforehand, hierarchical cluster analysis was chosen. Conceptually a limited number of clusters were expected, as available resources do not support a wide variety of holiday forms, the market being still in its developing phase. Cases were standardised, as response style should not bias results. The Ward Method was used to maximise within-cluster homogeneity. It produced a well interpretable solution and provided distinguishable segments, as confirmed by profiling. A four-cluster solution was best supported by the criterion of 'relative increase of agglomeration coefficient' (Hair *et al.* 1995). Scheffé *post-*

Table 18.2 Summary of PCA-solution

Factor	*Factor loading*	*Variance explained*	*Accumulated % variance explained*	*Cronbach alpha*
Factor 1: 'independent travel'				
• Ease in finding location	0.70613	25.9%	25.9%	0.7205
• Independence/flexibility	0.66009			
• Price	0.63824			
• Information	0.60673			
Factor 2: 'culture and tradition'				
• Traditional way of life	0.71669	10.6%	36.5%	0.7531
• Get to know culture and history	0.70023			
• Architecture/monuments	0.69017			
• Culture/folklore events	0.64870			
• Get to know rural life/ agriculture	0.56419			
Factor 3: 'social and active hedonism'				
• Entertainment/night life	0.78942	8.1%	44.6%	0.6851
• Opportunities for social-ising	0.70615			
• Opportunities for families with children	0.62599			
• Opportunities for sports	0.54235			
Factor 4: 'handicraft shopping'				
• Handicraft	0.62391	5.9%	50.5%	0.6497
• Accessibility	0.61920			
• Opportunities for shopping	0.53416			
• Professional service	0.51362			
Factor 5: 'calm and unpolluted environment'				
• Peaceful/quiet atmosphere	0.80328	5.4%	55.9%	0.7272
• Unpolluted environment	0.79613			
Factor 6: 'service/comfort'				
• Gastronomy	0.78082	4.7%	60.6%	0.5927
• Lodging	0.58471			
• Hospitality	0.50798			

hoc tests showed significant absolute differences between all clusters on the four factors. For segment profiling, differences on importance and satisfaction scores, demographic and behavioural characteristics and desired activities/opportunities for a countryside holiday were analysed (Tables 18.3 and 18.4), resulting in the following characterisation:

- Cluster 1 (25 per cent of sample) was most interested in 'socialising, fun and sports', next valued 'culture and tradition' and further conditions for 'independent travel'. However, compared to other clusters, members of this group did not appreciate particularly a 'calm and unpolluted environment'. This group, interested in a wide range of activities and opportunities, may be named *Want-it-all ruralists*. Members were mainly dissatisfied with opportunities for socialising, tourist information and signposting as well as opportunities for discovering culture and tradition, being mostly satisfied with hospitality and lodging. Their overall satisfaction level was relatively high. This group was young, dominated by Portuguese, followed by Dutch, and coming from scientific, liberal or student occupation groups. They travelled in larger groups and presented medium expenditure levels. This group, which generally took more holidays a year, stayed to an equal extent in the North and the Centre, mainly in TER and independent country houses, but also with friends and family. Personal recommendation or previous visits most influenced their destination choice.
- The *Independent ruralists* (24 per cent of sample) valued most 'independent travel' in a 'calm and unpolluted environment', with 'lodging' playing an important role. This group did not particularly value 'culture and tradition', nor 'social, active hedonism'. They were interested in 'walking' and 'eating out', apart from generally discovering a region by themselves while staying at a comfortable and peaceful place to relax. This group showed some dissatisfaction with tourist information, signposting and walking paths. Their overall average satisfaction level was comparatively low. Dominating nationalities were the British, German and Portuguese respectively. The average age of respondents was 42 years. Travel groups were mostly couples and mainly stayed in country houses in the North. This segment, although mainly foreign tourists, was rather independent of commercial distribution channels and influenced by personal recommendation and non-commercial communication, such as general literature and tourist guides. Their price consciousness may reflect a high quality orientation.
- Cluster 3 (30 per cent of sample) valued most 'culture and tradition' and a 'calm and unpolluted environment'. This group was not interested in 'socialising, fun and sports', nor in 'independent travel'. Members seemed to idealise the most traditional features of an unspoiled rural environment and may be called *Rural romantics*. Supply may only be improved through walking paths and typical restaurants, reflecting an

Table 18.3 Benefits and activities sought

	Want-it-all ruralists	*Independent ruralists*	*Rural romantics*	*Outdoor ruralists*
Benefit factors (Kruskal-Wallis mean ranks)				
Independent travel	107.32	137.89	66.48	66.82
Culture and tradition	115.7	56.31	131.39	57.64
Social, active hedonism	131.13	61.42	59.02	137.08
Calm and pure environment	42.7	98.71	112.12	124.33
Satisfaction levels (Kruskal-Wallis mean ranks)				
Overall weighted	97.26	76.23	106.66	77.97
Activities/opportunities sought (above 20%)				
	Typical restaurants (57%) Swimming (42%) Walking paths (40%) Horse riding (32%) Bars/pubs (26%) Excursions (25%) Cheap restaurants (21%) Museums (21%)	Walking paths (64%) Typical restaurants (49%) Horse riding (22%) Swimming (20%)	Walking paths (50%) Typical restaurants (27%) Vegetarian restaurants (21%)	Typical restaurants (64%) Walking paths (49%) Swimming (41%) Horse riding (28%) Cycling (23%) Tennis (23%) Theatre (23%)

Table 18.4 Profile of segments

	Want-it-all ruralists	*Independent ruralists*	*Rural romantics*	*Outdoor ruralists*
Average age	36	42	44	36
% married	49%	76%	79%	56%
Main nationalities (> 10%)	Portuguese (43%) Dutch (26%) German 13%) British (13%)	British (27%) German (24%) Portuguese (22%)	British (30%) Portuguese (22%) German (21%) French (11%)	Portuguese (62%) British (15%)
Main occupations (> 10%)	Scientific, liberal (36%) Students (15%) Clerical/service (11%) Retired (11%)	Scientific, liberal (31%) Clerical/service (16%) Upper management (13%) Housewives (11%)	Scientific, liberal (43%) Upper management (14%) Clerical/service (13%)	Scientific, liberal (39%) Clerical/service (23%) Students (10%)
Expenditures per person/ day (escudos)	18.000	16.400	23.150	12.300
Travel group	5.9 people	2.9 people	3.1 people	3.8 people
No holidays/year	2.6	2.1	2	1.7
No weekend breaks/year	2.9	2.3	2.6	2.8
Lodging type	TER (38%) Friends/family (19%) Indep. country houses (17%)	TER (56%) Indep. country houses (24%)	TER (63%) Hotel (16%)	TER (31%) Indep. country houses (28%) Friends/family (26%)
Destination choice influenced by (> 10%)	Pers. recommend. (51%) Catalogue (19%) Previous visit (15%) Newspaper ads (15%)	Pers. recommend. (38%) Catalogue (13%) Other literature (13%) Tourist guide (11%)	Catalogue (38%) Travel agency (23%) Pers. recommend. (20%) Tourist guide (18%) Previous visit (14%)	Pers. recommend. (38%) Previous visit (21%)
Region of stay	North (51%) Centre (49%)	North (77%)	North (89%)	North (61%) Centre (39%)

interest in preserving the countryside as it is. This group was the most satisfied, especially content with hospitality, lodging and price. Satisfaction levels may be increased, though, for the items 'ease in finding locations' and 'getting to know culture and tradition'. This segment contained the highest percentage of married tourists, was oldest on average (44 years), travelled mainly as couples, spent most and was the highest user of TER accommodation and hotels, staying mainly in the North. Main nationalities were the British, Portuguese and Germans. This group further relied heavily on commercial communications of tour operators and travel agents when selecting destinations. The significance of a previous visit shows some degree of loyalty towards the destination.

- Cluster 4 (21 per cent of sample) valued most a 'calm and unpolluted environment' as well as 'social, active hedonism', showing little interest in 'culture and tradition' or 'independent travelling'. A good name for this group would be *Outdoor ruralists*. They appreciated outdoor activities, like swimming, horse riding, cycling and tennis, and further demanded more typical restaurants and theatre. Their overall satisfaction level was relatively low. Especially the environment could be purer for this group, whereas hospitality satisfied most. The segment was relatively young (36 years), dominated by Portuguese and largely from scientific, liberal or clerical/service occupations. They travelled in larger groups, spent considerably less than the others, stayed in country houses or with friends and family, both in the North and Central Regions. They used to spend rural holidays in their home country and relied most on personal recommendation and a previous visit for destination choice.

It is clear that there are similarities between clusters 1 and 2, which both appreciate 'independent travel' and clusters 1 and 3 valuing 'culture and tradition', while clusters 1 and 4 both appreciate 'socialising, fun and sports'. However, cluster 1, showing interest in a broad range of items, differs from all the others in that it does not much appreciate a 'calm and unpolluted environment'. Despite different activities valued by the segments, more/better 'walking paths' are desired by all. Further, 'typical restaurants', 'swimming', 'horse riding' and museums evoke the interest of many. The low interest of segments 2 and 3 in improvements of the destinations through supplementary offerings may reflect a fear of commercialisation and modernisation/urbanisation of the countryside.

Conclusion

Before deriving any conclusions, the limitations of this study must be stressed. First, the sample was not representative, as this would require a random process of selecting respondents from a known population, which was impracticable. Emigrants and people visiting friends and relatives might be under-represented, such as tourists staying at hotels, pensions and the pousada, and also certain nationalities. Further, the selection of areas may be questioned. Another limitation lies in the eventually incomplete conception of the questionnaire, especially regarding the item-list on which benefit-segmentation was based. Finally, limitations are implicit in the multivariate statistical techniques, requiring subjective choices of methods and interpretations. However, results of this exploratory study may permit some interesting conclusions about the possible structure and driving forces of the summer tourism market in rural areas in the Portuguese North and Central Regions.

When comparing the analysed Portuguese market with rural tourism demand of other countries, the low significance of 'family holidays' and 'agro-tourism' is striking, eventually related to missing corresponding supply. There seems to be a market for speciality products, related to outdoors activities and cultural offerings. On the other hand, an important group of tourists appreciate most 'peace and quiet' and a traditional rural environment, advocating no changes at all. Therefore the introduction of new offerings must be prudent and well integrated in this environment. The preference of country houses in the Portuguese countryside illustrates the interest in a more personalised, small-scale accommodation, providing privacy, peace and quiet, closeness to nature and the 'authentic'. Also lodging forms similar to TER, but 'self-catering' seem to be successful and should eventually be integrated in the legislation and funding systems existing for TER, since 'independence' is an important benefit sought by some tourists.

It is interesting to note that foreign tourists outnumber Portuguese, which may be typical of TER accommodation, especially in the summer months (Edwards 1988; Moreira 1994). The often-suggested north–south tourist flow seems to be also visible in rural tourism. The main nationalities present in this sample are all top-ten tourist markets to Portugal, which suggests a 'spread' from main to secondary tourist destinations in the same country (INE 1993–95). The importance of tourists with high levels of education and corresponding professional status confirms findings in other countries. Further, most respondents are experienced travellers, also regarding rural tourism, which many spend abroad. Important competitors as international rural holiday destinations are France, Italy, Spain and Austria.

The findings suggest that the market can be divided into four tourist segments according to main benefits sought by tourists. The attractiveness of each segment from the point of view of the destination may be assessed in terms of expenditure levels and loyalty towards destination. From this angle, clusters 1 and 3 seem to be most interesting. A more detailed analysis of each cluster may reveal further potentials, strong and weak points, as well as possibilities of combining segments.

Cluster 3 (*Rural romantics*) shows, apart from highest expenditures, also highest satisfaction levels, and would require only small improvements. This group is concerned with the conservation of authenticity and desires only limited development. It may be considered an élite segment comprised of mainly foreign tourists, who frequently visit countryside destinations at home and abroad. Selected destinations may profit most from achieving these clients' loyalty. Important measures would include controlling tourist numbers, preserving heritage and improving its accessibility (walking paths, museums, handicraft expositions and information about local/regional culture and history). Avoiding pollution, improving signposts and offering vegetarian meals may further enhance satisfaction. Another important issue is this group's good accessibility through specialist tour operators. The consequent reduced control of distribution channels by the destination may be overcome by joint investments, relationship management, a 'destination developer', as suggested by Venema (1996), and the use of direct communication channels (Internet). Rural areas interested in this market should therefore improve their access to this market, apart from adapting offerings to these tourists' ideal of an 'authentic rural environment'.

Cluster 2 (*Independent ruralists*) seems compatible with cluster 3 in valuing most 'peace and quiet' and an 'unpolluted environment'. However, it is more demanding in terms of information, signposts and price and shows a general lower level of satisfaction and loyalty. Moreover, this group seems to be more difficult to reach, being more susceptible to non-commercial travel literature. Increasing this group's satisfaction should be a major concern, as personal recommendation is an important determinant of choice. This could be achieved by improving information, providing more walking paths, enhancing independent offerings (e.g. self-catering) and by avoiding urbanisation and tourist masses.

Cluster 4 (*Outdoor ruralists*) is similar to the above two in appreciating a peaceful, quiet and unpolluted environment. However, it requires more outdoor activities and opportunities for socialising, showing low overall satisfaction. However, offerings might be necessary which are incompatible with the desires of other clusters. If developed carefully, particularly in limited areas and in the off-season, these offerings may be successfully integrated in the region's supply. Unfortunately, this group is very price-sensitive which may not justify necessary investments. However, offerings also attractive to other segments and further available as leisure facilities for locals may be reasonable (walking paths, swimming pools, horse riding, cycling, excursions, tennis courts, typical restaurants, museums, bars, etc.). The low spending power of this segment may be related to its young age. On the other hand, travel groups of this segment are generally larger, which increases their actual impact.

Finally, cluster 1 (*Want-it-all ruralists*) seems to be the most different from the rest as it does not really value a quiet and unpolluted environment and misses most opportunities for socialising (e.g. discos). It seems to be a segment that would not mind 'urbanisation' of parts of the countryside as respondents miss particularly entertainment/nightlife, activities and infrastructures. Some interests of this segment may be satisfied together with those of others without degrading of the rural destination in the eyes of the most sensitive 'countryside idealists' and 'environmentalists'. Spatial and temporal differentiation of offerings may be even more adequate for this segment, since different behaviours may conflict when brought too closely together (see Grove and Fisk 1996). Cultural offerings may attract this group together with cluster 3. Activity and socialising both interest clusters 1 and 4. These seem to be most similar considering concrete improvement wishes. Both stay further to a larger extent in the Central Region and include many Portuguese, travelling in larger groups. However, expenditure levels are higher for cluster 1.

The importance of the domestic market in clusters 1 and 4 deserves some attention. It is typically the most important market for rural tourism in many countries, should be more stable, less seasonal and less sensitive to international competition, as well as easier to 'understand' and 'satisfy' by local suppliers. Furthermore, a 'healthy' domestic market may reduce the negative image of an 'overwhelming tourism industry' in the eyes of those who look for the authentic and do not want to feel like being in a 'foreign colony'. Finally, the domestic tourist market is far from being fully developed (only 30 per cent of Portuguese spend holidays away from home) mainly due to economic constraints.

Apart from the identified differences, the general demand for 'walking paths' should be highlighted, as this interest should be easy to respond to, with little investment and few negative impacts. Also general dissatisfaction with signposting and information should be taken seriously. The importance of an overall destination image, depending on items like 'landscape' and 'environment', calls for an integrated 'destination marketing approach'. 'Destination image' may be assessed by different approaches, as demonstrated by Edwards *et al.* (Chapter 19) who studied affective responses to the Alto Minho, using semantic differential scales. Also more qualitative approaches should be useful here, as the construct is too complex to be embraced by rating procedures. However, its analysis is extremely important for understanding choices and satisfactions, distinguishing segments and defining positioning strategies.

The North and the Central regions seem to attract different tourists, looking for distinct benefits. The North appears to be generally more attractive, especially to foreign tourists, which may be due to a lack of knowledge about and market access to the Central region. However, different areas' specialisation on distinct types of tourism and corresponding target markets plus positioning strategies may be reasonable. A specific, geographically limited tourism development strategy may provide a framework, in which each supplier decides how specialised or general an offer should be, as long as incompatibilities are avoided. The aim would be attracting a market whose satisfaction is achievable with existing resources, desirable from the perspective of the resident population and considering natural and cultural heritage preservation. Accordingly, the optimal marketing mix can be defined. The success of this strategy can be tested by continuous research on tourist satisfaction, trends in benefits sought and images of competitive destinations, reaction towards concrete elements of the marketing mix, as well as by monitoring economic, social-cultural and environmental impacts.

Notes

1. The terms 'rural area' and 'rurality' are also debatable. The OECD (1994) defined areas with fewer than 150 inhabitants per km^2 as rural and regions as essentially or sensitively rural with a certain percentage of rural areas (> 50 per cent versus 15–50 per cent). But also other features are important, such as agriculture, cultural heritage, traditions and life-style (Badouin 1982).
2. TER stands for *turismo no espaço rural* and refers to family-owned country houses which serve as tourist accommodation, which are specially regulated, registered and classified by the national tourism entity DGT. TER combines heritage concerns with the ideal of personalised, high-quality tourism supply. DGT distinguishes: TH (*turismo de habitação*), the most 'noble' form of accommodation in manor houses with high architectural value and quality decoration and equipment; TR (*turismo rural*), good-quality accommodation in typical rustic family houses located in a rural setting; AT (*agro-turismo*), accommodation in country or manor houses integrated in a functioning farm.
3. The chi-square test of cross-tabulation is only used with a minimum expected cell count of 5, so that 'a chi-square distribution can be used to determine an approximate critical value to specify the rejection region' (Sinich 1992: 1003, as cited by Ryan 1995).
4. Low communalities, low factor loadings, low MSAs, negative effect on Cronbach alpha of resulting factor constellation, conceptual weakness, low contribution to distinguishing factors (several similar, low loadings) and deletion yields a better overall factor solution.

References

Badouin, R. (1982) *Socio-Economia do Ordenamento Rural*, Porto: RES.

Crosby, Daries, Fernandez, Luengo, Galán, García, Sastre, Mendoza (1993) *Desarrollo Turístico Sostenible en el Medio Rural*, Madrid: European Centre for Environmental and Touristic Education.

Edwards, J. (1988) *A survey of the proprietors of houses belonging to the 'Turihab' organisation of Northern Portugal and a selection of their guests during 1987*, Bournemouth: Department of Tourism/Dorset Institute.

Grove, S. and Fisk, R. (1996) *The Impact of Others upon Customers' Service Experiences: A Critical Incident Examination of 'Getting Along'*, unpublished paper, Porto: ISEE/University of Porto.

Hair, J., Anderson, R., Tatham, R. and Black, W. (1995) *Multivariate Data Analysis with Readings*, Englewood Cliffs, NJ: Prentice Hall, 4th edn.

Herzberg, F. (1991) 'One more time: how do you motivate employees' in Gabarro (ed.) *Managing People and Organizations*, Boston: Harvard Business School, 159–78.

INE (Instituto Nacional de Estatísticas) (1993–95) *Estatísticas do Turismo 1992–1994*, Lisbon: INE.

Iso-Ahola (1987) 'Psychological Nature of Leisure and Tourism Experience', *Annals of Tourism Research* 14: 314–31.

Keane, M. (1992) 'Rural Tourism and Rural Development', in Briassoulis *et al.* (eds), *Tourism and the Environment*, Dordrecht: Kluwer, 43–55.

Leite, A. (1990) 'Que turismo rural no Alto Minho?' in *1o Encontro: Minho-Identidade e Mudança* (conference proceedings) Braga: Universidade do Minho, 396–402.

Moreira, F. (1994) *O Turismo em Espaço Rural: Enquadramento e Expressão Geográfica em Portugal*, Lisbon: Universidade de Lisboa.

OECD (1994) *Politique du Tourisme et Tourisme International dans les pays de l'OECD 1991–1992*, Paris: OECD.

Ryan, C. (1995) *Researching Tourist Satisfaction – Issues, Concepts, Problems*, London and New York: Routledge.

Sinich, T. (1992) *Business Statistics by Example*, New York: Macmillan, 4th edn.

Venema, M. (1996) 'Marketing destinations by tour operators', paper presented at the International Conference *Marketing de Destinos Turísticos*, Faro: Universidade do Algarve.

19 Tourism brand attributes of the Alto Minho, Portugal

Jonathan Edwards, Carlos Fernandes, Julian Fox and Roger Vaughan

Introduction

Tourism is a business and, along with other businesses, the tourism industry of an area includes amongst its most important potential assets intangible factors such as the symbols and slogans used, the perceived quality and the customer base together with any brand name and underlying associations. These are the assets which Aaker (1996) argues comprise brand equity and are the primary source of competitive strategic advantage. Therefore a brand should have a clear and distinct image that truly differentiates it from its competitors. To do this Aaker argues it is necessary to create a broad brand vision that recognises a brand as something greater than a simple set of physical attributes. This broad brand vision is based on brand associations made up of, for example, associations with quality and with a particular way of relating to the customer. Whilst Aaker developed this argument in relation to companies it may equally be applicable to tourist destinations, particularly as the associations can be shown to be more enduring and more resistant to imitation by competitors than a specified set of product attributes which can be imitated and surpassed. Aaker argues that if this is the case it is much more difficult to duplicate the unique values, peoples and programmes of an organisation than to copy a product.

This chapter examines the potential for the development of a brand image for the Alto Minho region in the north-west of mainland Portugal through an exploration of the perceptions of this destination by visitors. It has four main parts. First, there is a brief review of the literature on image and on modelling the way people form the images they hold. Second, there is a brief description of the research on which the results presented are based. Third, some results from a survey conducted in the Alto Minho in 1996 which explores the perceptions of visitors to the area about the nature of the area as a place for a holiday, are presented. Fourth, a brief review is given of the factors which underpin the physical characteristics of the images reported on.

Image of an area

Basis of image

Whilst few tourist destinations are currently branded, many do have images, which they may or may not seek to manage, although it is by no means always clear what various authors, agencies and individuals are referring to when they use the term image. However, a useful definition is that provided by Dobni and Zinkan (1990). They define image as consisting of the perception of a consumer based on reasoned (cognitive) and emotional (affective) interpretation.

Mayo and Jarvis (1981) suggest that such cognitive (attribute-based beliefs describing the area) and affective (the feelings aroused in the individual about the area) perceptions are the way people make sense of the world. It is these perceptions that influence behaviour. In the case of tourism, such influence will extend to the choice of whether or not to visit a destination.

Cognitive images of an area

Echtner and Ritchie (1991, 1993) argue, based upon an extensive literature review, that destination images can only be measured by recognising the complexity of the various components and measuring these using a variety of methods. They suggest that the majority of studies have been confined to attempting to measure only the physical attributes of the destination as a basis for measuring destination images. They suggest that a better conceptual framework can be envisaged as a three-dimensional matrix (Figure 19.1), with the three continua being:

Physical attribute	to	Holistic imagery
Functional characteristic	to	Psychological characteristic
Common	to	Unique

The first two of these dimensions result in four quadrants. The top two quadrants define the physical characteristics of the area. In the top left quadrant (quadrant 1) the broad components of the area are defined (for example, the broad components of the area such as land use, people, etc.). In the top right quadrant (quadrant 2) are the more detailed characteristics of those broad components (for example, the main characteristic feature of a rural area may be that it consists of small fields). The bottom two quadrants consist of the translation of those physical attributes by the individual into either specific or holistic images. The bottom left quadrant (quadrant 3) consists of the description of the specific physical characteristics (for example, the countryside is lush). The bottom right quadrant (quadrant 4) presents the overall image (for example, that it is a 'traditional' area).

The third continuum, Common–Unique, reflects the findings of Pearce (1988) and McCannell (1989) which suggest that perceptions of destination image can be determined both by their unique features and equally by their

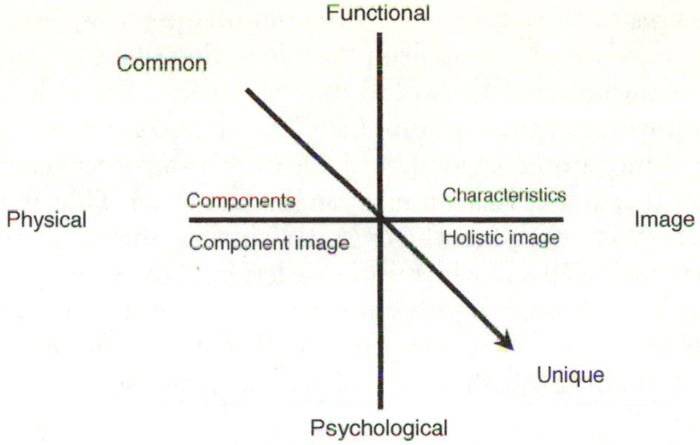

Source: Based on Echtner and Richie,1993

Figure 19.1 Cognitive model: component parts.

more common characteristics. Thus, while all areas are likely to have the common components of quadrant 1, the uniqueness of any area will be determined by the detailed characteristics of those components (quadrant 2) and the physical image they present (quadrant 3). The uniqueness is summed up in the holistic image of quadrant 4.

Affective images of an area

Further analysis of the holistic image quadrant (quadrant 4) is possible by following Baloglu and Brinberg (1997). They propose that in addition to

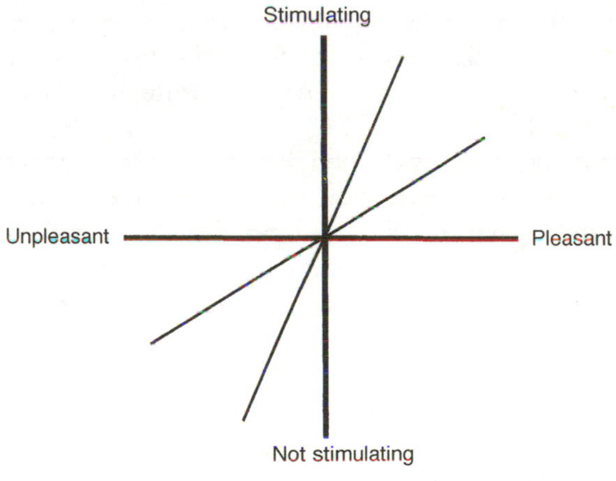

Source: Based on Baloglu and Brinberg, 1997

Figure 19.2 Affective model: overall evaluation.

describing the area in these cognitive (holistic) image terms, based on the physical attributes of the area, it is also possible to describe the area in terms of the reaction of the individual visitor to the image they have of the area. The analysis of this affective image is accomplished using two continua: one based on the level of stimulation provided by the area and the other reflecting the pleasantness or otherwise of that stimulation (Figure 19.2). Thus the Baloglu and Brinberg approach translates the cognitive image (the image of what is being observed) into an affective image (the reaction of the visitor) consisting of four alternative states: unpleasantly stimulated as when scared (quadrant 1), pleasantly stimulated as when intrigued (quadrant 2), unpleasantly not stimulated as when bored (quadrant 3) and pleasantly not stimulated as when relaxed (quadrant 4).

Rural tourism in the Alto Minho

By studying visitor perceptions of a destination image it should be possible to identify those characteristics which are contributing to, or detracting from, the formation of a positive and strong image of the area. The two conceptual frameworks described above have been used in the research reported on in this chapter to examine the cognitive image(s) held by, and the affective responses of, visitors to the Alto Minho region of northern Portugal.

The Alto Minho is a predominantly rural area with a landscape dominated by fertile river valleys, the Lima and the Minho, running westward from the higher ground of the Peneda Geres national park to the Atlantic coast. The tourism product comprises restricted developments on the coastal plain and more recently the development of 'rural' tourism in the interior. It is with this 'rural' tourism that this chapter is concerned.

The term 'rural' tourism is only partly described by objective, spatial, geographic criteria. More accurately the term reflects the prevailing culture and as such may be recognised as a particular form of cultural tourism, in that it is certain aspects and outcomes of an active rural culture that attracts tourists to these areas (Edwards 1991).

Whilst 'culture' may be variously defined, Ritchie and Zins, in their seminal paper discussing culture as a determinant of the attractiveness of a region, adopted the definition given by Kluckholm and Kelly (Ritchie and Zins 1978: 254). Essentially, this definition is that culture is a system of living, with both implicit and explicit elements, which is shared by all members of the group at that one time. Building on this definition Ritchie and Zins proposed a conceptual framework which sought to recognise the variables which influence the attractiveness of a region and which incorporates socio-cultural elements (Table 19.1).

Subsequently, various authorities (Curado 1996), including the National Tourist Board in Portugal, have adopted this analysis both in preparing itineraries of cultural resources and as a basis for marketing and promotional activity. The emphasis in this approach is on the development of selected

Table 19.1 Variables influencing the attractiveness of a region

General factors	*Socio-cultural elements*
Natural beauty	Work
Climate	Dress
Socio-cultural characteristics*	Architecture*
Accessibility	Handicrafts
Attitudes towards tourists	History
Infrastructure	Language
Price levels	Religion
Shopping/commercial facilities	Education
Sport/recreation facilities*	Traditions
Family and friends	Leisure activities
	Art/music
	Gastronomy*
	Countryside*
	Built areas*
	Local population*

Note
* Characteristics used in the Alto Minho study.
Source: Ritchie and Zins (1978).

manifestations of culture, which are readily visible to the visitor, through the identification of priority areas for investments in cultural tourism. This has included the modernisation of tourist accommodation (including tourist accommodation in historic buildings), authentication of the cultural heritage, training, and museums and 'other' cultural resources.

The more diffuse, but equal, manifestation of culture in terms of landscapes which result from patterns of cultivation characteristic of rural areas generally, and particularly of intensively cultivated areas such as the Alto Minho, are not defined as a significant cultural variable although they encapsulate the definition of culture proposed by Kluckholm and Kelly which was cited above. A similar argument can be made not simply for individual examples of architecture but for the totality of the built areas. Equally the local population going about their everyday life are as much a manifestation of the culture as the 'special event' displays of folk dancing or handicraft sales. In going about their everyday life they act effectively as the unpaid attractions and workforce of the tourism industry.

Method

The broader more holistic view of culture in the context of rural tourism, referred to above, informed the approach taken in determining the reason for visiting and the perceptions of the Alto Minho by visitors.

The data were collected in the Alto Minho during the late spring and summer of 1996. The survey was based on self-completion questionnaires which were distributed to respondents at selected locations throughout the Alto Minho. Those locations were accommodation units and covered the range of accommodation on offer to visitors. A total of 141 questionnaires were completed.

The self completion questionnaires which were used incorporated a range of questions determining the:

1. reasons for visiting: for example nature, source of information, attractions
2. activities during visit
3. perceptions of the Alto Minho: countryside, built areas, local people, gastronomy, and overall
4. future visit intentions
5. good and bad points about the Alto Minho.

The majority of questions, that is those relating to points 1,2,4 and 5 were closed and, in some instances, required respondents to rank alternatives. However, the questions which explored perceptions of the Alto Minho, point 3 above, used semantic differentials based on bipolar opposites. For example, was the location 'varied' or 'the same'. Specific questions were directed at the perceptions of visitors of the countryside, the built environment, the region's gastronomy and the host population. It is the information generated by these responses upon which this chapter focuses.

Perceptions of the Alto Minho

Basis of the results

The results presented in this section are in two parts. The first part considers the overall perceptions of the area by all visitors. The second part considers whether first-time visitors hold different perceptions from repeat visitors. The results presented, however, are only a selection of the perceptional descriptors used.

The results are based on the mean scores of the answers given by the visitors. These scores are based on the individual being asked to indicate where, on a five-point scale between two sets of descriptors with opposite meanings, they perceived the area to be. The closer to the word the respondent placed their description the stronger the perception. This is, therefore, an application of semantic differentials based on bi-polar opposites. In Table 19.2 the lower the number (average score) the closer the perception was to the first descriptor given.

Comparisons are made later in this section between first-time and repeat visitors. Where the results for the two groups are similar (not shown to be different through a t-test at the 95 per cent level of confidence) the area can be considered to have the same image for both groups, thus it is a consistent

Table 19.2 Cognitive perceptions of the Alto Minho

Component	*Alternative descriptors*	*First-time visitors*	*Repeat visitors*	*All visitors*
		Average score*		
Countryside				
	small fields–wide open spaces	2.67	2.55	2.62
	lush–arid	2.52	2.00	2.27
	wild–agricultural	2.97	3.43	3.17
Towns and villages				
	stone built–block built	2.15	2.41	2.26
	open–enclosed	3.25	3.62	3.42
	traditional–modern	2.03	1.71	1.88
People				
	helpful–unhelpful	1.85	1.58	1.72
	knowledgeable–uninformed	2.43	2.58	2.50
	friendly–unfriendly	1.29	1.27	1.28
Food				
	meat based–vegetarian	2.18	1.88	2.04
	rich–plain	3.17	2.20	2.73
	distinctive–international	2.39	1.89	2.17
Overall				
	exciting–boring	2.53	1.73	2.15
	safe–risky	1.67	1.59	1.63
	relaxed–hectic	1.66	1.62	1.64

Note
* Average of scores that could be between 1 to 5 inclusive.

image. However, where they are shown to be different, then there may be a problem with any brand image developed as, while it may be 'strong' for one group, it may be less so or even the opposite for the other. (The term 'strong' in this context is not used in the same way as 'strong' in branding theory which relates to recognition and recall.)

Overall image

This sub-section translates the findings of the survey into the cognitive model developed by Echtner and Ritchie (1993) and affective model developed by Baloglu and Brinberg (1997). These are shown in Figures 19.3 and 19.4 respectively. The results on which these graphical presentations of the models are based are shown in Table 19.2.

The cognitive model starts with the broad physical components (quadrant 1). These components (countryside, 'urban' areas, food, people) make up the physical nature of the area and form the underlying basis of the attraction of the area. Each area has different types and combinations of these broad components; therefore each will contribute differently to the attraction of the area for the visitor.

These broad components are then described in terms of detailed physical (quadrant 2) or psychological (quadrant 3) descriptors. Thus, for example, the countryside is characterised by small fields (quadrant 2) and as being lush (quadrant 3) while the built environment is characterised as being constructed of stone and as being somewhat enclosed. These characteristics are the specific details that will vary between different destinations and which give each area its 'personality'.

Finally, in terms of the cognitive model, the holistic psychological descriptors are identified (quadrant 4). Thus the area is perceived as consisting of a predominantly agricultural, rather than wild, landscape. The buildings within the area are thought to have retained a traditional style. The people are considered to be friendly. The gastronomy is perceived to be distinctive and set apart from the 'internationalised' gastronomy of many destinations.

The final questions relating to the perceptions of visitors to the area were designed to gather their overall holistic perception using descriptors that give a personal dimension to the cognitive image quadrant. That is, following Baloglu and Brinberg (1997), they give a description of the feelings that are evoked in the visitor by the cognitive images they hold. It is apparent from Figure 19.4 and Table 19.2 that visitors find the Alto Minho an exciting, safe and relaxing destination.

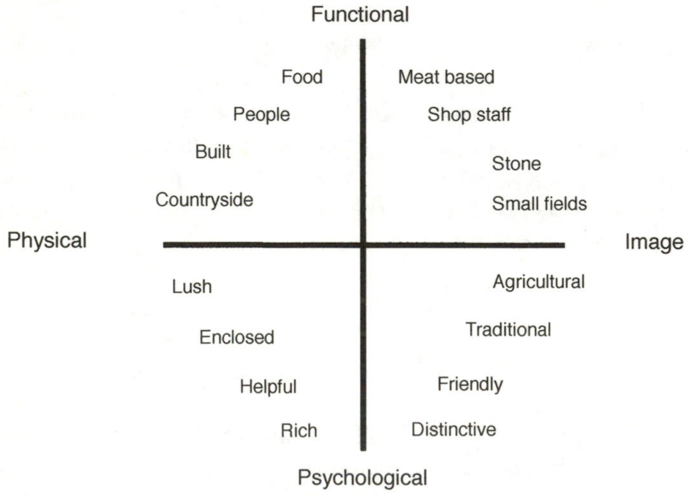

Source: Based on Echtner and Richie, 1993

Figure 19.3 Applied cognitive model.

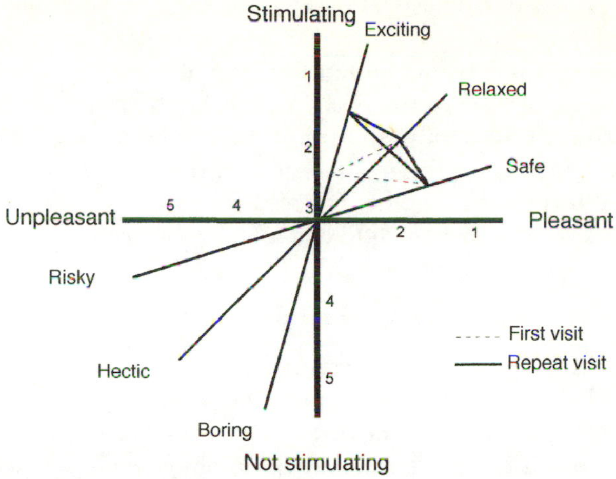

Source: Based on Baloglu and Brinberg, 1997

Figure 19.4 Applied affective model.

Images held by different visitor types

While the above analysis suggests that brand attributes can be described both in cognitive and affective terms, a question remains as to whether such perceptions are held by different visitor types to the same degree.

Refining the earlier analysis with respect to the perceptions of first-time and repeat visitors revealed (Table 19.2) that:

- There was little difference in the perceptions of the countryside held by the first-time and repeat visitors; both groups perceived it as being characterised by small lush/green fields. However, the repeat visitors perceived the area as being more lush and more agricultural than the first-time visitors.
- The built areas were perceived similarly by both groups of respondents; the buildings were perceived as being constructed of stone and as being enclosed. However, the repeat visitors more strongly perceived the built environment to be traditional than the first-time visitors.
- Both first-time and repeat visitors perceived, from their contacts, the people of the Alto Minho to be helpful, knowledgeable and to behave in a friendly manner.
- The gastronomy of the Alto Minho elicited significantly different responses from the two groups. Repeat visitors had an enhanced appreciation of the gastronomy and had become aware of the area's variety and potential excitement. They more strongly perceived the food as richer and more distinctive.

In the points above the differences highlighted are those that were found to be statistically significant.

Finally, it is apparent from Figure 19.4 that, while both types of visitor find the area exciting, repeat visitors find the Alto Minho more exciting than the first-time visitors. Both groups share similar perceptions of a safe and relaxing destination. The axes on Figure 19.4 go from 1 to 5 with the intersection being at 3. When joined together the results suggest that the first-time and repeat visitors occupy slightly different 'affective' spaces.

Underpinnings of the potential brand

The Alto Minho contains an agriculturally based society set in a particular landscape, with particular ways of social behaviour and with a distinctive gastronomy. These characteristics are key determinants of how the visitors perceive the area. It is the cultural underpinnings of these characteristics that underpin the development and implementation of a clear brand identity for the Alto Minho based on the types of images presented in the results of the research (as presented above).

The method of cultivating the land results in particular landscapes and gastronomies. The fields or minifundia are small and the total plot holding is often under three hectares with this total being made up of seven to ten different and scattered 'gardens'. With overpopulation a reality for several centuries, cultivation has evolved to maximise land availability. One of the most obvious manifestations of this is the 1.5–2.0m high trellises around the margins of individual plots supporting vines. This pattern of vine cultivation in turn is partly responsible for the distinctive character of the region's 'green' wines, first imported into England in the twelfth century. Cultivation has until relatively recently been carried out by hand, mainly by women, with cattle being used to plough and draw carts.

In the lowland river valleys cattle are generally 'stall' fed, being kept for milk production and as draught animals. In the upland regions the soil, unless terraced, is too poor for cultivation and consequently beef breeds of cattle and goats are allowed to roam extensively, grazing on the tough natural vegetation. The meat from both cattle (Barroso beef) and goats is of good quality and contributes to the distinctiveness of the Alto Minho's gastronomy. Also pork meat and virtually all parts of the pig are traditionally used in the gastronomy. Traditional pork recipes include *Rojoes con Sarrabulho* (a variety of cuts of pork including liver and various sausages are served with rice that has been cooked in the animal's blood).

The built environment is perceived as being traditional and this indicates that the widespread use of granite as a building material for public and religious buildings, bridges and many houses built before 1960 dominates the built environment. Many of the area's most striking buildings were built in the eighteenth and nineteenth centuries reflecting the wealth that some families acquired principally from investments in Brazil.

While families of the aristocracy often benefited from the Brazilian adventure, for many of the peasant classes Brazil represented an opportunity for an emigrant to acquire wealth and status which effectively did not exist in the Alto Minho. As such emigration, and return migration to and from Brazil, marked the beginning of what became a distinguishing feature of society in the Alto Minho: male migration and return. A current manifestation of this migratory practice is reflected in the large numbers of villagers even in remote areas who speak French, France having been a major destination country for migrants in the 1960s and 1970s. As a result of this migratory phenomenon when, unusually in comparison to other societies, the intention of many male emigrants was to return, the responsibility for maintaining society fell to women and consequently society became demonstrably matriarchal. For example, in the absence of male members of the family the land was worked by women and even when the men returned the women continued to undertake the majority of the agricultural tasks.

The perception of visitors of the local population is highly significant to all developing destinations. In the Alto Minho we find a people who have travelled but who, in many cases, prefer to bring up a family in the Alto Minho rather than California. It is a society held together and moulded for centuries by the women who waited for husbands and sons to return, a society where the Catholic Church remains effectively unchallenged in the religious sphere and a people who are perceived by first-time as well as repeat visitors as friendly, helpful and reasonably knowledgeable.

Conclusion

For a brand to be successful it should have a clear and distinct image that truly differentiates it from its competitors. From the research it appears (given the dangers of generalising from a small sample) that the visitors to the Alto Minho perceive the possible components of the brand very clearly; an area that is agricultural, with a distinctive gastronomy, clear traditions and a friendly population. It also appears to be true that visitors find the area exciting, safe and distinctive. Whether these are sufficiently different to form the basis of a branding exercise would require comparative research to be undertaken (although any review of literature published about the Alto Minho would demonstrate a clear brand equity in that visiting writers rejoice in the old ways of living and the organic 'sustainable' farming practices while the relatively few tour companies who operate in the Alto Minho stress aspects of the environment that will appeal to allocentrics). However, the basis of those perceptions could be undermined by the very development of the brand if the underlying key determinant, the socio-economic structures of the local community, are undermined.

References

Aaker, D. (1996) *Building Strong Brands*. New York, The Free Press.

Baloglu, S. and Brinberg, D. (1997) 'Affective images of tourist destinations'. *Journal of Travel Research* 35, 4: 11–15.

Curado, C. H. (1996) 'Cultural tourism in Portugal' in G. Richards (ed.) *Cultural Tourism in Europe*. Wallingford, CAB International: 249–66.

Dobni, D. and Zinkan, G. M. (1990) 'In search of brand image: a foundation analysis'. *Advances in Consumer Research* 17: 110–19.

Echtner, C. M. and Brent Ritchie, J. R. (1991) 'Meaning and measurement of destination image'. *Journal of Tourism Studies* 2, 2: 2–12.

Echtner, C. M. and Brent Ritchie, J. R. (1993) 'The measurement of destination image and empirical assesment'. *Journal of Travel Research* 31: 3–13.

Edwards, J. R. (1991) 'Guest–host perceptions of rural tourism in England and Portugal' in M. T. Sinclair and M. J. Stabler (eds) *The Tourism Industry: An International Analysis*. Wallingford, CAB: 143–64.

McCannell, D. (1989) *The Tourist*. New York, Schoeken Books.

Mayo, E. and Jarvis, L. (1981) *The Psychology of Leisure Travel*. London, CBI.

Pearce, P. L. (1988) *The Ulysses Factor*. New York, Springer Verlag.

Ritchie, B. and Zins, M. (1978) 'Culture as determinant of the attractiveness of a tourism region'. *Annals of Tourism Research* 5, 2: 252–67.

20 Conclusions

Greg Richards and Derek Hall

In this concluding chapter we briefly examine and summarise the evidence presented by the authors in this volume to identify critical success factors in sustainable community development. Finally, some potential future research agendas are outlined.

Conceptualising the community

One of the major problems facing those analysing 'communities' is to find adequate definitions of the concept of community. The increasing popularity of community-based approaches to tourism development has led to a more widespread and in many cases relatively vague use of the concept in social and/or spatial terms. As Frank Howie points out in Chapter 7, 'community' has almost become a shibboleth. It has tended to become idealised, as if 'communities' are inherently good and external influences inevitably bad. It is as if the fashionable adoption of 'community' thinking has locked itself onto some kind of set of hegelian binary oppositions. But of course, as the range of case studies presented in this volume have indicated, there is a wide range of types of 'communities' and groups within 'communities' ('sub-communities'?), all of whom can respond differently to tourism.

The communities analysed in this volume have included the entrepreneurial community as depicted by Khoon Koh (Chapter 13), agricultural communities, urban inner city communities, and communities dominated by the mafia. This diversity of real and imagined communities suggests that to implement successfully community-based strategies for tourism development, more varied concepts and models are required. Joachim Kappert noted in Chapter 17 that the community cannot be isolated from its locality or geographic context. Yet there exists the aspatial 'community of interest' which has tended to receive relatively little attention in the tourism literature, despite its relevance. Certainly more attention needs to be paid to social and cultural aspects of 'community' and communities.

Involving the community

The range of case studies presented in this volume illustrates that the problem of involving the community effectively in the sustainable development of tourism is a global issue. Despite the community-orientated rhetoric of much sustainable tourism policy, it remains problematic to find ways and means of ensuring that all sectors of the community participate in tourism development and that conflicts surrounding the use of community resources are resolved or at least minimised.

As Jayne Stocks pointed out in Chapter 15, community development is not a new phenomenon, but, as a spatially targeted finite social policy, it has been pursued for several decades. Many of the issues which were appropriate to community development ideals and processes 30 years ago are, ironically, still relevant today. Indeed, in many cases they have been brought into sharper relief by the speed and scale of the social, cultural and economic change characteristic of late modernity.

Participation is often a problem of power relationships within the community, and empowerment practices, such as bottom-up planning strategies, are not matched by empowering philosophies. Unequal distribution of power and uneven flows of information can disenfranchise members of the community when decisions are taken about tourism development. This point was made very clearly in the analyses of tourism development in Indonesia by Heidi Dahles in Chapter 10 and Theo Kamsma and Karin Bras in Chapter 11. In the most extreme cases state power is used to further the interests of developers to the detriment of small local entrepreneurs, stifling grass-roots approaches to development. They also point out that policies of developing quality tourism may privilege external over local interests.

Bill Bramwell and Angela Sharman in Chapter 2 provided an analytical framework of issues which affect approaches to community participation in tourism planning (Table 2.4) and suggested possible solutions through encouraging a wide range of participants in decision-making processes which were representative of all relevant stakeholders in the community. This is easier to achieve in the developed world, where wealth and power tend to be more evenly distributed.

Frank Howie (Chapter 7) pointed out that disadvantaged local communities may not be able to identify with tourism development because they view it as an exogenous development with benefits that largely accrue to outsiders. Inability to identify with tourism development in turn may lead either to resistance or indifference. In the UK, Guy Jackson and Nigel Morpeth (Chapter 8), suggested that Local Agenda 21 (LA21) does offer the potential for marginalised communities to be involved in tourism development, but this is often frustrated by lack of awareness or divided responsibilities in local authorities. Policies relating to different aspects of sustainability are not often integrated (e.g. tourism and transport – see Hall

1999), hampering the exchange of information within policy-making bodies, and impeding the flow of information between different policy levels.

As Derek Hall underlines in Chapter 4, the lack of stability in communities in transition may undermine participative models. Ironically, however, conditions of instability may also provide windows of opportunity for community involvement which may not be apparent in more stable communities.

Informing the community

Involving the community and allowing members of the community to make informed decisions about the course of tourism development requires a free flow of appropriate intelligible and useable information within that community. The flow of information is crucial, since the concept of bottom-up development presupposes that all sections of the community are adequately informed about the nature and consequences of tourism development. This is also necessary for the sustainable development of future generations. Yet the implementation of sustainable tourism development may be hampered by the lack of suitable tools, as was indicated in the chapters of Part 2.

Sustainability as an objective is often conflated with the natural environment, and policies for environmental sustainability may often be based on technocratically derived 'scientific' criteria, such as the measurement of air, water and soil quality, as indicated by Brian Goodall and Mike Stabler in Chapter 5. This approach can exclude and disenfranchise many segments of the community, who are not, or who perceive themselves not to be, equipped to analyse or counter the 'scientific' arguments and measures of technocratic élites.

Sustaining the community

A major issue for communities is the economic sustainability of tourism, which can provide a means to stimulate local employment, inhibit out-migration (and encourage in-migration) and prevent the emergence of a local old-age demographic structure. Because the business of tourism consists predominantly of small and medium enterprises (SMEs) and is dependent on innovation for the creation of new products, the role of the entrepreneur can be critical. Entrepreneurs form an important element in the economy of most communities, and can represent a crucial alternative to the growing power of transnational companies. In order for small, local entrepreneurs to survive in the face of growing global competition, however, they need to be supported in a number of ways, as Graeme Evans and Robert Cleverdon emphasised in Chapter 9 when they argued for a global 'fair trade' network to help small businesses in tourism in the developing world.

As well as developing global networks, the construction and maintenance of local networks is also crucial to the development of local enterprise, and as the chapters in this volume have demonstrated, the support of local enterprise poses specific problems in different environments, depending on the level of development and location. Heidi Dahles pointed out in Chapter 10 that local enterprise systems in the developing world depend on the activities of distinct types of entrepreneurs: 'patrons' and 'brokers'. The patrons provide the capital and resources necessary for tourism development to occur, but the brokers represent an essential interface between patrons and tourists, anticipating, translating and satisfying tourist needs through innovation and ingenuity. Dahles illustrated for Indonesia that small-scale tourism enterprises have the potential to contribute a vital force in the community transforming local resources into tourist products and services. This type of bottom-up product generation, particularly where there is increasing demand for flexibility in tourism supply by the state, can form an integral part of the industry rather than being marginalised through excessive regulation. But governments can easily stifle such local inventiveness, however, as Karin Bras and Theo Kamsma graphically illustrated in case studies from the same country in Chapter 11.

Local production systems in the developing world can also be supported through direct links with consumer markets in the developed world. Local producer networks can be supported not only by global fair trade distribution systems, but also by consumer organisations and producer networks in the generating markets. The sale of 'community friendly' tourism products could be supported by appropriate branding and consumer information. Network linkages to internal and external actors are also important for providing mutual support and the dissemination of innovation and good practice in developed economies, as Janine Caalders emphasised in the case of Friesland (Chapter 12). Khoon Koh demonstrated in Chapter 13 the way in which such effective networks can provide the conditions necessary for entrepreneurialism to flourish, in the case of small Texan communities. This process can be assisted by regional, national and supranational (e.g. EU) policies and frameworks. In Friesland, the public sector has played a leading role in developing and supporting network development, providing financial subsidies for community initiatives and integrating different policy levels.

Financial subsidies alone, however, are insufficient to guarantee involvement of all actors in the community, particularly if information flows are uneven. The ability of the public sector to support entrepreneurial development is also dependent upon reliable information about the birth and death of tourism enterprises in the community (Chapter 13).

Rural communities, with their relative isolation, lack of access to capital and fragile environments represent a particular problem in the development of sustainable tourism. The chapters in Part 4 of this volume have provided a variety of examples of the issues facing rural communities in developing tourism sustainably.

The intensification of agriculture and subsequent erosion of the economic base of rural communities also poses problems for tourism development. As Jan van der Straaten (Chapter 14) illustrated in the case of France, small rural communities often find it difficult to afford the infrastructure costs of developing low-impact tourism, and may be forced to opt for larger-scale development which will change the nature of the community. The concept of sustainable tourism is likely to be able to increase the potential of regions more or less favourably located.

Joachim Kappert (Chapter 17) argued that although tourism development in rural Portugal may stimulate demand for agricultural products, it also creates labour shortages and raises land prices, directly damaging the agricultural sector. The development of community-based co-operatives may provide a solution to these problems, extending the available resource base and increasing the diversity of agricultural production to reduce over-dependence on tourism. But the success of such ventures may depend heavily upon characteristics of the local culture. Northern Portugal does not have a strong tradition of co-operative development, unlike the more collectivist cultures evident in the south of the country.

Marketing the community

One of the consequences of modernisation is that the communities visited by tourists are also consumed by them as commodities. Communities are literally sold as part of the tourist product, and they are often expected to conform to the tourist's image of the idealised community, particularly in rural areas, where modernisation is not expected to impinge on rustic tranquillity. This commodification of community is evidenced by programmes aimed at restoring or conserving the look of localities for the tourist gaze. The Irish Tourist Board, for example, has issued guidelines to farmers on how to make their farms more 'traditional' and 'rural' for visiting tourists (Carrol 1995). Local residents may come to resent the images they are supposed to represent, however, as Howie (Chapter 7) demonstrates in the case of *Trainspotting* tourism in Edinburgh.

The re-imaging of rural areas (Butler and Hall 1998a), arising out of a combination of the commodification of the countryside (Urry 1990) and processes of economic restructuring (Gannon 1994), is of critical concern to rural dwellers, tourists and tourism promoters. Yet traditional, 'idyllic' images of timeless sustainability are still idealised, and their power may well be sufficient to stifle the articulation of real, local identity.

The way of life of the local community is an essential part of the marketing and branding of the tourist product. As Jonathan Edwards *et al.* (Chapter 19) showed in the Alto Minho region of Portugal, visitors recognise the unique features of the community and the landscape it has forged as important aspects of the image of the region. These images are in turn utilised by tourism marketeers to attract tourists to the region, but it is essential that the way of life on which the attractiveness of the region is founded is not damaged in the process.

Elisabeth Kastenholz (Chapter 18) pointed out that the demand for different aspects of the way of life of the community is highly segmented. By matching the nature of the local community more carefully with the needs of different tourist segments, therefore, it may be possible to minimise the mismatch between tourist and resident expectations, and to develop a more sustainable form of rural tourism.

Research issues

The chapters contained in this volume have illustrated many of the challenges facing communities in developing tourism sustainably, and has pointed to some possible solutions. It is clear, however, that many questions remain about the relationship between tourism and the community. This section outlines some of the areas for possible future research which have been highlighted by this book.

Definitions of community

The term 'community' is used by authors in this volume in a number of different ways, reflecting a wider inconsistency and lack of rigour over the definition and use of the term within the social sciences. 'Community' is a word which we think we understand, but the imprecise nature of the term makes misunderstanding very easy. In the tourism literature 'community' is usually assumed to be equated with the presence of a set of common social characteristics and goals held by a population residing in a local area. But there exist many aspatial 'communities of interest', and, for example, as early as 1951 the 'community of limited liability' (Janowitz 1951) was recognised, whereby residents take only a limited interest in their area of residence because the main focus of their activities – work or recreation – lies elsewhere (Hall 1982) or perhaps because there are negative elements of their local area which cause disaffection.

Internal and external perceptions of 'community', the nature of spatial and social boundaries and their articulation in local social and economic cohesion or disruption are important factors in the tourism role, and its sustainability, of local communities, each of which will exhibit different configurations.

Power and empowerment

Communities themselves are unequal, and the fashionability of research into social exclusion has re-emphasised the inequalities within and between self-recognised groups (e.g. Cloke and Little 1997; Milbourne 1997). Yet the 'otherness' of race, gender, sexuality, age and ableism and their role as determinants of tourism power dimensions is only beginning to be addressed (Morgan and Pritchard 1998).

As Mowforth and Munt (1998) have emphasised in the context of the Third World, the analysis of power relationships is crucial to an understanding of the impact of tourism on the community. Not all members of the community, however defined, are equally able to influence the decisions which affect them. Recent studies of communities have tended to emphasise the need for 'empowerment', or the devolution of power to the grass roots of the community (Wilson 1996). In many cases, however, empowerment remains at the level of rhetoric, an ideal still waiting to be realised.

Jan van der Straaten (Chapter 14) emphasised the power that the metropolitan core of Europe is able to exercise over the rural periphery, for example. Solutions for one community may create problems for another. This suggests that research on tourism and the community needs to analyse the external relationships of the community, as well as its internal power structures. This may be particularly important in Europe, where the creation of a 'Europe of the Regions' is producing new alliances and relationships between communities across national boundaries. Examples of new forms of networks are represented in this volume in the context of fair trade networks, and charity-based initiatives such as the SUSTRANS project (Chapter 8). Without more empirical research (e.g. Cope *et al.* 1998), the success of these organisations is likely to remain largely a matter of assertion. The types of questions that need to be addressed include: what impacts do these initiatives have within the community? Are they more responsive to 'community-friendly' or sustainable tourism products?

Networks – development of the third sector?

One potential solution to the problems of participation and empowerment is seen in the growth of the third sector. The global growth of NGOs and local associations and community groups has led some commentators to view them as a resource which can be employed to redress some of the power imbalances in modern society, as argued in this volume by Graeme Evans and Robert Cleverdon (Chapter 9), through the creation of 'fair trade' networks.

In some cases the disparate nature of the third sector may hamper efforts to generate community-based initiatives. In particular, the tension between pragmatic and radical approaches to community development may prevent co-operation between third-sector groups, and more radical groups will often be wary of working with commercial interests – invariably an important issue in tourism.

The growth of the third sector is a direct consequence of the dialectic tensions between globalisation and localisation. While globalisation may arguably threaten to homogenise local communities, it also creates resistance in the form of localisation, which may be used creatively to sustain the community. More research is certainly needed on the processes of commodification and decommodification which are so crucial to the interplay of tourism and community in the modern world. To what extent can local communities influence their relationships with global processes, and stimulate decommodification?

To date, the effectiveness of NGOs and other third-sector organisations in challenging existing power relations remains a grey area. There are some notable successes in the fields of environmental protection and consumer protection (e.g. the UK Consumer Association – founded by Michael Young, one of the pioneers of UK community studies in the 1950s – and its *Which?* series of publications, including *Holiday Which?*), but it is unclear just what impact these organisations have had in the tourism field. Are governments and commercial organisations beginning to take more notice? Are consumers becoming better organised?

Research is hampered by the fragmented and often *ad hoc* nature of the organisations involved. The information about their activities is limited, often through lack of funds or access to established communication channels (e.g. Friends of the Earth, pursuing autonomous action by national and local groups, at the expense of co-operation with or against transnational companies).

We also need to understand how this 'community' of interest itself functions. Indeed, tourists themselves comprise a global community (or many communities?) of interest.

Globalisation and localisation

Power relationships may shift radically as a result of the forces of globalisation. However, new communication media may offer effective means for local groups to communicate to a global audience. The information technology associated with globalisation, for example, can, via internet and e-mail, make it possible for local organisations, even those based in remote rural regions or the developing world, to reach a global audience. This is already happening in some areas (e.g. the Scottish islands – Hall *et al.* 1999). However, the effectiveness of these strategies for local communities may depend on the way in which access to, and control of the Internet develops. Further, multiple websites, for example, can also convey conflicting and confusing images of the same local area.

There need to be mechanisms for democratic control. The trend towards consumer-driven development has removed or weakened these controls in some areas, with the balance of power shifting from local citizens towards global consumers. This process is already evident in the working of the EU *Package Holiday Directive*, which has placed tour operators and their clients in a much stronger position relative to local communities. Because local residents are normally not consumers of the tourism product, they may have little say in a *laissez-faire*, consumer-oriented tourism market. Participation may become limited to playing a role in the production and reproduction of the tourist experience, without much control over its effects. Such a situation may be tolerable for those working in the tourism industry, but may not be so acceptable for other members of the community.

Embedding tourism

In many parts of the world there remains a major policy implementation gap between the ideal of sustainable community-based (tourism) development and its application (Wheeller 1993). A major error which policy makers have often made with respect to tourism is to treat the industry (industries?) in isolation from the other factors which constitute the social, environmental and economic fabric of communities, however defined and conceived (Butler and Hall 1998b). A major task for tourism researchers at all scalar levels from the local to the global, is to provide analytical and policy frameworks which firmly embed tourism within dynamic development processes and which can identify and inform appropriate sustainable pathways.

References

Butler, R. W. and Hall, C. M. (1998a) 'Image and reimaging of rural areas' in R. Butler, C. M. Hall and J. Jenkins (eds) *Tourism and recreation in rural areas*, John Wiley & Sons, Chichester and New York, pp. 115–22.

Butler, R. W. and Hall, C. M. (1998b) 'Tourism and recreation in rural areas: myth and reality' in D. Hall and L. O'Hanlon (eds) *Rural tourism management: sustainable options*, The Scottish Agricultural College, Auchincruive, pp. 97–108.

Carroll, C. (1995) 'Tourism: Cultural Construction of the Countryside' MA Thesis, Programme in European Leisure Studies, Tilburg University.

Cloke, P. and Little, J. (eds) (1997) *Contested countryside cultures: otherness, marginalisation and rurality*, Routledge, London.

Cope, A. M., Doxford, D. and Hill, T. (1998) 'Monitoring tourism on the UK's first long-distance cycle route' *Journal of Sustainable Tourism*, 6(3), 210–23.

Gannon, A. (1994) 'Rural tourism as a factor in rural community economic development for economies in transition' *Journal of Sustainable Tourism*, 2(1/2), 51–60.

Hall, D. (1982) 'Valued environments and the planning process: community consciousness and urban structure' in J. R. Gold and J. Burgess (eds) *Valued environments*, George Allen & Unwin, London, pp. 172–88.

Hall, D. (1999) 'Conceptualising tourism transport: inequality and externality issues' *Journal of Transport Geography*, 7(4).

Hall, D., Gallagher, C. and Boyne, S. (1999) 'Restructuring peripherality: the reconfiguring of Bute' in F. Brown and D. Hall (eds) *Peripheral area tourism in Europe: case studies*, Bornholm Research Centre, Bornholm, Denmark.

Janowitz, M. (1951) *The community press in an urban setting*, Free Press, Glencoe.

Milbourne, P. (ed.) (1997) *Revealing rural 'others': representation, power and identity in the British countryside*, Pinter, London and Washington.

Morgan, N. and Pritchard, A. (1998) *Tourism, promotion and power: creating images,*

creating identities, John Wiley & Sons, Chichester and New York.
Mowforth, M. and Munt, I. (1998) *Tourism and sustainability: new tourism in the Third World*, Routledge, London.
Urry, J. (1990) *The tourist gaze: leisure and travel in contemporary societies*, Sage, London.
Wheeller, B. (1993) 'Sustaining the ego' *Journal of Sustainable Tourism*, 1(2), 121–9.
Wilson, P. A. (1996) 'Empowerment: Community Economic Development from the Inside Out' *Urban Studies*, 33(4–5), 617–30.

Index

Aaker, D. 285
access projects 88
accommodation 144–5, 237; Portugal 263–4, 268–9, 270, 280, 283; small-scale in Indonesia 161–2, 173, 175–6, 179–80, 180–1
activity-based accommodation 237
affective images 287–8, 292–3
Agenda 21 9, 102, 119–20; and fair trade in tourism 149–50; and the tourism industry 69–70; *see also* Local Agenda 21
agricultural support centre 198, 201–2
agriculture 230; Friesland 197–9, 200, 201–2; intensification in Europe 225; Portugal 259–60, 262, 265–6, 294
agritourism 199, 201; Bangunkerto 11, 242–57
aid: development 146; project assistance 192, 200–1
Albania 8, 48–59; post-communist transformation 49–51; rural tourism 8, 53–5, 55–6; tourism as development sector 51–2
Alberta 245
Algarve Coast 264
Alps 224–5; GTA 227–8
Alto Minho 12, 285–96; perceptions of 290–4; rural tourism 288–9; underpinnings of the potential brand 294–5
Amsterdam 110
Andersson, A. 3
anti-participatory attitude 158
Archer, B. 244
Ashworth, G.J. 103, 106
Association of Independent Tour Operators (AITO) 143
Athens, Texas 209–14
Atlantis cycle route 129
attitudes: to agritourism 249–53; tourists in Portugal 270–3; towards tourism 40–1
Auckland 110
authenticity 105–6; staged 4

Balaton, Lake 8, 36–47; attitudes towards tourism 40–1; impacts of tourism development 42–4; reactions of residents to tourist activities 45; tourists–residents relationship 41–2
Bali 161, 176, 178–9
Bali Tourist Development Corporation 157–8
Baloglu, S. 287–8, 292, 293
Bangunkerto, Indonesia 11, 242–57; agritourism site 245–8; community aspects 249; visitors 248–9
Basri Group 174–5
Bätzing, W. 221, 224
Begg, D. 108
Beltane Festival 113
benchmarking 77; fair trade activity 147–9
benefit-segmentation 12, 268–84
best available technology (BAT) 77
Bird, B.J. 207
Bjorklund, E.M. 45
Boissevain, J. 159
Boniface, P. 102
Bonneval in Maurienne 228–9
Boorstin, D. 105, 107
Bord Failte (Irish Tourism Board) 235, 236, 238
Borobudur 248
bottom-up approaches 67–9, 79–80
Bramwell, B. 5, 6, 32
brand 285–96; underpinnings of potential brand 294–5
bridge actors 188, 201–2

Brinberg, D. 287–8, 292, 293
Brohman, J. 171, 243–4
brokers 159–60, 162–5, 166
Brückl, S. 226
Bryson, B. 101, 109
built environment 293, 294
Bupati of West Lombok 174
businesses, monitoring 94
Butcher, H. 121–2
Butler, R.W. 45, 178
Bygrave, W.D. 207

C2C Route 130
CaféDirect 140–1
calm and unpolluted environment 274, 275
Camagni, R. 187–8, 189
Canon of Lek Dukajin 54
capacity management 24–6
car parking 31–2
car trail 85, 87, 96
Castleton 19, 30–1, 31–2
Castleton Chamber of Trade 32
Catholic Institute for International Relations (CIIR) 147, 148
Cavaco, C. 245
CESCAFE co-operative 141
cities 102–3; sustainable city 104–5; *see also* Edinburgh
coconut plantations 173
codes of conduct: corporate 147, 148; in tourism industry 70
coffee 140–1
cognitive images 286–7, 291–2
Cohen, E. 101, 105
Coker, C. 77
co-management 48
commissions 163, 164
Common Agricultural Policy (CAP) 230, 266
Common Ground 107
Commonwealth Heads of Government Meeting (CHOGM) 111–12
communication 197–8
community: approaches to tourism in the Gaeltacht 236–41; cities and 105–6; communities in conflict 6–7; concept 1–2, 2–5, 297; definitions 302; and development 48–9; informing 299; Local Agenda 21 and 121–3; marketing 301–2; research issues 302–5; salak project and 249, 251; and sustainable environment 79; sustaining 299–301
community benefits from tourism 24, 25
community control 243, 253–4
community culture 243, 253–4
community development 242–3; and rural tourism in Albania 53–5; small entrepreneurs and 156; tourism as initiative for 243–5
Community Forum 90, 99
community participation 49, 231, 244, 298–9; approaches to 26–32; Gili Trawangan 10, 170–84; levels of 243; Local Agenda 21 126; techniques 29–30; TTTMP 90–2
community tourism entrepreneurism 11, 205–17
conflict, communities in 6–7
consensus 49; degree of 28–9, 31–2
consumer demand *see* demand
consumer markets 300
consumer product 144
co-operation: between entrepreneurs 162, 193, 201; resistance to in Albania 54; stimulation of co-operation agreements 193, 201
Cooperativa Agrícola 266
co-operatives 301
corporate codes of conduct 147, 148
crafts 149, 240
Craigmillar 114–15
Craik, J. 105
cultivation of land 294
cultural tourism: fair trade 149; Ireland 11, 233–41
culture: Alto Minho 288–9, 294–5; impacts of tourism 8, 36–47; institutional culture 20, 24, 54–5; perceived differences 41–2; political culture 20, 24; and tradition 274, 275
cycle network, national 128–31

Dargie, T. 93
degree of consensus 28–9, 31–2
demand 223–4; for fair trade 145–6
demographic profile 270–3
demolition of businesses and homes 170, 175
demonstration effect 37, 51
Department of Tourism and Trade 235
deregulation policy 154
design of planning processes 193–4, 201–2
destinations *see* tourist destinations
developing countries 138, 139; *see also* fair trade
development: community and 48–9; *see also*

community development; regional development; rural development
development aid 146
dialogue 28, 30
disenfranchised communities 102, 114–15
dislocation 52
distributive empowerment 7
distrust 54–5
diversification 264
Dobni, D. 286
domestic market 282
Donegan, L. 101
dual economy 157–9
Durrell, L. 107

Earth Summit 102, 119–20
Echtner, C.M. 286–7, 291, 292
Ecological Main Structure 197–8, 200
ecological sustainability 243, 253–4
ecology 103–4
economic considerations 103–4
economic monitoring 94
economic self reliance 243, 253
Edinburgh 9, 101–18; city and its tourism industry 107–8; local festivals 112–14; Old Town 110–11; tourism at neighbourhood level 109–12; villages 114–15
Edinburgh Exchange 111
Edinburgh International Conference Centre (EICC) 111–12
embedding tourism 305
emigration 49–50
employment 50, 52, 53, 141
empowerment 7, 302–3
endogenous development 186
endogenous linkages 187–8
English Tourist Board 6
entrepreneurship 10–11, 299–300; community tourism entrepreneurism in Texas 11, 205–17; Gili Trawangan 10, 170–84; small entrepreneurs and community development in Indonesia 10, 154–69; tourism enterprise birth rates 210–13; types of small entrepreneurs 159–65
environment 102, 238; calm and unpolluted 274, 275; Hope Valley 21; impact of tourism 65–7; improvements and TTTMP 89; monitoring 92–3; Portugal 264; sustainability 5–6, 20, 223
environment-led tourism 21, 23, 26
environmental indicators 73
environmental manuals 70–1
Environmental Performance Standards (EPSs) 9, 67–9, 71–2
Environmental Quality Standards (EQSs) 9, 67–9, 71–2
environmental school 206
environmental standards 8–9, 63–82; evaluating improvements in performance 76–8; measurement of environmental performance 72–6; top-down and bottom-up approaches 67–9
ethics 103–4
Europe 224–31
Europe of the Regions 303
European Fair Trade Association (EFTA) 140
European Union (EU) 120–1, 225; CAP 230, 266; Package Holiday Directive 304; Travel and Tourism Directive 138
exogenous development 186
exogenous linkages 187–8
external production units 192, 200
fair trade 9–10, 137–53; Agenda 21 and 149–50; consumer demand 145–6; crafts and cultural tourism 149; features of 140–1; vs free trade 138–40; future for 150–1; monitoring and benchmarking 147–9; in tourism services 142–5
Fair Trade Foundation 147
Federation of Nature and National Parks in Europe (FNNPE) 123
festivals, local 105, 112–14
finance 237–8
first-order resources 159–60
first-time visitors 290–1, 293–4
focusing on the most promising parts of the area 192–3, 201
folk museum 238
food/cuisine 265, 293, 294
'foot-loose' industry 187
footpaths, long-distance 227–8
fossil fuels 223
Fowler, P.J. 102
France 226, 227–8, 229, 295
free trade 138–40
Friesland 10–11, 185–204; centre for agricultural support 198; evaluation of WCL policy 199–202; history of WCL policy 196–9; WCL policy 194–6
funding 98
Fyall, A. 20

Gaasterlân-Sleat 194–5

Gaeltacht 11, 233–41; community approach to tourism 236–8; Glencolumbkille 238–40; tourism strategy 235–6, 237–8
Gambia, the 150
Garrod, B. 20
gastronomy 265, 293, 294
gatekeepers 188, 202
General Agreement on Tariffs and Trade (GATT) 138
Generasi Jaya, Pt. 173, 174
generative power 7
geographical polarisation 225–7
Gili Trawangan 10, 170–84; Basri Group vs local people 174–5; future scenario 180–1; Lombok 176–8; present-day 175–6
Glencolumbkille 238–40
global trade exhibitions 146
globalisation 304–5
Go, F.M. 102–3
Gorgie-Dalry 114
Gothenburg 126
Gouldson, A. 78
government support for investors 175, 180–1
GR 5 227–8
Graburn, N.H.H. 106
Grande Traversata delle Alpi (GTA) 227–8
greenhouse effect 223
Grenoble, France 226
Gresse-en-Vercors 229
grey-area tourism 102, 109–12
grey economies 50, 52
guides: informal 163–5; student 249

Hague, The 126
Hall, C.M. 27, 103, 244
handicraft shopping 274, 275
Hawaii 186
Healey, P. 28
Herzberg, F. 271
Hinch, T.D. 105
historic-cultural tourist cities 102–3; *see also* Edinburgh
Hjalager, A. 245
Hogmanay Festival 113–14
homestays 161–2, 173, 175–6, 179–80
Hope Valley 8, 17–35; community participation 29–32; sustainable tourism 21–6; visitor-management plan 24–6
Hope Valley Visitor Management Plan Working Group 18, 21, 24, 29
Howie, F. 103
Hughes, G. 88, 91, 95, 97, 104, 127, 132
human capital, redundant 52, 55
Hungary *see* Balaton, Lake
Hunter, C. 21, 22–3, 26
hygienic factors 271–2

illegal enterprises 160–1
images 282, 286–8; of the Alto Minho 290–4; impact of tourism on 40–1
imagined communities 3–4
importance ratings 271–2, 273–4, 275
independent patrons 159–60, 160–2, 164–5, 166
independent ruralists 276, 277, 278, 281
independent travel 274, 275
India 150
individual needs 243, 253–4
individualism 122
Indonesia 10, 154–84; agritourism in Bangunkerto 11, 242–57; dual economy 157–9; Gili Trawangan 10, 170–84; tourism policy 157–9, 172, 178–81
informal guides 163–5
informal sector 155, 156
information 89, 299
innovative milieus 187–8; network management and planning 189–90
inputs, environmental 65–9
instability 56
institutional culture 20, 24; Albania 54–5
integrated rural development 264–6
integrated tourism resort 177–8
integration 199; policy integration 191–2, 199–200
integrative school 207
intensification of agriculture 225, 230
intensity of community participation 28, 30–1
intermediaries 143
International Union for the Conservation of Nature (IUCN) 222
intervention policies 192, 200–1
investment: climate (Q-factors) 207–14; pyramid schemes 50–1; *see also* outside investors
Ireland 11, 233–41
Irish language 233–4, 237, 240, 241
Italy 227–8

Jakarta 178–9
Jarvis, L. 286
Jenkins, J.M. 103

Kalimantan 179
Keszthely–Hévíz region 39–45
Kinder, R. 110
Koh, K.Y. 207–8
Kosovar Albanians 50
Krippendorf, K. 103–4
Kusuma, Wasita 173, 174
Kuta 177, 178, 179

Lake Balaton *see* Balaton, Lake
land: cultivation in Portugal 294; disputes in Gili Trawangan 170–1, 173, 174–5
language, Irish 233–4, 237, 240, 241
large tourist resorts 158–9
Lash, C. 2
Leith 109
Leslie, D. 127, 132
Lewis, N. 101
Li, T. 243, 254
Limburg 196
linkages 187–8; *see also* networks
Local Agenda 21 (LA21) 9, 119–34; and community 121–3; national cycle network 128–31; relevance to tourism 120–1; significance to tourism in UK 123–8
local economic development 144
local festivals 105, 112–14
local government 122, 124–5, 131–2
Local Government Management Board (LGMB) 124, 129, 133
localisation 304–5
locality, community and 4
Loch Lomond and Trossachs National Park 97
Lombok 149, 176–8, 179
Lombok Tourism Development Corporation (LTDC) 177–8
Long, V.H. 244
long-distance footpaths 227–8
Lord Provost's Commission on Sustainability 108
losmen (homestays) 161–2, 173, 175–6, 179–80
Lynch, K. 107

MacCannell, D. 105
McDyer, Father 238
McIntosh, R.W. 244
MacLellan, R. 93–4, 97
Malik, Nellie Adam 174
Manado 179
market development 230–1
market forces 124–5
marketing 228, 230; agritourism and salak marketing 251–3; the community 301–2; Gaeltacht 237; Northern Portugal 262–3
Mary Queen of Scots 115
matriarchal society 295
Mayo, E. 286
Meithal Forbartha na Gaeltachta (MFG) 234–5, 235, 236, 237, 238
Merapi, Mount 249
migration: Albania 49–50; male migration and return, Alto Minho 295
Millenium Link canal project 115
Molt, W. 226
monitoring: fair trade activity 147–9; TTTMP 92–4, 98
most promising parts of the area, focusing on 192–3, 201
motivational factors 271–2
Mount Pleasant, Texas 209–14
Murphy, P. 4, 106, 244
mutual dependence 166

nation state 3
National Cycle Network 128–31
National Landscape Foundation 199
National Landscape Policy 195–6, 197; Project Office 196, 199, 200–1
needs, individual 243, 253–4
Neil, A. 108
neotenous tourism 21, 23
neotribalism 4
Netherlands: Friesland 10–11, 185–204; WCL policy 194
network management 189–90
network specialists 159–60, 162–5, 166
networks 300; development of third sector 303–4; Indonesia 162, 166; tourism in Friesland 10–11, 185–204
Niass islands 179
Nord Zee Route 129
Nozick, M. 243, 253
Nusa Dua 158, 179

Office of Fair Trading (OFT) 140
Oideas Gael 240
Oporto Metropolitan Area 262
ordinary 101, 105–6; importance of 106–7
organised crime 52, 54–5
outdoor ruralists 277, 278, 279, 281
outputs, environmental 65–9

outside investors 10, 170–84; government support 175, 180–1
Oxfam 140

P-factors 207–14
Parc Naturel Régional de Vercors (PNR Vercors) 229
partial consensus 28–9, 32
participation: community *see* community participation
partnerships 83–100
patrons 159–60, 160–2, 164–5, 166
Peak Park Joint Planning Board 24, 32
Peak Tourism Partnership (PTP) 24, 29, 31
Pearce, D. 215
Peneda-Gerês National Park 264
people school 206
performance measurement 8–9, 63–82; environmental performance 72–6; evaluating improvements in environmental performance 76–8
Philbrick, A.K. 45
pilot projects 126
planning, regional 189–90
planning process, design of 193–4, 201–2
Plog, S.C. 101
polarisation, geographical 225–7
police raids 164–5
policy integration 191–2, 199–200
political culture 20, 24
polluting sectors 222, 223, 230
Ponte de Lima 269
Portobello 114
Portugal 245; accommodation structure 263–4; Alto Minho 12, 285–96; community and rural development in Northern region 12, 258–67; demographic profile and tourist behaviour 270–3; market for rural tourism in North and Central Portugal 12, 268–84; planning for integrated rural development 264–6; tourism in Northern region 260–1; tourism in rural areas 261–3; tourist segments 273–9, 280–2
possessive individualism 122
post-communist transition 49–51
post-modernism 105–6
pousadas 263
power 244, 302–3
premium pricing 145
Prentice, R. 28, 104, 106
pressure–state–response concept 73–4
Pretty, J. 243
pricing 145
Princes Street, Edinburgh 112
principal components analysis 273–4, 275, 279
product-led tourism 21, 22
propensity to enterprise (P-factors) 207–14
provincial government 158, 175; *see also* local government
Putri Nyale Resort 171, 177–8
pyramid investment schemes 50–1

Q-factors (quality of investment climate) 207–14
Qeparo 53, 54
quality: tourism in the Gaeltacht 237–8; tourism in Indonesia 179–80; of visitor experience in the Hope Valley 24, 25
quality of life: Edinburgh 102, 109–12; sustainable city 104–5

Raban, J. 105, 106, 109
Ranck, S. 244
Read, S.E. 103
'REAL' tourism 103
redundant human capital 52, 55
Reed, M. 244
refuge sector 52
regional development: elements for regional planning 189; theories 187–8; tourism and 258–60
regionalism 4
registration 165
regulation 154–5, 165
repeat visitors 290–1, 293–4
residents: attitudes towards tourism 40–1; reactions to tourist activities 45; relationship with tourists 41–2
resort life cycle model 178–9
resort tourism 158–9
resources 159–60
Reynolds, P.D. 208
Riley, M. 52
Ritchie, J.R.B. 286–7, 288, 289, 291, 292
Rudini, Interior Minister 175
rural communities 11–12, 300–1; Bangunkerto 11, 242–57; Gaeltacht 11, 233–41; sustainable tourism and rural regions in Europe 11, 221–32
rural development: Europe 224–7; integrated 264–6
rural romantics 276–9, 280–1
rural tourism: Albania 8, 53–5, 55–6; Alto

Minho 12, 285–96; market for in North and Central Portugal 12, 268–84; Northern Portugal 12, 258–67; principles for development 190–4
Ryan, C. 110

St Lucia 145
salak agritourism project 246–53
Sammeng, A.M. 159
Sanur 179
satisfaction ratings 272–3
savings 50–1
scope of participation by community 27–8, 29–30
Scotland: Edinburgh 9, 101–18; Trossachs 9, 83–100
second-order resources 159–60
segments, tourist 273–9, 280–2
self reliance, economic 243, 253
Senggigi 171, 177, 181
service and comfort 274, 275
services sector 138, 139
sex tourism 109–10
Shannon Development 235
Sharpley, J. 245
Sharpley, R. 245
Shaw, G. 206
Shkoder 53
Shoard, M. 106
Simmons, D.G. 106
Simms, D.M. 206
small entrepreneurs: Albania 50, 52; Indonesia 10, 154–84; types of 159–65
small and medium-sized (SME) enterprise networking 143–4
small-scale accommodation 161–2, 173, 175–6, 179–80, 180–1
small-scale tourism projects: Albania 53, 56; Indonesia 159
Smith, V.L. 101
Soajo 269
social and active hedonism 274, 275
social impacts of tourism 94; Lake Balaton 8, 36–47
Spain 186
spatial planning concept 175
staged authenticity 4
stakeholders 20; fair trade in tourism 151; Hope Valley 21–4, 29; representation 27–8, 29
Stevenson, R.L. 109
street guides 163–5
student guides 249
subsidies 261
sustainability 1, 5–6, 299; and community 299–301; ecological 243, 253–4; indicators 96, 98; positions 18–20; TTTMP 94–6
sustainable city 104–5
sustainable development 17, 121; 'columns' 103–4; initiatives and Local Agenda 21 125–6
Sustainable Rural Tourism 126, 128
'Sustainable Themes' 129–30
sustainable tourism 17, 121; approach in the Hope Valley 21–6; approaches to 18–26; dimensions 32; principles of management 5, 6; tourism sector and 222–4
Sustrans 128–31
Sutowo, Ponco 174
Szivas, E. 52
Sztompka, P. 54

Tapeis Gael 240
taxes 165
Taylor, G. 106
TER (*turismo no espaço rural*) 268–9, 280, 283
Texas 11, 205–17; measurement of P- and Q-factors 208–14
Texas Fishing Hall of Fame 210
third sector 7, 303–4
tips and gifts 164
top-down approaches 67–9, 80
Tour Indo 248
tourism: embedding 305; fair trade in tourism services 142–5; Local Agenda 21's significance to 123–8; nature of 190–1; policy in Indonesia 157–9, 172, 178–81; and regional development 258–60; relevance of Local Agenda 21 120–1; socio-cultural impacts 36–9; strategy in the Gaeltacht 235–6, 237–8; use as a community development initiative 243–5
tourism enterprise birth rates 210–13
Tourism and the Environment Task Force (TETF) 24, 91, 94, 95, 96, 99, 126; aims 84
tourism imperative 21, 22
tourism industry: Agenda 21 and 69–70; codes of conduct 70; development sector in Albania 51–2; environmental manuals 70–1; and environmental problems 69–71; impacts of development 37, 42–4, 45; and sustainable tourism 222–4

Tourism Management Programmes (TMPs) 84–5; *see also* Trossachs
tourism priority zones 51
Tourism Youth Association 247, 249
tourist behaviour 270–3
tourist destinations: features of 20–1; images 40–1, 282, 286–8, 290–4
Tourist Development Corporations 157–8
tourist–host relationship 37, 40–2, 45
tourist segments 273–9, 280–2
towns 102–3, 265
Trainspotting 108, 109, 110
transition, post-communist 49–51
transparency 151
Trans-Pennine cycle route 129
transport 96, 144–5
Trossachs 9, 83–100; community participation 90–2; future 96–7; monitoring 92–4; sustainability 94–6; Trossachs Trail Tourism Management Programme (TTTMP) 85–90; TTTMP projects 88–9
Trossachs Discovery Centre 97
TUI 148–9
Tunbridge, J.E. 103, 106
Turner, R.K. 18–20, 65, 66
Udaras na Gaeltachta 235–6, 240
United Kingdom (UK) 123–33; Local Agenda 21's significance to tourism 123–8; national cycle network 128–31
United Nations Conference on Environment and Development 48, 119–20
urban tourism 102–3; *see also* Edinburgh
Urry, J. 2, 5

Van de Ven, A.H. 207
Van der Borg, J. 113
Venice 113
villages 114–15
visitors: attractions and facilities for 89; monitoring 92–3; salak agritourism project 248–9; visitor types and images 290–1, 293–4; *see also* tourist behaviour; tourist–host relationship

want-it-all ruralists 276, 277, 278, 281–2
Waterland 197
WCL (*Waardevol Cultuurlandschap*) policy 193–202
weaving 240
Welford, R. 78
Welsh, I. 109
Wester Hailes 115
wildlife tourism 96
Williams, A.M. 206
women 56
World Commission on Environment and Development (WCED) 222, 223; *Our Common Future* 17, 18
World Trade Organisation (WTO) 138
World Wide Web 146

Zinkan, G.M. 286
Zins, M. 288, 289

[redacted]